# Introduction of Biotechnology in India's Agriculture

CMA Publication no. 247

Vasant P. Gandhi • Dinesh Jain

# Introduction of Biotechnology in India's Agriculture

## Impact, Performance and Economics

Springer

विद्याविनियोगाद्विकासः
IIM
AHMEDABAD

Vasant P. Gandhi
Centre for Management in Agriculture
Indian Institute of Management
Ahmedabad (IIMA)
Ahmedabad, India

Dinesh Jain
Entrepreneurship Development
Institute of India
Ahmedabad, India

ISBN 978-981-10-9327-2 ISBN 978-981-10-1091-0 (eBook)
DOI 10.1007/978-981-10-1091-0



Printed on acid-free paper

This Springer imprint is published by Springer Nature
The registered company is Springer Science+Business Media Singapore Pte Ltd.

# Foreword

The Centre for Management in Agriculture (CMA), Indian Institute of Management, Ahmedabad, is actively engaged in applied and problem-solving research on agriculture, food and agribusiness management, towards achieving the major goals of agricultural and rural development in the country and the world. As a result, over the years, CMA has developed considerable expertise in a large spectrum of areas of agriculture and agribusiness, including the management of agricultural inputs, agro-processing, agri-food marketing, rural infrastructure, grass-roots innovations, appropriate technologies for arid and semiarid regions, international agricultural trade and WTO issues, global competitiveness, commodity markets, food safety and quality including organic food, food supermarkets and food value chains. CMA undertakes research of this kind, especially for the Ministry of Agriculture, Government of India, and on its own as well as at the request of other client groups.

The context of the present research is the urgent need for new breakthroughs in agricultural technology and India's experience in the introduction of biotechnology in cotton, a major commercial crop in India. Scientific advances in biological sciences are leading to major breakthroughs in the recent years, and one of the important outcomes is the development of Bt cotton. The study, undertaken at the request of the Ministry of Agriculture, Government of India, has sought to objectively examine the experience of Bt cotton vs non-Bt cotton in the major cotton states of Andhra Pradesh, Gujarat, Maharashtra and Tamil Nadu which together account for about 70 % of the cotton production. With a sample of 694 farm households, the study reports on the yields, pesticide costs, seed costs, overall cost of production and profits. It also reports on the environmental impacts, the satisfaction with the technology and ways of improving its effectiveness.

I am sure the study will be found useful by policymakers, academicians, managers and others interested in issues of agriculture, new technology and the welfare of the farmers in the country.

Indian Institute of Management
Ahmedabad, India

Ashish Nanda

# Preface

The book is dedicated to the memory of Professor D.K. Desai, the founder of the Centre for Management in Agriculture and the pioneer of agribusiness education in India, and Professor Gunvant Desai, the outstanding professor and researcher from whom the authors learnt so much. The book is a result of research studies conducted by the authors and associated researchers to objectively examine the performance of the recently introduced Bt cotton varieties in India. The study was undertaken at the request of the Ministry of Agriculture, government of India, as a coordinated research project covering the major cotton-growing states of Andhra Pradesh, Gujarat, Maharashtra and Tamil Nadu, to objectively evaluate its performance. The study was conducted under the leadership of the Centre for Management in Agriculture (CMA), Indian Institute of Management, Ahmedabad (IIMA), and in cooperation with the Agro-Economic Research Centres (AERCs) in these states, associated with the ministry. While the study in Maharashtra and the consolidated study were carried out by CMA, the other studies were largely undertaken by the respective AERCs in these states in coordination with CMA. This book manuscript consolidates the following research studies:

1. Economics of Bt Cotton vis-à-vis traditional Cotton Varieties – Study in Andhra Pradesh, by N. Ramgopal, Agro-Economic Research Centre, Andhra University, Visakhapatnam, 2006.
2. Returns to Bt Cotton vis-à-vis Traditional Cotton Varieties in Gujarat State, by V.D. Shah, Agro-Economic Research Centre, Sardar Patel University, Vallabh Vidyanagar, 2007.
3. Returns and Economics of Bt Cotton vis-à-vis Traditional Cotton Varieties in the State of Maharashtra in India, by Vasant P. Gandhi and N.V. Namboodiri, Centre for Management in Agriculture, Indian Institute of Management, Ahmedabad, 2009.
4. Returns to Bt Cotton vis-à-vis Traditional Cotton in Tamil Nadu, by A. Pushpavalli, Agro-Economic Research Centre, University of Madras, Chennai, 2007.

The authors very gratefully acknowledge the contributions of N. Ramgopal, Agro-Economic Research Centre (AERC), Visakhapatnam, Andhra Pradesh;

V.D. Shah, AERC, Vallabh Vidyanagar, Gujarat; S.S. Kalamkar, AERC, Pune, Maharashtra; A. Pushpavalli, AERC, Chennai, Tamil Nadu; and N.V. Namboodiri, CMA-IIMA, as well as the heads and staff of all the centres. The authors wish to sincerely thank Darshan Ajudia, Varsha Khandker, Chandrabhushan, Aashish Argade, PV Sethumadhavan, Barathi Ramachandran, Mini Nair and Dipali Solanki for their assistance. Support of the Ministry of Agriculture, Government of India, and CMA-IIMA is gratefully acknowledged.

Ahmedabad, India

Vasant P. Gandhi
Dinesh Jain

# Contents

# List of Figures

# List of Tables

# About the Authors

**Vasant P. Gandhi** is currently NABARD chair professor at the Indian Institute of Management, Ahmedabad (IIMA) in India. He did his PhD from Stanford University, Masters from IIMA, and Bachelors in Agriculture from Pantnagar, India. He has been Chairman of IIMA's Centre for Management in Agriculture, and a member of the Board of Governors of IIMA. He has worked with the World Bank, and the International Food Policy Research Institute (IFPRI) in Washington, and has been a Visiting Professor at the University of Sydney and the James Cook University, Australia. He has been Consultant/Advisor to FAO, the World Bank, Government of India, various state governments, Agriculture Insurance Company of India, McKinsey & Company, various high level government committees, and is on several company boards. His areas of research in which he has published extensively include international food and agriculture policy, technology for agricultural growth, water resource management and institutions, agricultural markets and agribusiness.

**Dinesh Jain** obtained his doctorate from Indian Institute of Management, Ahmedabad (IIMA) and is associated with the Entrepreneurship Development Institute of India (EDI) as Associate Faculty. He did the doctoral Fellow Programme in Management (FPM) with specialisation in Agribusiness from IIMA, and prior to this, his Masters in Business Administration (Agribusiness Management) and Bachelors in Agricultural Engineering. At EDI, he is the Programme Coordinator for EDI's Post Graduate Diploma in Management-Development Studies. He has done extensive research through projects of Australian Centre for International Agriculture Research (ACIAR), International Food Policy Research Institute (IFPRI), Global Development Network (GDN), and Ministry of Agriculture – Govt. of India. His areas of research include institutional innovations and design, especially in water management, agri-entrepreneurship, new technology adoption in agriculture, food security, and impact assessment of development initiatives.

# Part I
# Introduction

# Chapter 1
# Introduction and Objectives of the Study

## 1.1 Introduction

Cotton is the most important cash crop in India and the country ranks first in cotton area and second in cotton production in the world. About 15 million farmers in the country spread across 10 states are engaged in cotton production and grow cotton on an area of about 11 million hectares. India also holds a prominent position in cotton textile industry in the world, manufacturing cotton textile products for a large number of end uses in India and abroad. Despite being one of the top most cotton countries in the world, the cotton yields in India are one of the lowest. A major reason for the low yields is the susceptibility to severe insect pest attacks which cause extensive crop damage. The major cotton producing states in the country are Maharashtra, Gujarat, Andhra Pradesh, Punjab and Tamil Nadu, and among them Maharashtra alone accounts for over one-third of the cotton area in the country.

Following a long history of cultivation of traditional varieties, hybrid cotton was introduced in India for the first time in 1970. This was in the state of Gujarat, and by virtue of its high yield potential it became extremely popular. A large number of hybrids were released. However, it was soon realized that the hybrids were highly susceptible to insect pests which cause severe damage. This became a huge problem especially from 1993 to 1994 onwards, leading to frequent crop failures as well as fluctuating and declining yields. Over 150 different insect pests' species are reported to attack cotton at various stages of its growth causing severe reduction in yields. This drives massive pesticide use by farmers and high cost of cultivation. It is estimated that over 55 % of the pesticides sold in the country are used on cotton. As a result of this, cotton farming was in serious trouble and farmers were looking for a solution. It was at this juncture that the transgenic Bt cotton varieties arrived on the world stage.

It was after much hesitation and delay that the Government of India allowed the cultivation of three genetically modified Bt cotton hybrids in April 2002 for a period


V.P. Gandhi, D. Jain, *Introduction of Biotechnology in India's Agriculture*, DOI 10.1007/978-981-10-1091-0_1

of 3 years. This was preceded by the controversial unauthorized introduction of Bt cotton hybrids in some areas of the country. Analysis from several years of Indian trial data had demonstrated the superiority of Bt technology in terms of reduced pesticides' application and increase in effective yield. The impact assessment commissioned by Mahyco-Monsanto Biotech claimed sizable benefits for Bt adopters (Nielson and ORG MARG Study 2004). However anti-biotechnology activists declared the technology as a complete failure (e.g. Shiva and Jafri 2003). Even though the performance of Bt cotton has been projected to be satisfactory in government circles, there is great discontent in different quarters with Bt cotton. Strong views both for and against Bt technology has surfaced. The major advantages claimed for Bt Cotton include reduction in the use of pesticides by almost 50 %, reduction in the harmful effect of pesticides on the environment, good quality of cotton fiber at par with that of non-Bt cotton, better yield per unit of input use, and lesser residue of pesticides in the fiber resulting in reduced harmful effects such as allergic reactions. However, the voices against Bt Cotton indicate that the gene may spread and its impact on the eco-system is not known, Bt Cotton seed would be very expensive compared to Non-Bt seed for the farmers, some companies may have a monopoly on Bt seed, Bt cotton farmers may still need to use pesticides, Bt cotton seed cake will cause harm to animals, Bt may enter the human food chain and cause harm, transgenic varieties will lead to disappearance of native varieties and biodiversity in the country, and insects will soon become resistant to Bt Cotton making the pest control even more difficult in the near future.

Despite these concerns, Bt cotton cultivation has spread quite rapidly in India and elsewhere and farmers in developing countries are willing to adopt this technology. In view of the above mentioned diverse views on Bt cotton and considering the importance of cotton in Indian agriculture, it seemed important to undertake a comprehensive and systematic review to study the economic returns and other related aspects of the cultivation of Bt cotton as opposed to non-Bt cotton in major cotton producing states in the country.

This study was undertaken at the request of Ministry of Agriculture, Government of India, as a coordinated study in four states namely Andhra Pradesh, Gujarat, Maharashtra and Tamil Nadu, under the leadership of the Centre for Management in Agriculture (CMA) Indian Institute of Management, Ahmedabad, along with the Agro-Economic Research Centers (AERC) in these states. While the study in Maharashtra as well as the consolidated study was carried out by CMA, the other studies were done by the respective AERCs in these states in coordination with CMA.

## 1.2 Objectives and Methodology of the Study

The main objectives of the study are as follows:

- To examine the development and the advantages and disadvantages of Bt cotton as a pest-resistant variety under both irrigated and unirrigated conditions in selected states

- To assess the cost of cultivation and economics of Bt cotton as compared to other cotton varieties grown by farmers
- To assess the net returns to Bt cotton as compared to other cotton varieties
- To examine the observations of farmers on Bt cotton with respect to the technology, crop performance, pest incidence, and impact on other crops and the environment
- To comment on the usefulness of Bt cotton technology under Indian conditions and indicate ways for further improvement.

This study is based on the primary data collected from four states namely Andhra Pradesh, Gujarat, Maharashtra and Tamil Nadu. Effort was made to adopt similar methodology, content and survey instruments in all these states, as far as possible. On determining the farmer sample, a multi-stage stratified random sampling design was used. The first stage of sampling was the selection of districts, the second stage was the selection of villages in the selected district and the third stage was the selection of sample farmer households of different types (Bt and Non-Bt, farm sizes, irrigated-unirrigated) in the selected villages. Effort was made to have nearly an equal number of Bt and Non-Bt farmers in the sample. Effort was also made through stratification to cover both irrigated and unirrigated farms under Bt and non-Bt cotton, as well as small, medium and large farmers. The sampling methodology is discussed in greater details in the respective state studies. The primary data collected pertains to the agricultural year 2004–2005.

Since the major objective of the study is the economics of Bt Cotton vis-à-vis non-Bt cotton, a comparative study of them on important parameters has been undertaken. Wherever possible the comparative performance of Bt and non-Bt farmer households under different farm size and irrigation status has been studied. Tabulation as well as basic statistical tools and regression analysis have been used to study the differences and impact of Bt cotton as compared to non-Bt cotton, measure some of the differences and check their statistical significance. The following chapters give a brief overview of cotton in India and present the results emerging from the study.

## References

Nielson, A. C., & ORG MARG Study. (2004). *Benefits of Bollgard cotton*. Mumbai: Mahyco and Monsanto Biotech.

Shiva, V., & Jafri, A. (2003, May 31). Failure of the GMO's in India. Retrieved September 2011, from Mindfully website: http://www.mindfully.org/GE/2003/India-GMO-Failure-Shiva31may03.htm

# Chapter 2
# An Overview of Cotton in India

## 2.1 Introduction

In this chapter we briefly describe the position of India in world cotton economy, India's recent performance in cotton production and also the performance of cotton production in the four selected states.

## 2.2 World Scenario

The world annual production of cotton is estimated at about 100 million bales (one bale equals 480 lbs.) (Table 2.1). China occupies the top position with a share of 29 % of the global production, followed by India with a share of 21.7 %, and USA has a share of 12.8 %. The other two countries with a share of over 5 % of the world cotton production are Pakistan and Brazil. Although India occupies the top position in terms of area under cotton, its share is low in global production due to low yields. The cotton yields in the country is hardly one-third that of China and little over 40 % that of USA (Table 2.2). The yield in India is less than one-fourth of that of some of the smaller cotton producing countries such as Australia, Syria and Greece.

## 2.3 Cotton Cultivation in India: Recent Performance

This section examines the trends in cotton area, production, and yield in India since 1950–1951, with particular emphasis on its performance since 1990–1991. The performance during the period 1990–1991 to 2010–2011 has been scrutinized in two periods viz., 1990–1991 to 2001–2002 and 2002–2003 to 2010–2011. This is to see the likely impact of adoption of Bt cotton in the country initiated in 2002–2003.


V.P. Gandhi, D. Jain, *Introduction of Biotechnology in India's Agriculture*, DOI 10.1007/978-981-10-1091-0_2

**Table 2.1** World cotton production: average for 2008–2009 to 2011–2012

| | Countries | Million bales of 480 lbs | Percentage to world total |
|---|---|---|---|
| 1. | China | 33.175 | 28.96 |
| 2. | India | 24.825 | 21.67 |
| 3. | United States | 14.675 | 12.81 |
| 4. | Pakistan | 9.225 | 8.05 |
| 5. | Brazil | 7.175 | 6.26 |
| 6. | Uzbekistan | 4.2 | 3.67 |
| 7. | Australia | 3.1 | 2.71 |
| 8. | African Franc Zone | 2.3 | 2.01 |
| 9. | Turkey | 2.275 | 1.99 |
| 10. | Turkmenistan | 1.575 | 1.37 |
| 11. | EU-27 | 1.275 | 1.11 |
| 12. | Greece | 1.075 | 0.94 |
| 13. | Mexico | 0.75 | 0.65 |
| 14. | Burkina | 0.75 | 0.65 |
| 15. | Mali | 0.55 | 0.48 |
| 16. | Others | 7.625 | 6.66 |
| | World total | 114.55 | 100 |

Source: Derived from Ministry of Textiles, Govt. of India

**Table 2.2** Area, production and yield of cotton in major cotton producing countries: 2010

| | Area (000HA) | Production ('000 tonnes) | Yield (kg/ha) |
|---|---|---|---|
| China | 4,849 | 17,910 | 3,693.5 |
| India | 11,142 | 17,760 | 1,594.0 |
| United States | 4,329.66 | 9,473.8 | 2,188.1 |
| Pakistan | 2,689.1 | 5,613.928 | 2,087.7 |
| Uzbekistan | 1,342.5 | 3,442.8 | 2,564.5 |
| Brazil | 829.753 | 2,949.845 | 3,555.1 |
| Turkey | 480.439 | 2,150 | 4,475.1 |
| Australia | 208.3 | 939 | 4,507.9 |
| Argentina | 440.911 | 753.524 | 1,709.0 |
| Turkmenistan | 550 | 684 | 1,243.6 |

Source: http://faostat.fao.org/

The following observations on the impact can be made from the data and results presented in the Table below.

During the period 1990–1991 to 2001–2002 the lowest and highest levels of cotton production in the country were 117 and 178 lakh bales respectively, which indicates a difference of about 60 lakh bales, over a span of 12 years (Table 2.3 and Figs. 2.1 and 2.2). However, the lowest and highest levels of production achieved during 2002–2003 and 2010–2011, i.e., in years, were 136 and 330 lakh bales respectively, which at 194 lakh bales is higher. The production level during 2010–2011was greater by 150 lakh bales than even the previous peak of 178 lakh bales

**Table 2.3** Growth in production, area and yield of cotton: all India

| | Production in lakh bales | Area in lakh hectare | Yield: in kg per ha |
|---|---|---|---|
| 1950–1951 | 32.8 | 58.8 | 95 |
| 1960–1961 | 56.8 | 76.1 | 127 |
| 1970–1971 | 53.5 | 76.1 | 120 |
| 1980–1981 | 78 | 78.2 | 170 |
| 1990–1991 | 117 | 73.9 | 269 |
| 1991–1992 | 118 | 73.8 | 271 |
| 1992–1993 | 138 | 75.4 | 311 |
| 1993–1994 | 121.5 | 74.4 | 278 |
| 1994–1995 | 138.5 | 78.6 | 300 |
| 1995–1996 | 170.7 | 90.6 | 320 |
| 1996–1997 | 177.9 | 91.7 | 330 |
| 1997–1998 | 158 | 88.3 | 307 |
| 1998–1999 | 165 | 92.9 | 302 |
| 1999–2000 | 156 | 87.3 | 304 |
| 2000–2001 | 140 | 85.8 | 278 |
| 2001–2002 | 158 | 87.3 | 308 |
| 2002–2003 | 136 | 76.7 | 302 |
| 2003–2004 | 177 | 77.9 | 387 |
| 2004–2005 | 213 | 89.7 | 404 |
| 2005–2006 | 185 | 86.8 | 362 |
| 2006–2007 | 226 | 91.4 | 421 |
| 2007–2008 | 258.8 | 94.1 | 467 |
| 2008–2009 | 228.8 | 94.1 | 413 |
| 2009–2010 | 240.2 | 101.3 | 403 |
| 2010–2011 | 330 | 112.4 | 499 |
| Coefficient of variation | | | |
| 1990–1991 to 2001–2002 | 14.19 | 9.02 | 6.62 |
| 2002–2003 to 2010–2011 | 24.86 | 12.06 | 13.96 |
| Annual rate of growth | | | |
| 1990–1991 to 2001–2002 | 2.68 | 1.99 | 0.72 |
| 2002–2003 to 2010–2011 | 8.23 | 4.16 | 4.05 |

achieved during 1996–1997. The variability in production during the second period was also somewhat greater than the first period – the coefficient of variation increased from 14.19 to 24.86. This was mainly due to the crash in cotton production to 136 lakh bales in 2002–2003. The annual rate of growth of cotton production during the period 2002–2003 to 2010–2011 was at 8.23 % as compared to 2.68 % during 1990–1991 to 2001–2002. The growth in cotton production during the former period was mainly due to growth in area rather than growth in yield witnessed during 2002–2003 to 2010–2011. The average cotton yield during 1990–1991 to 2001–2002 was about 300 kg per hectare and it increased substantially to over 400 kg per hectare during 2002–2003 to 2010–2011.

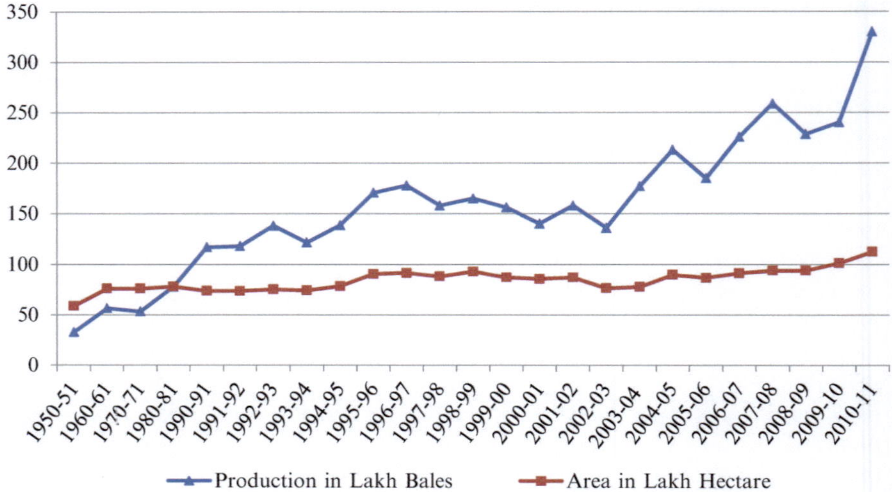


**Fig. 2.1** Production and area of cotton in India

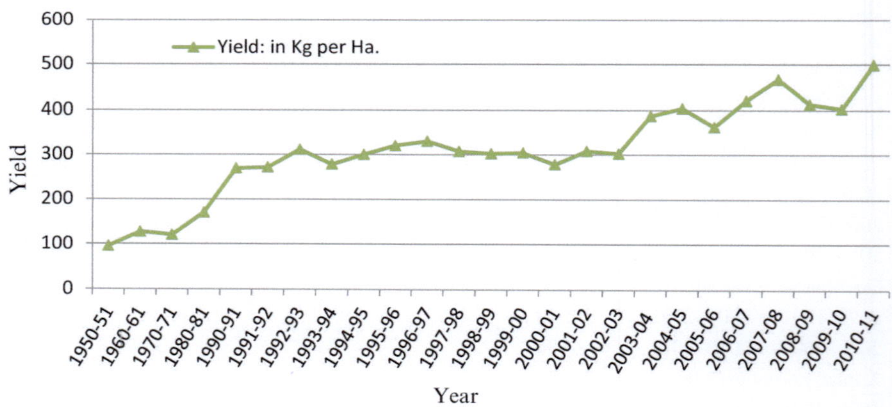


**Fig. 2.2** Yield of cotton in India (kg/ha)

## 2.4 Geographical Distribution of Area and Production of Cotton

Based on cotton production during the triennium ending 2007–2008, Gujarat ranks at the top with a share of 36 %, followed by Maharashtra with 17.8 % and Andhra Pradesh with 13.2 % (Table 2.4). Tamil Nadu has a share of only 1.86 % in the national production. Together, Gujarat, Maharashtra, Andhra Pradesh and Tamil Nadu accounted for 69 % of the cotton production in India in the triennium ending 2007–2008. In terms of area under cotton, Maharashtra occupies the top position

**Table 2.4** Area, production and yield of cotton in major states of India: triennium ending 2007–2008

| | State | Production (lakh bales of 170 kg) | Area in lakh Ha | Yield in kilogram,/ha |
|---|---|---|---|---|
| 1 | Maharashtra | 49.33 | 30.50 | 273.28 |
| | | (17.75) | (33.22) | (9) |
| 2 | Gujarat | 100.00 | 23.38 | 729.82 |
| | | (35.97) | (25.36) | (1) |
| 3 | Andhra Pradesh | 36.67 | 10.35 | 590.27 |
| | | (13.19) | (11.27) | (4) |
| 4 | Madhya Pradesh | 19.00 | 56.34 | 509.33 |
| | | (6.83) | (6.90) | (5) |
| 5 | Haryana | 15.00 | 5.30 | 496.74 |
| | | (5.40) | (5.78) | (6) |
| 6 | Punjab | 23.33 | 6.04 | 666.31 |
| | | (8.39) | (6.58) | (3) |
| 7 | Karnataka | 6.83 | 3.99 | 294.85 |
| | | (2.46) | (4.34) | (8) |
| 8 | Rajasthan | 9.33 | 3.97 | 400.40 |
| | | (3.36) | (4.32) | (7) |
| 9 | Tamil Nadu | 5.17 | 1.27 | 691.83 |
| | | (1.86) | (1.38) | (2) |
| 10 | Others | 1.00 | 0.64 | 273.15 |
| | | (0.48) | (0.76) | (8) |
| | Total | 298.37 | 91.80 | 514.01* |
| | | (100) | (100) | |

Figures in brackets under Production and Area represent the percentage to total production and area respectively. Figures in brackets under Yield shows their respective ranks in yield. States have been arranged in descending order of total area under cotton

with a share of 33.2 % in the 9.2 million hectares of area under cotton cultivation in the country, followed by Gujarat with 25.36 % and Andhra Pradesh with 11.3 % during triennium ending 2007–2008. The average yield of cotton is highest in the state of Gujarat with 729.82 kg per hectare followed by Tamil Nadu with 691.83 kg per hectare. However, the average yield of cotton in Maharashtra at 273 kg per hectare is one of the lowest as against 514 kg per hectare for the country as a whole.

## 2.5 Recent Performance of Cotton Cultivation in the Selected States

Andhra Pradesh shows a huge acceleration in production rate of growth from 3.76 % in 1990–1991 to 2001–2002 to 13.66 % in 2002–2003 to 2010–2011. This shows a huge impact of Bt cotton technology introduction. The growth rate during 2002–2003 to 2010–2011 on account of area and yield were at 8.83 % and 4.81 % respectively

**Table 2.5** Growth in production, area and yield of cotton in Andhra Pradesh

| Year | Area (lakh hectares) | Production (lakh bales) | Yield (kg per hectare) |
|---|---|---|---|
| 1990–1991 | 6.55 | 11.1 | 288 |
| 1991–1992 | 7.06 | 12.99 | 313 |
| 1992–1993 | 8.05 | 11.47 | 242 |
| 1993–1994 | 7.28 | 13.49 | 315 |
| 1994–1995 | 8.45 | 14.26 | 287 |
| 1995–1996 | 10.59 | 16.1 | 258 |
| 1995–1997 | 10.15 | 18.78 | 315 |
| 1997–1998 | 9.06 | 13.2 | 248 |
| 1998–1999 | 12.81 | 15.22 | 202 |
| 1999–2000 | 10.46 | 15.79 | 257 |
| 2000–2001 | 10.22 | 16.63 | 277 |
| 2001–2002 | 11.08 | 18.77 | 288 |
| 2002–2003 | 8.03 | 10.86 | 230 |
| 2003–2004 | 8.37 | 18.9 | 384 |
| 2004–2005 | 11.78 | 21.9 | 316 |
| 2005–2006 | 10.33 | 21.08 | 347 |
| 2006–2007 | 9.72 | 21.82 | 382 |
| 2007–2008 | 11.34 | 34.91 | 523 |
| 2008–2009 | 13.99 | 35.69 | 434 |
| 2009–2010 | 14.68 | 32.32 | 374 |
| 2010–2011 | 17.76 | 38.9 | 372 |
| Coefficient of variation | | | |
| 1990–1991 to 2001–2002 | 20.24 | 17.08 | 12.48 |
| 2002–2003 to 2010–2011 | 27.06 | 36.05 | 21.33 |
| Annual compound growth rate (%) | | | |
| 1990–1991 to 2001–2002 | 4.87 | 3.76 | −1.10 |
| 2002–2003 to 2010–2011 | 8.83 | 13.66 | 4.81 |

to achieve the overall production growth rate of over 13.66 % (Table 2.5 and Figs. 2.3, 2.4, and 2.5).

Gujarat has experienced high variability in cotton production during the 1990s largely due to yield fluctuation (Table 2.6 and Figs. 2.3, 2.4, and 2.5). The state experienced a decline in the growth rate (−3.62) in yield during 1990–1991 to 2001–2002. Since 2002–2003, the growth in cotton production in the state was been phenomenal, and the state's annual rate of growth of production shot up from 1.54 % to 17.14 % comparing 1990–1991 to 2001–2002 and 2002–2003 to 2010–2011. This shows a huge impact of Bt cotton technology introduction.

Maharashtra accounts for the largest share of area under cotton in the country but with lowest levels of cotton yield in the country. Despite low levels of yield the area, under cotton in the state has been the largest over the years. The state achieved 3.6 % growth rate in cotton production during 1990–1991 to 2002–2003, but during the

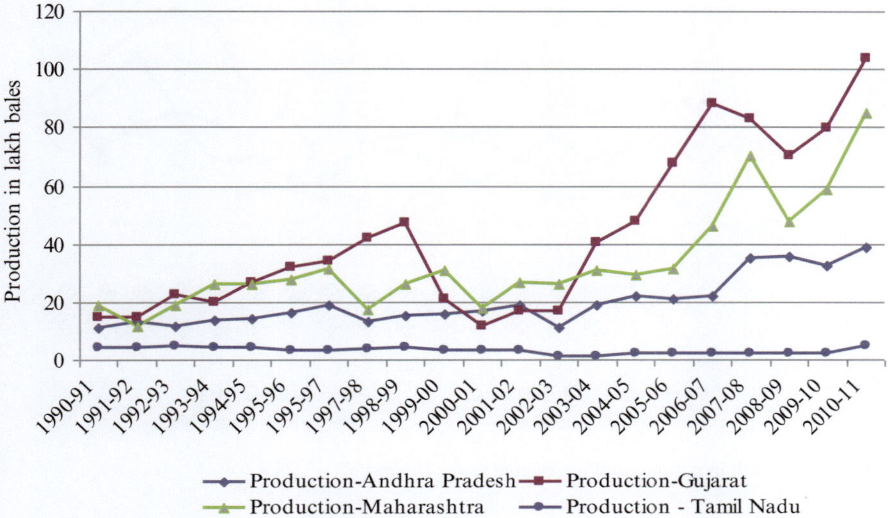


**Fig. 2.3** Growth in cotton production in selected states (lakh bales)

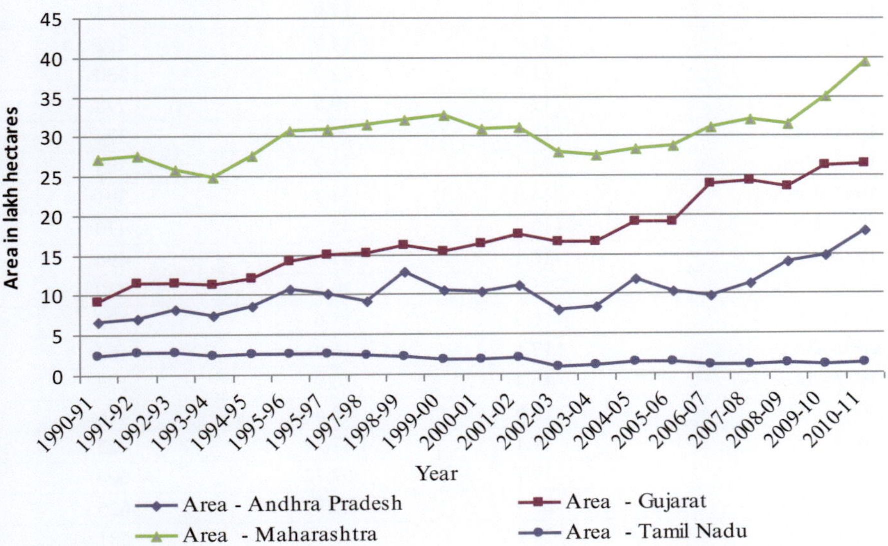


**Fig. 2.4** Growth in area under cotton in selected states (lakh hectares)

period 2002–2003 to 2010–2011 the state achieved an annual growth in cotton production of 14.04 % largely due to growth in yield (Table 2.7 and Figs. 2.3, 2.4, and 2.5). This shows a huge impact of Bt cotton technology introduction.

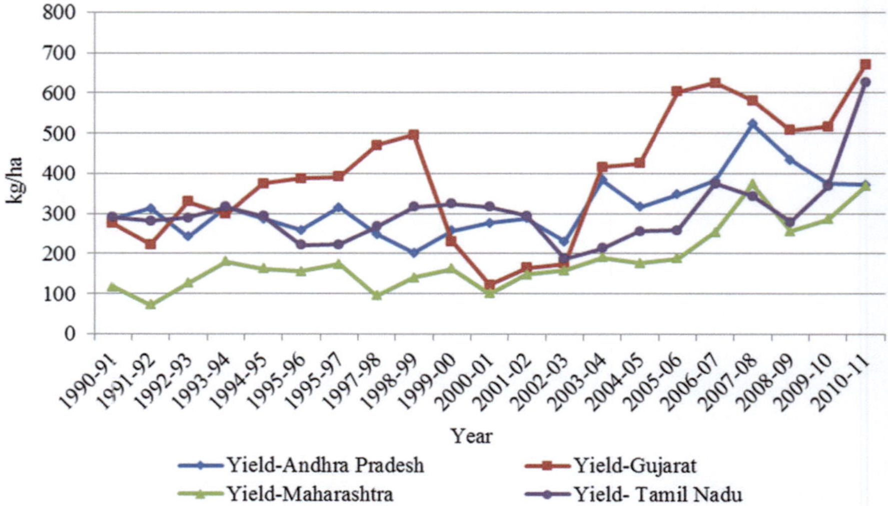


**Fig. 2.5** Growth in cotton yield in selected states (kg/ha)

**Table 2.6** Growth in production, area and yield of cotton in Gujarat

| Year | Area (lakh hectares) | Production (lakh bales) | Yield (kg per hectare) |
|---|---|---|---|
| 1990–1991 | 9.2 | 14.9 | 275 |
| 1991–1992 | 11.4 | 14.9 | 222 |
| 1992–1993 | 11.5 | 22.3 | 330 |
| 1993–1994 | 11.3 | 19.8 | 298 |
| 1994–1995 | 12.1 | 26.6 | 374 |
| 1995–1996 | 14.1 | 32.2 | 388 |
| 1995–1997 | 14.9 | 34.3 | 391 |
| 1997–1998 | 15.2 | 42 | 470 |
| 1998–1999 | 16.1 | 47 | 496 |
| 1999–2000 | 15.4 | 20.8 | 230 |
| 2000–2001 | 16.2 | 11.6 | 122 |
| 2001–2002 | 17.5 | 16.9 | 164 |
| 2002–2003 | 16.4 | 16.9 | 175 |
| 2003–2004 | 16.5 | 40.4 | 416 |
| 2004–2005 | 19.1 | 47.72 | 425 |
| 2005–2006 | 19.1 | 67.72 | 603 |
| 2006–2007 | 23.9 | 87.87 | 625 |
| 2007–2008 | 24.22 | 82.76 | 581 |
| 2008–2009 | 23.53 | 70.13 | 507 |
| 2009–2010 | 26.25 | 79.86 | 517 |
| 2010–2011 | 26.33 | 104 | 671 |
| Coefficient of variation | | | |
| 1990–1991 to 2001–2002 | 18.60 | 44.86 | 37.19 |
| 2002–2003 to 2010–2011 | 18.24 | 40.62 | 29.92 |
| Annual compound growth rate (%) | | | |
| 1990–1991 to 2001–2002 | 5.17 | 1.54 | −3.62 |
| 2002–2003 to 2010–2011 | 6.57 | 17.14 | 10.57 |

**Table 2.7** Growth in production, area and yield of cotton in Maharashtra

| | Area (lakh hectares) | Production (lakh bales) | Yield (kgper hectare) |
|---|---|---|---|
| 1990–1991 | 27.21 | 18.75 | 117 |
| 1991–1992 | 27.59 | 11.56 | 71 |
| 1992–1993 | 25.74 | 18.9 | 125 |
| 1993–1994 | 24.81 | 26.26 | 180 |
| 1994–1995 | 27.6 | 26.25 | 162 |
| 1995–1996 | 30.64 | 27.99 | 155 |
| 1996–1997 | 30.85 | 31.43 | 173 |
| 1997–1998 | 31.39 | 17.53 | 95 |
| 1998–1999 | 31.99 | 26.19 | 139 |
| 1999–2000 | 32.54 | 30.99 | 162 |
| 2000–2001 | 30.8 | 18 | 99 |
| 2001–2002 | 31 | 26.9 | 148 |
| 2002–2003 | 28 | 26 | 158 |
| 2003–2004 | 27.6 | 30.8 | 190 |
| 2004–2005 | 28.4 | 29.4 | 176 |
| 2005–2006 | 28.8 | 31.6 | 187 |
| 2006–2007 | 31.1 | 46.2 | 253 |
| 2007–2008 | 32 | 70.2 | 373 |
| 2008–2009 | 31.5 | 47.5 | 256 |
| 2009–2010 | 35 | 58.6 | 285 |
| 2010–2011 | 39.3 | 85 | 368 |
| Coefficient of variation (%) | | | |
| 1990–1991 to 2001–2002 | 8.86 | 26.57 | 25.30 |
| 2002–2003 to 2010–2011 | 12.24 | 43.42 | 32.25 |
| Annual compound growth rate (%) | | | |
| 1990–1991 to 2001–2002 | 2.00 | 3.60 | 1.61 |
| 2002–2003 to 2010–2011 | 3.97 | 14.04 | 10.06 |

Tamil Nadu had a negative growth rate in production and area during 1990–1991 to 2001–2001 but achieved a 13.09 % growth rate in 2002–2003 to 2010–2011 positive growth in cotton production mainly due to yield growth (Table 2.8 and Figs. 2.3, 2.4, and 2.5). This again shows a huge impact of Bt cotton technology introduction.

**Table 2.8** Growth in production, area and yield of cotton in Tamil Nadu

| Year | Area (lakh hectares) | Production (lakh bales) | Yield (kg per hectare) |
|---|---|---|---|
| 1990–1991 | 2.39 | 4.09 | 290 |
| 1991–1992 | 2.64 | 4.37 | 281 |
| 1992–1993 | 2.67 | 4.54 | 289 |
| 1993–1994 | 2.29 | 4.26 | 316 |
| 1994–1995 | 2.55 | 4.40 | 293 |
| 1995–1996 | 2.61 | 3.39 | 221 |
| 1996–1997 | 2.52 | 3.30 | 222 |
| 1997–1998 | 2.28 | 3.58 | 267 |
| 1998–1999 | 2.19 | 4.06 | 316 |
| 1999–2000 | 1.78 | 3.40 | 324 |
| 2000–2001 | 1.70 | 3.17 | 317 |
| 2001–2002 | 1.88 | 3.26 | 295 |
| 2002–2003 | 0.76 | 0.84 | 188 |
| 2003–2004 | 0.98 | 1.23 | 213 |
| 2004–2005 | 1.29 | 1.95 | 256 |
| 2005–2006 | 1.41 | 2.13 | 258 |
| 2006–2007 | 1.00 | 2.21 | 374 |
| 2007–2008 | 0.99 | 2.01 | 344 |
| 2008–2009 | 1.15 | 1.88 | 279 |
| 2009–2010 | 1.04 | 2.25 | 368 |
| 2010–2011 | 1.22 | 4.50 | 627 |
| Coefficient of variation | | | |
| 1990–1991 to 2001–2002 | 14.92 | 13.46 | 12.05 |
| 2002–2003 to 2010–2011 | 17.84 | 48.01 | 40.73 |
| Annual compound growth rate (%) | | | |
| 1990–1991 to 2001–2002 | –3.51 | –2.90 | 0.63 |
| 2002–2003 to 2010–2011 | 2.48 | 13.09 | 11.53 |

# Chapter 3
# Development and Adoption of Bt Cotton

## 3.1 Introduction

Bt cotton gets its name from a bacterium called *Bacillus thuringiensis*. Bt cotton contains a foreign gene obtained from *Bacillus thuringiensis,* which is an aerobic bacterium, a natural enemy of boll worms, characterized by its ability to produce crystalline inclusions during sporulation. This bacterium was first discovered by a Japanese bacteriologist in 1901 and subsequently in 1915 a German scientist isolated crystal toxin in Thuringen region of Germany. *B. thuringiensis* was registered as a microbial pest control agent in 1961 under federal Insecticide and Rodenticide Act in the US. In India, Bt formulations have been registered under Pesticides Act, 1968. With the advent of biotechnology, the bacterial gene was introduced genetically into the cotton genome, and it protects the plants from bollworms, the major pest of cotton. The worms feeding on the leaves of a Bt cotton plant become lethargic and sleepy, and are finally eliminated.

The first Bt cotton varieties were introduced commercially through a licensing agreement between the gene discoverer, Monsanto, and the leading American cotton germplasm firm, Delta and Pine Land Company (D&PL). These varieties contain the Cry1Ac gene and are commercialized under the trade name Bollgard®. Varieties with transgenes for insect resistance and herbicide tolerance (Bt/HT) stacked together were introduced in the United States in 1997. Further, Monsanto received regulatory approval in some markets for a new product that incorporates two Bt genes, Cry1Ac and Cry2Ab2. This product, known as Bollgard II®, was commercialized in 2003. The incorporation of two Bt genes is believed to improve the effectiveness of the product and delay the development of resistant pests.

The chronological progress of field trials and the adoption of Bt cotton across countries are given in Tables 3.1 and 3.2. The commercial cultivation of Bt cotton has been on in the United States of America, Australia and Mexico since 1996 and in China and South Africa with a lag of 1 year. Countries such as India, Indonesia




V.P. Gandhi, D. Jain, *Introduction of Biotechnology in India's Agriculture*, DOI 10.1007/978-981-10-1091-0_3

**Table 3.1** Progress of field trials and commercial releases of Bt cotton around the world

| | |
|---|---|
| Argentina | Approved in 2001. By 2005–2006 sown on around 13 % of the total cotton area. The level of adoption rose to 85 % by 2011–2012 |
| Australia | Introduced in 1996. By 2002–2003 accounts for around 30 % of total cotton crop. This increases to 80 % in 2004–2005 with the release of Monsanto's Bollgard II variety. By 2011–2012, the level of adoption has reached 90 % |
| Brazil | Field trials approved in March 2005. Smuggling of Bt cotton seeds from Argentina and Paraguay is widespread. At least 5 % of the 1.3 million tons produced in the 2005–2006 season comes from "black market" Bt varieties. Brazil started commercial cultivation and production of Bt Cotton on wide scale in 2006–2007 |
| Burkina Faso | About 8500 ha of area was planted for first time with Bollgard ® II Cotton in 2008. Acreage increased to 115,000 and 247,000 ha in 2009 and 2010 respectively |
| China | Released in 1997. Now planted on well over half of the national cotton area |
| Colombia | Imported by Monsanto in 2002, without environmental clearance. Legal action results in the suspension of the authorization. Commercial cultivation finally started in 2003–2004 and by 2011–2012 its level of adoption had already reached 85 % |
| Costa Rica | Monsanto began field trials without regulatory oversight in 1992. By 2004, 638 ha area has been planted, mainly for the export of seeds |
| Egypt | Monsanto and Egypt's Agriculture Genetic Engineering Research Institute collaborated in field trials of Bt cotton. They claim commercial introduction could take place as early as 2006 |
| India | Commercial introduction in 2002. By 2004, Bt cotton accounts for 6 % of total cotton area and is only permitted for cultivation in six states. In 2006–2007, Monsanto begins sales of Bollgard II. By year 2010–2011, over 800 varieties of BT cotton have been approved for cultivation in India. By 2011–2012, 11 million hectares of land (85 % of total cotton area) BT Cotton, with, grown with BT cotton, which accounts for 47 % of the world total |
| Indonesia | Introduced in South Sulawesi province in 2001. Two years later it is withdrawn after its failure to perform triggers farmer protests |
| Kenya | Monsanto imports Bt cotton into Kenya in 2004 for field trials |
| Myanmar | Official Production stated in the year 2010–2011 |
| Mexico | Bt cotton introduced in 1996. Government subsidizes purchase of Bt cotton seeds. In2002–2003, 25 % of the national cotton area planted to Bt cotton, slightly less than the percentage in 2000 |
| Pakistan | In May 2005 the Pakistan Atomic Energy Commission provides 40,000 kg of Bt cotton seed to farmers in the Punjab. Official production stated in year 2009–2010 |
| Paraguay | In July 2005, The Minister of Agriculture announces that it will approve Monsanto's GM cotton as part of a joint project with the company |
| Philippines | In January 2005, the Cotton Development Authority signs a memorandum of agreement with the Philippine Rice Research Institute to begin field trials of Bt cotton |
| Senegal | Irregular field trials later abandoned |

(continued)

**Table 3.1** (continued)

| | |
|---|---|
| South Africa | Bt cotton approved for commercial planting in 1997. Adoption very rapid and by 2002, an estimated 75 % of national cotton area planted to GM cotton. In 200304 only 35,700 ha of cotton was planted, an 80 % reduction since 2000, ascribed to low world prices and droughts. In 2004–2005 the area planted was 21,700 ha, an extraordinary 40 % drop in area planted to cotton in 1 year |
| Thailand | Field tests in 1997. Abandoned after mass protests |
| USA | Over 80 % of the cotton area in the US is Bt cotton. Studies show reduction in pesticide use since Bt cotton introduced in 1996, but now secondary pests are becoming an increasing problem |

Derived from International Cotton Advisory Committee (2012) & http://grain.org/go/btcotton

**Table 3.2** Adoption of Bt cotton in major cotton growing countries

| Country | 1996 | 1997 | 1998 | 1999 | 2000 | 2001 | 2002 | 2003 | 2004 | 2005 | 2006 | 2008 | 2010 |
|---|---|---|---|---|---|---|---|---|---|---|---|---|---|
| USA | √ | √ | √ | √ | √ | √ | √ | √ | √ | √ | √ | √ | √ |
| Australia | √ | √ | √ | √ | √ | √ | √ | √ | √ | √ | √ | √ | √ |
| Brazil | | | | | | | | | | | √ | √ | √ |
| Burkina Faso | | | | | | | | | | | | √ | √ |
| China | | √ | √ | √ | √ | √ | √ | √ | √ | √ | √ | √ | √ |
| India | | | | | | | √ | √ | √ | √ | √ | √ | √ |
| Indonesia | | | | | | | √ | √ | | | | | |
| Myanmar | | | | | | | | | | | | | √ |
| Mexico | √ | √ | √ | √ | √ | √ | √ | √ | √ | √ | √ | √ | √ |
| Argentina | | | √ | √ | √ | √ | √ | √ | √ | √ | √ | √ | √ |
| Colombia | | | | | | | √ | √ | | | √ | √ | √ |
| South Africa | | √ | √ | √ | √ | √ | √ | √ | √ | √ | √ | √ | √ |
| Costa Rica | | | | | | | | | √ | √ | √ | √ | √ |
| Pakistan | | | | | | | | | | √ | √ | √ | √ |
| Paraguay | | | | | | | | | | √ | √ | √ | √ |

Source: International Cotton Advisory Committee (2012) & www.grain.org/research 2

and Colombia have taken up its commercial cultivation since 2002. The area under Bt cotton, including Bt with herbicides tolerance, has increased from mere 0.8 million hectares during 1996 to over six million hectares by the year 2004 (Tables 3.1, 3.2, and 3.3) (Fig. 3.1).

## 3.2 Review of Literature

The reported potential advantages of Bt cotton include agronomic, economic and environmental. The major agronomic attributes of Bt cotton are improved pest control and yield advantage compared to conventional cotton varieties. The major economic benefits envisaged are reduced use of pesticides and effective yield

**Table 3.3** Global adoption of Bt and HT cotton (million hectares)

| Year | Bt Cotton | Bt and HT Cotton | Total |
|---|---|---|---|
| 1996 | 0.8 | 0.0 | 0.8 |
| 1997 | 1.1 | 0.1 | 1.2 |
| 1998 | 1.4 | 0.1 | 1.5 |
| 1999 | 1.3 | 0.8 | 2.1 |
| 2000 | 1.5 | 1.7 | 3.2 |
| 2001 | 1.9 | 2.4 | 4.3 |
| 2002 | 2.4 | 2.2 | 4.6 |
| 2003 | 3.1 | 2.6 | 5.7 |
| 2004 | 3.3 | 3.0 | 6.3 |
| 2010 | 16.1 | 4.9 | 21.0 |
| 2011 | 17.9 | 6.7 | 24.6 |

Source: James C (2003, 2011). HT stands for herbicide tolerant, also developed through biotechnology

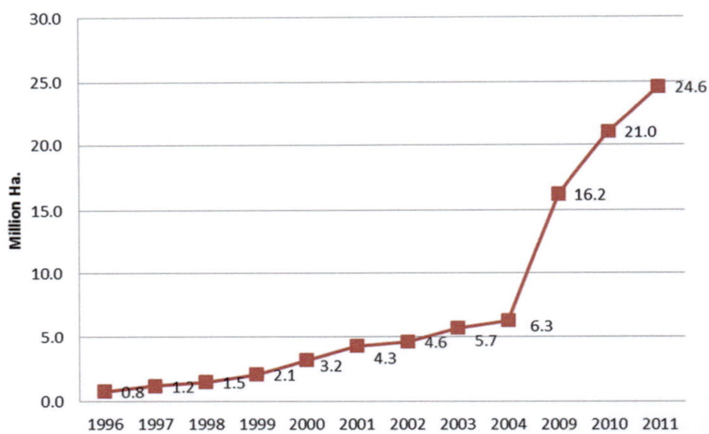


**Fig. 3.1** Global adoption of Bt/HT cotton

superiority over non Bt cotton. Major environmental benefits include reduction in number of insecticides spray, less insecticide in soils and aquifers, less exposure to pesticides for human beings and animals, and increase in the population of beneficial insects. These issues are reviewed below based on various studies conducted in India and elsewhere in the world.

### *3.2.1 Pesticide Use*

A major agronomic attribute of Bt cotton over conventional cotton is its high level of resistance to the bollworm complex. As a result, the need to use insecticides gets greatly reduced since the use of insecticides against bollworms is very high for the conventional cotton hybrids/ varieties. But there are conflicting views on these counts. Data based on field trials from a number of countries indicate that Bt cotton reduces the need of pesticides from seven sprays to two or three sprays (James 2002). Survey of Bt cotton in China during 1999–2001 period showed that on an average the incidence of insecticides poisoning in farmers using Bt cotton is four times less than farmers using conventional varieties (Pray et al. 2001). Growers in the US are reported to have reduced insecticides use by 1.9 million pounds of active ingredient per year since 2001 (Gianessi et al.). It is reported that in China the insecticide application in cotton was reduced by 67 % (Huang et al. 2002).

But the field level observations from various parts of India are mixed. Some observe that since Bt cotton does not offer protection against pink boll worm, it is essential to spray pesticides at almost the same level as for non-Bt (Sahai and Rahman 2003). However, quite a few studies have found that there was significant reduction in the use of pesticides on Bt cotton as compared to non-Bt cotton (Sharma et al. 2002). A study carried out in four states of India during the first season of Bt cotton adoption shows that the Bt technology leads to significant pesticides reduction (Naik et al. 2005). Around 70 % of the farmers in Andhra Pradesh who have used Bt Cotton varieties responded favorably to it indicating that the variety is resistant to pests. The Indian Council of Agricultural Research (ICAR) indicates that about 65 % of the insecticide used in cotton production is to tackle the menace of bollworms, and if the genetically modified (GM) varieties are resistant to the pest, their cultivation must be encouraged (Statement of ICAR Director General quoted in Indian Express 2003).

Others indicate that the variety is susceptible to the bollworm and the yield is below par. A study (Venkateshwarlu 2002) conducted in 11 villages of Warangal district in AP, indicates that non-Bt cotton produced 30 % more and there is only a marginal difference in the pesticides use. It found that farmers sprayed pesticides four to six times in Bt, and five to seven times on non-Bt cotton. Bt farmer had to pay Rs.1, 150 more towards the purchase of seed. Besides, the labour charges are stated to be about Rs. 150 more for picking Bt cotton. The price of Bt cotton was reported to be 10 % less in the local market (Business Line 2002). The study indicated that Bt Cotton has failed on many counts and the claims made by the company were wrong. It neither improved yield through better plant protection nor reduced the pesticide usage and the returns were less since the pods were small, seeds were more, lint and the staple length were less (Venkateshwarlu 2002).

In some cases, it was reported that the new pests and diseases emerged, and Bt cotton failed to prevent even the boll worm attack. The economics that was worked out by the Indian Council of Agricultural Research (ICAR), Genetic Engineering Approval Committee and Monsanto-Mahyco are questioned. Bt cotton was also afflicted with the 'leaf curl virus' in the northern states of India. In Maharashtra, the Bt cotton crop in Vidarbha was been badly affected by the root-rot disease. In Gujarat heavy infestation of bollworm on the Bt cotton was reported in the districts of Bhavanagar, Surendranagar and Rajkot. Some reports indicated that initially Bt Cotton showed resistance to boll worms but as soon as the formation of bolls started, the worms started attacking them (RFSTE 2002). The above literature indicates that the opinions in the context of the resistance of Bt cotton to pests are divergent and require investigation.

### 3.2.2 Cost of Production and Yield

It was generally believed that significant decline in the use of pesticides would reduce the total cost of cultivation. But it may not be so mainly due to high cost of Bt seed (Iyengar and Lalitha 2002). Besides, the use of yield increasing inputs is relatively high in Bt cotton and so the total cost of cultivation is found to be relatively high not only in India but also elsewhere in the world (Financial Express 2003). However, a study in China for the years 1999–2001 showed that even though the cost of seed was greater for Bt cotton, this was offset by a reduction in pesticides cost and a reduction in labour cost because Bt cotton farmers do not have to spend as much time spraying pesticides (Pray et al. 2002). The positive impacts of Bt cotton on yield were reported from various parts of the world (Chaturvedi 2002; Pray et al. 2001). Significant yield gains by Bt cotton were reported from Maharashtra, Karnataka and Andhra Pradesh in India during the year 2002 (Naik et al. 2005). The net benefit of Bt cotton over non-Bt cotton was found to be around Rs. 7000 per acre mainly due to increase in yield (Thomas 2002; www.kisanwatch.org). The gross margin for Bt cotton was substantially higher in case of Bt cotton in Maharashtra, Karnataka and Tamil Nadu (Naik et al. 2005). The net benefit from Bt cotton were reported to be higher in US, China, and South Africa (Pray et al. 2001; Dong et al. 2004). Thus majority of the studies mentioned here are by and large of the opinion that Bt cotton does have effective yield and profit superiority compared to non-Bt cotton.

### 3.2.3 Bt Cotton Seed Price

The prices of Bt seeds were almost three times that of non-Bt seeds and this has been a major issue in several parts of the country. The governments of Andhra Pradesh, Maharashtra and Gujarat, which constitute the cotton belt of India, have recently directed the sub-licensees of Monsanto not to charge more than Rs 750 per 450 g packet of Bt cotton seed. After the Andhra Pradesh State Government referred the matter to the Monopolies and Restrictive Trade Practices Commission (MRTPC), the seed prices have been slashed to Rs. 750 per packet. The Mahyco-Monsanto Biotech India Limited (MMBL) after discussing with the seed companies brought down the trait value for Bt to Rs. 150 in Kharif 2006. The MMBL and seed companies decided to amend the relevant clauses of the agreement to avoid any issue vis-à-vis the MRTP Act. The intervention by the three states led to an injunction passed by MRTPC on May 11, 2006 directing Monsanto to reduce its trait value of Rs 900 per 450 g of Bt cotton seed to the level it charges in China, which works out to as little as Rs 40 for the same quantity. The price fixed by the states for the 2006–2007 season was at least 50 % less than the price at which the same seed was sold in the previous year – on account of the technology fee or "trait value" charged by Monsanto.

### 3.2.4 Environmental Considerations

Significant decrease in the number of insecticides sprays for the control of the major Lepidopteron insect pests – the bollworm – should substantially reduce the environmental hazards due to high toxicity of the insecticides. Lesser farmer exposure to insecticides would reduce health implications. The reduction in the use of insecticides will also reduce the risks to mammals, birds, bees, fish and other organisms (USEPA 2001). No systematic study has reported any direct adverse impact of Bt cotton on the environment.

### 3.2.5 Voices for and Against Bt Cotton

Many voices have been raised for and against Bt cotton in India since the mid-2000s.

Advantages of Bt Cotton have been reported to be:

- Reduction in the use of insecticides by almost 50 %.
- Reduction in the use of insecticides and hence reduction in the harmful effect on the environment, including soil, water, atmosphere and life.
- The quality of cotton fibre is at par with that of non-Bt cotton.
- Better yield per unit of input use.
- Reduction in the use of insecticides favours building up of population of beneficial insect pests.

  Lesser residue of pesticides in the fibre produced which reduces the chances of harmful effects such as allergic reactions and so on.

The criticisms and fears that have been voiced against Bt Cotton include:

- The gene may spread and its impact in the eco-system is not known
- Transgenic crop varieties will lead to the destruction of the native crop of the country (Varieties as used in this study include varieties and hybrids).
- The Bt Cotton seed will be very expensive compared to Non-Bt seeds for the farmers and some companies may have a monopoly in seed multiplication and sales
- Even on Bt cotton the farmers may require to use insecticides same as that of non-Bt cotton
- The Bt cotton seed cake produced may cause harm to the animals and the toxin may enter the human food chain
- Farmers will have to purchase Bt cotton seeds every year
- High likelihood that pests may soon develop resistance to Bt gene and hence the control of boll worm will become even more difficulty in the near future

In view of these diverse views and considering the importance of cotton in Indian agriculture, it was considered worthwhile to undertake a systematic study examining the returns and other significant aspects of Bt cotton as opposed to non-Bt cotton.

## 3.3 Adoption of Bt Cotton in India

After much hesitation and delay, Bt cotton was approved by the Government of India for commercial cultivation in India in 2002. After the unauthorized appearance of Bt Cotton in Gujarat in 2001, in March 2002 the Genetic Engineering Approval Committee (GEAC), the regulatory authority of the Government of India for transgenic crops approved the commercial cultivation of three Bt cotton varieties: Bt Mech 12, Bt Mech 162 and Bt Mech 184. These varieties were developed by Monsanto in collaboration with its Indian partner the Maharashtra Hybrid Seeds Company (MAHYCO) and were meant for commercial cultivation in central and southern India. Later, the GEAC approved large scale field trials and seed production of 12 more varieties of Bt cotton in 2005. While MAHYCO is Monsanto's partner in India, Rasi Seeds and Ankur Seeds are sub-licensees of Monsanto. Ankur Seeds was given the green signal to conduct large scale field trials and seed production of Ankur 651 Bt and Ankur 2354 Bt in North India, and Ankur 651 Bt and Ankur 09 Bt in Central India. In 2005, RCH 2 Bt became the fourth transgenic cotton variety to be approved for commercial cultivation in the country. By year 2010–2011, over 800 varieties of BT cotton have been approved for cultivation in India.

Gujarat and Maharashtra were the early adopters of Bt cotton in the country that commenced in 2002 followed by Andhra Pradesh and Karnataka (Table 3.4). By official estimates, the area under Bt cotton in India was about 1 million hectare, or about 11 % of the total area under cotton in the country in 2005 which has increased to about 112 million hectare by 2010–2011, or about 58% of the total area under cotton in the country. As of 2005, as per official statistics, the share of area under Bt cotton to total area under cotton was over 27 % in Madhya Pradesh, about18 % in Maharashtra. These statistics do not include the substantial un-reported area in the

**Table 3.4** Growth in area under Bt cotton in India ('000 ha)

| | | | | | Area under Bt cotton as percent of total area under cotton | | | |
|---|---|---|---|---|---|---|---|---|
| State | 2003 | 2004 | 2005 | 2010 | 2003 | 2004 | 2005 | 2010 |
| Andhra Pradesh | 5.46 | 71.22 | 90.41 | 1701 | 0.65 | 6.07 | 9.30 | 95.35 |
| Madhya Pradesh | 13.35 | 86.12 | 136.21 | 600 | 2.26 | 14.95 | 21.45 | 92.31 |
| Gujarat | 41.68 | 125.92 | 149.25 | 2088 | 2.53 | 6.61 | 7.19 | 79.30 |
| Maharashtra | 21.85 | 161.47 | 508.67 | 3543 | 0.79 | 5.42 | 17.61 | 90.11 |
| Karnataka | 3.04 | 34.30 | 29.34 | 373 | 0.97 | 6.70 | 8.08 | 68.44 |
| Tamil Nadu | 7.69 | 11.99 | 17.02 | 62 | 7.46 | 8.45 | 11.34 | 50.82 |
| Punjab | Neg*. | Neg. | 70.42 | 510 | Neg. | Neg. | 12.14 | 96.23 |
| Haryana | Neg. | Neg. | 10.77 | 418 | Neg. | Neg. | 1.80 | 84.96 |
| Rajasthan | Neg. | Neg. | 2.31 | 260 | Neg. | Neg. | 0.51 | 77.61 |
| Total | 93.08 | 491.02 | 1014.40 | 9555 | 1.22 | 5.50 | 11.51 | 85.01 |

Based on official statistics

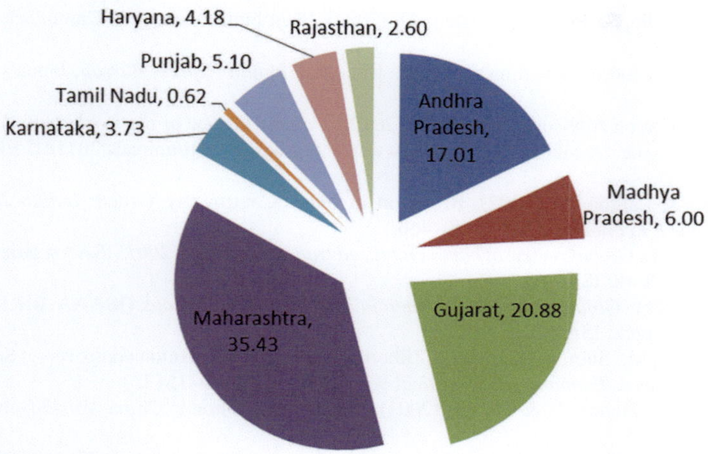


**Fig. 3.2** Distribution of area under Bt cotton in various states – 2010

**Table 3.5** Commercial cultivation of Bt cotton hybrids in India, 2002 (hectares)

| State | MECH-12 | MECH-162 | MECH-184 | Total |
|---|---|---|---|---|
| Maharashtra | 112 | 9300 | 5334 | 14,746 |
| Madhya Pradesh | 60 | 404 | 1756 | 2220 |
| Karnataka | 0 | 3828 | 80 | 3908 |
| Andhra Pradesh | 44 | 5564 | 0 | 5608 |
| Gujarat | 76 | 4136 | 4642 | 8854 |
| Tamil Nadu | 0 | 2042 | 660 | 2702 |
| Total | 292 | 25,274 | 12,472 | 38,038 |

Source: R.B. Barwale, V.R. Gadwal, Usha Zehr, and Brent Zehr, Prospects for Bt Cotton Technology in India, AgBioForum, Vol. 7, 2004

state of Gujarat. Official statistics also suggest that by 2010–2011, over 85 % of total cotton area in India, over 90 % area under cotton in states of Andhra Pradesh, Madhya Pradesh, Maharashtra and Punjab, and more than 80 % in Gujarat is under Bt cotton cultivation (Fig. 3.2 and Table 3.5).

## References

Chaturvedi, S. (2002). Agricultural biotechnology and new trends in IPR regime: Challenges before developing countries. *Economic and Political Weekly, 37*(13), 1212–1222.

Dong, H., Li, W., Tang, W., & Zhang, D. (2004). Development of hybrid Bt cotton in China – A successful integration of transgenic technology and conventional techniques. *Current Science, 86*(6), 778–782.

Financial Express. (2003). Regulator says no to Bt cotton cultivation. *Financial Express*, 25 April 2003.

Huang, J., Rozelle, S., Pray, C., & Wang, Q. (2002). Plant biotechnology in China. *Science, 295*, 674–677.
Indian Express. (2003). Government defends Bt cotton, *Indian Express Bureau*, January 15, New Delhi.
International Cotton Advisory Committee. (2012). *Cotton: Review of the world situation*. 65(5), May-July Issue. Available at https://www.icac.org/wp-content/uploads/2011/05/ERev_june.pdf
Iyengar, S., & Lalitha, N. (2002). Bt cotton in India: Controversy visited. *Indian Journal of Agricultural Economics, 57*(3), 459–466.
James, C. (2002). *Global review of commercialized transgenic crops: 2001* (ISAAA Brief No. 26). Ithaca/New York: ISAAA.
James, C. (2011). *Global status of commercialized biotech/GM crops* (ISAAA Brief No. 43). Ithaca/New York: ISAAA.
Naik, G., Qaim, M., Subramanian, A., & Zilberman, D. (2005). Bt cotton controversy: Some paradoxes explained. *Economic and Political Weekly, 40*(15), 1514–1517.
Pray, C., Ma, D., Huang, J., & Qiao, F. (2001). Impact of Bt cotton in China. *World Development, 29*(5), 813–825.
Pray, C. E., Huang, J., Hu, R., & Rozelle, S. (2002). Five years of Bt cotton in China – The benefits continue. *The Plant Journal, 31*, 423–430.
RFSTE. (2002). Failure of Bt cotton in India. *Research foundation for science, technology and ecology*.
Sahai, S., & Rahman, S. (2003). Performance of Bt cotton: Data from first commercial crop. *Economic and Political Weekly, 38*(30), 3139–3141.
Sharma, H. C., Crouch, J. H., Sharma, K. K., Seetharama, N., & Hash, C. T. (2002). Applications of biotechnology for crop improvement: Prospects and constraints. *Plant Science, 163*(2002), 381–395.
The Hindu Business Line Bureau. (2002, December 9). Bt cotton lets down AP farmers: Study. *The Hindu Business Line*.
Thomas, E. C. (2002). India adopts Bt cotton. *Yojana, 46*(6), 14–17.
USEPA. (2001. October 15). *Bt plant-incorporated protectants – Biopesticides registration action document*. Washington, DC: U.S. Environmental Protection Agency.
Venkateshwarlu, K. (2002). *The Hindu*, India, December 30, 2002.

# Part II
# Bt Cotton vis-a-vis Non-Bt Cotton Overall in India

# Chapter 4
# Sampling and Methodology of the Study Across the States

## 4.1 Sample Design

The target sample size for each state was 180 cotton farmers comprising of 90 Bt and 90 Non-Bt farmers. Maharashtra proved to be an exception where the coverage was 85 Bt and 69 Non-Bt farmers (Tables 4.1 and 4.2). The Gujarat sample does not have unirrigated cotton as a sufficient number of such farmers were not available in the sample districts. The number of sample farmer households under unirrigated Bt cotton was relatively less in Andhra Pradesh and Maharashtra, but were relatively more in Tamil Nadu. The study had an overall sample size of 694 farm households.

The average size of operational holding did not differ markedly between Bt and Non-Bt but was found to be marginally higher under Bt cotton compared to Non-Bt cotton in Andhra Pradesh, both irrigated and unirrigated (Table 4.3). The same was true for Gujarat and Tamil Nadu. In Maharashtra, the average operated area for both irrigated and unirrigated Bt cotton farmers was significantly greater than that of non-Bt farmers.


V.P. Gandhi, D. Jain, *Introduction of Biotechnology in India's Agriculture*, DOI 10.1007/978-981-10-1091-0_4

**Table 4.1** Sample size

| | Bt cotton | | | Non-Bt cotton | | |
|---|---|---|---|---|---|---|
| Farm size in hectare | Irrigated | Unirrigated | Total | Irrigated | Unirrigated | Total |
| Andhra Pradesh | | | | | | |
| Small (Below 2 Ha) | 24 | 18 | 42 | 24 | 18 | 42 |
| Medium (2–4 Ha) | 18 | 12 | 30 | 18 | 12 | 30 |
| Large (Above 4 Ha) | 12 | 6 | 18 | 12 | 6 | 18 |
| Total | 54 | 36 | 90 | 54 | 36 | 90 |
| Gujarat | | | | | | |
| Small (Below 2 Ha) | 42 | 0 | 42 | 42 | 0 | 42 |
| Medium (2–4 Ha) | 30 | 0 | 30 | 30 | 0 | 30 |
| Large (Above 4 Ha) | 18 | 0 | 18 | 18 | 0 | 18 |
| Total | 90 | 0 | 90 | 90 | 0 | 90 |
| Maharashtra | | | | | | |
| Small (Below 2 Ha) | 8 | 9 | 17 | 9 | 14 | 23 |
| Medium (2–4 Ha) | 14 | 9 | 23 | 5 | 13 | 18 |
| Large (Above 4 Ha) | 26 | 19 | 45 | 7 | 21 | 28 |
| Total | 48 | 37 | 85 | 21 | 48 | 69 |
| Tamil Nadu | | | | | | |
| Small (Below 2 Ha) | 12 | 30 | 42 | 12 | 30 | 42 |
| Medium (2–4 Ha) | 11 | 19 | 30 | 11 | 19 | 30 |
| Large (Above 4 Ha) | 9 | 9 | 18 | 9 | 9 | 18 |
| Total | 32 | 58 | 90 | 32 | 58 | 90 |
| Overall | | | | | | |
| Small (Below 2 Ha) | 86 | 57 | 143 | 87 | 62 | 149 |
| Medium (2–4 Ha) | 73 | 40 | 113 | 64 | 44 | 108 |
| Large (Above 4 Ha) | 65 | 34 | 99 | 46 | 36 | 82 |
| Total | 224 | 131 | 355 | 197 | 142 | 339 |
| Grand total | | | | | | 694 |

**Table 4.2** Sample farm size distribution (percentage)

| Farm size in hectare | Bt cotton | | | Non-Bt cotton | | |
|---|---|---|---|---|---|---|
| | Irrigated | Unirrigated | Total | Irrigated | Unirrigated | Total |
| Andhra Pradesh | | | | | | |
| Small (Below 2 Ha) | 44.44 | 50.00 | 46.67 | 44.44 | 50.00 | 46.67 |
| Medium (2–4 Ha) | 33.33 | 33.33 | 33.33 | 33.33 | 33.33 | 33.33 |
| Large (Above 4 Ha) | 22.22 | 16.67 | 20.00 | 22.22 | 16.67 | 20.00 |
| Total | 100 | 100 | 100 | 100 | 100 | 100 |
| Gujarat | | | | | | |
| Small (Below 2 Ha) | 46.67 | – | 46.67 | 46.67 | – | 46.67 |
| Medium (2–4 Ha) | 33.33 | – | 33.33 | 33.33 | – | 33.33 |
| Large (Above 4 Ha) | 20.00 | – | 20.00 | 20.00 | – | 20.00 |
| Total | 100 | – | 100 | 100 | – | 100 |
| Maharashtra | | | | | | |
| Small (Below 2 Ha) | 16.67 | 24.32 | 20.00 | 42.86 | 29.17 | 33.33 |
| Medium (2–4 Ha) | 29.17 | 24.32 | 27.06 | 23.81 | 27.08 | 26.09 |
| Large (Above 4 Ha) | 54.17 | 51.35 | 52.94 | 33.33 | 43.75 | 40.58 |
| Total | 100 | 100 | 100 | 100 | 100 | 100 |
| Tamil Nadu | | | | | | |
| Small (Below 2 Ha) | 37.50 | 51.72 | 46.67 | 37.50 | 51.72 | 46.67 |
| Medium (2–4 Ha) | 34.38 | 32.76 | 33.33 | 34.38 | 32.76 | 33.33 |
| Large (Above 4 Ha) | 28.13 | 15.52 | 20.00 | 28.13 | 15.52 | 20.00 |
| Total | 100 | 100 | 100 | 100 | 100 | 100 |
| Overall | | | | | | |
| Small (Below 2 Ha) | 38.39 | 43.51 | 40.28 | 44.16 | 43.66 | 43.95 |
| Medium (2–4 Ha) | 32.59 | 30.53 | 31.83 | 32.49 | 30.99 | 31.86 |
| Large (Above 4 Ha) | 29.02 | 25.95 | 27.89 | 23.35 | 25.35 | 24.19 |
| Total | 100 | 100 | 100 | 100 | 100 | 100 |

**Table 4.3** Average operational holding size of Bt and non-Bt cotton farmer households (area in hectares)

| | Bt cotton | | Non-Bt cotton | |
|---|---|---|---|---|
| | I | UI | I | UI |
| Andhra Pradesh | | | | |
| Small | 1.36 | 1.19 | 1.39 | 1.03 |
| Medium | 3.10 | 1.67 | 2.83 | 1.90 |
| Large | 6.52 | 3.88 | 5.13 | 4.27 |
| Overall | 2.92 | 2.59 | 2.62 | 2.49 |
| Gujarat | | | | |
| Small | 1.58 | 0 | 1.28 | 0 |
| Medium | 3.15 | 0 | 3.09 | 0 |
| Large | 8.33 | 0 | 7.49 | 0 |
| Overall | 3.45 | 0 | 3.13 | 0 |
| Maharashtra | | | | |
| Small | 1.38 | 0.64 | 1.62 | 0.52 |
| Medium | 3.45 | 2.13 | 3.40 | 0.72 |
| Large | 9.16 | 3.98 | 8.66 | 2.51 |
| Overall | 6.06 | 2.81 | 4.94 | 1.38 |
| Tamil Nadu | | | | |
| Small | 1.15 | 0.65 | 0.99 | 0.44 |
| Medium | 2.83 | 1.74 | 2.69 | 1.12 |
| Large | 5.05 | 2.04 | 2.83 | 1.70 |
| Overall | 2.49 | 1.52 | 2.30 | 1.00 |

# Chapter 5
# Nature and Performance of Bt Cotton vs. Non-Bt Cotton Across the States

## 5.1 Introduction

In Andhra Pradesh, cotton is a very important crop and stands at the third place in terms of area cultivated. Over the past two and half decades there had been a significant increase in the area under cotton in the state. Cotton has been grown under both irrigated and unirrigated conditions and the latter dominates. Many improved and hybrid varieties of cotton are used by the cultivators. Bt cotton was introduced in the state in 2002. The GEAC permitted Mahyco-Monsanto the commercial release of three Bt cotton varieties namely MECH-12, MECH-184 and RCH-20.

In Gujarat, cotton shares one of the two top positions in terms of cropped area. The total area under cotton in Gujarat during 2005–2006 was about 2.08 million hectares which is roughly about 23 % of the total area under cotton in the country. Almost two-thirds of the area under cotton in the state is under hybrid varieties. The Bt varieties have been spreading fast in the state, both approved and non-approved varieties of Bt cotton.

In Maharashtra, cotton has traditionally been one of the most important crops. Area under cotton in Maharashtra is about 3 million hectares and the state accounts for about 30 % of the cotton area in the country. As against its share in total area, the share in cotton production in the country continued to be low due to its relatively low yields. Although cotton is not a major crop in Tamil Nadu, it plays an important role in the economy of the selected districts in the state.

## 5.2 Cropping Pattern

The cropping pattern of the sample farm households during the survey year is given in Tables 5.1, 5.2, 5.3, and 5.4. The cropping pattern for the Bt and Non-Bt sample households was not collected separately in Andhra Pradesh. Among the sample households, cotton is the major crop followed by maize, chillies and paddy. In




V.P. Gandhi, D. Jain, *Introduction of Biotechnology in India's Agriculture*, DOI 10.1007/978-981-10-1091-0_5

**Table 5.1** Cropping pattern among sample households: Andhra Pradesh

| Sr. no. | Crop | Average area in hectares | Percentage to total cropped area |
|---|---|---|---|
| 1 | Bt cotton | 0.69 | 26.14 |
| 2 | Non-Bt cotton | 0.77 | 29.17 |
| 3 | Paddy | 0.27 | 10.23 |
| 4 | Maize | 0.35 | 13.26 |
| 5 | Chillies | 0.3 | 11.36 |
| 6 | Turmeric | 0.07 | 2.65 |
| 7 | Others | 0.19 | 7.20 |
| | All crops | 2.64 | 100 |

**Table 5.2** Cropping pattern among sample households: Gujarat

| Sr. no. | Crops | Area in hectares | | Percentage to total cropped area | |
|---|---|---|---|---|---|
| | | Bt cotton farmers | Non-Bt cotton farmers | Bt cotton farmers | Non-Bt cotton farmers |
| 1 | Cotton-NC[a] | 1.02 | 0.07 | 26.43 | 1.89 |
| 2 | Cotton-G[b] | 0.86 | 0.02 | 20.92 | 0.57 |
| 3 | Cotton-non Bt | 0.11 | 1.65 | 2.62 | 42.21 |
| 4 | Groundnut | 0.62 | 0.76 | 15.11 | 19.96 |
| 5 | Wheat | 0.23 | 0.19 | 5.66 | 5.02 |
| 6 | Sugarcane | 0.22 | 0.21 | 5.32 | 5.60 |
| 7 | Tur (pigeon pea) | 0.21 | 0.21 | 5.17 | 5.62 |
| 8 | Fodder | 0.39 | 0.25 | 9.45 | 6.60 |
| 9 | Vegetables | 0.08 | 0.12 | 1.96 | 3.15 |
| 10 | Garlic | 0.06 | 0.17 | 1.54 | 4.60 |
| 11 | Other | 0.24 | 0.18 | 5.81 | 4.79 |
| | Total cropped area | 4.08 | 3.79 | 100 | 100 |

[a]Cotton-NC stands for non confirmed Bt cotton
[b]Cotton-G stands for genuine Bt cotton or registered Bt cotton

Gujarat cotton accounted for 50 % of the cropped area for Bt cotton farmers and 45 % for the non-Bt farmers. Groundnut was the other major crop grown by both Bt and Non-Bt sample households followed by wheat, sugarcane and pulses. In Maharashtra cotton is one of the most important crops in cropping pattern of the sample households. The other major crops grown by the sample households were jowar and pulses. There was no major difference in cropping pattern between Bt and non-Bt farmers. Although cotton is not a major crop in Tamil Nadu, it accounted for about 31 % of the total cropped area for Bt farmers and 42.1 % of the cropped area for non-Bt farmers. Cholam (jowar), maize and pulses were the other major crops grown by the sample farm households.

**Table 5.3** Cropping pattern among sample households: Maharashtra

| Sr. no | Crops | Area in hectare | | Percentage to total cropped area | |
|---|---|---|---|---|---|
| | | Bt cotton | Non-Bt cotton | Bt cotton | Non-Bt cotton |
| 1 | Cotton | 1.45 | 30.33 | 1.35 | 32.93 |
| 2 | Jowar | 0.59 | 12.36 | 0.51 | 12.54 |
| 3 | Maize | 0.20 | 4.15 | 0.20 | 4.94 |
| 4 | Soybean | 0.25 | 5.16 | 0.24 | 5.82 |
| 5 | Urad+Mung | 0.35 | 7.28 | 0.37 | 9.08 |
| 6 | Banana | 0.22 | 4.66 | 0.28 | 6.91 |
| 7 | Sugarcane | 0.23 | 4.74 | 0.02 | 0.39 |
| 8 | Tur | 0.30 | 6.26 | 0.21 | 5.13 |
| 9 | Chilli | 0.17 | 3.64 | 0.20 | 4.84 |
| 10 | Mango | 0.13 | 2.62 | 0.02 | 0.49 |
| 11 | Wheat | 0.31 | 6.43 | 0.38 | 9.18 |
| 12 | Gram | 0.29 | 6.10 | 0.07 | 1.78 |
| 13 | Sunflower | 0.08 | 1.61 | 0.04 | 1.09 |
| 14 | Jowar | 0.18 | 3.73 | 0.13 | 3.16 |
| 15 | Groundnut | 0.02 | 0.51 | 0.03 | 0.79 |
| 16 | Onion | 0.02 | 0.51 | 0.04 | 0.89 |
| | Total | 4.78 | 100 | 4.10 | 100 |

**Table 5.4** Cropping pattern among sample households: Tamil Nadu

| Sr. no | Crop | Area in hectare | | Percentage to total cropped area | |
|---|---|---|---|---|---|
| | | Bt cotton | Non-Bt cotton | Bt cotton | Non-Bt cotton |
| 1 | Cotton | 0.86 | 1.10 | 30.71 | 42.15 |
| 2 | Groundnut | 0.37 | 0.37 | 13.21 | 14.18 |
| 3 | Cholam | 0.3 | 0.32 | 10.71 | 12.26 |
| 4 | Paddy | 0.3 | 0.18 | 10.71 | 6.90 |
| 5 | Maize | 0.25 | 0.06 | 8.93 | 2.30 |
| 6 | Pulses | 0.19 | 0.11 | 6.79 | 4.21 |
| 7 | Cumbu | 0.13 | 0.1 | 4.64 | 3.83 |
| 8 | Sugarcane | 0.11 | 0.02 | 3.93 | 0.77 |
| 9 | Vegetables | 0.11 | 0 | 3.93 | 0.00 |
| 10 | Sunflower | 0.07 | 0 | 2.50 | 0.00 |
| 11 | Maize | 0.04 | 0.33 | 1.43 | 12.64 |
| 12 | Sunflower | 0.04 | 0.01 | 1.43 | 0.38 |
| 13 | Tapioca | 0.02 | 0 | 0.71 | 0.00 |
| 14 | Turmeric | 0.01 | 0 | 0.36 | 0.00 |
| 15 | Chillies | 0 | 0.01 | 0.00 | 0.38 |
| | Total cropped area | 2.80 | 2.61 | 100 | 100 |

**Table 5.5** Average area under cotton among sample households (hectares/per farm)

| | BT | | | Non-Bt | | |
|---|---|---|---|---|---|---|
| | I | UI | Total | I | UI | Total |
| Andhra Pradesh | | | | | | |
| Small | 0.41 | 0.35 | 0.39 | 0.24 | 0.39 | 0.30 |
| Medium | 0.63 | 0.80 | 0.70 | 0.78 | 0.82 | 0.80 |
| Large | 1.34 | 1.64 | 1.47 | 1.22 | 1.42 | 1.28 |
| Overall | 0.65 | 0.76 | 0.69 | 0.62 | 0.74 | 0.66 |
| Gujarat | | | | | | |
| Small | 1.03 | 0 | 1.03 | 0.87 | 0 | 0.87 |
| Medium | 2.05 | 0 | 2.05 | 1.59 | 0 | 1.59 |
| Large | 3.57 | 0 | 3.57 | 3.56 | 0 | 3.56 |
| Overall | 1.88 | 0 | 1.88 | 1.65 | 0 | 1.65 |
| Maharashtra | | | | | | |
| Small | 0.86 | 0.63 | 0.74 | 0.85 | 1.07 | 0.92 |
| Medium | 1.55 | 1.21 | 1.42 | 1.38 | 1.18 | 1.24 |
| Large | 1.79 | 1.66 | 1.74 | 2.02 | 1.58 | 1.69 |
| Overall | 1.56 | 1.3 | 1.45 | 1.37 | 1.32 | 1.35 |
| Tamil Nadu | | | | | | |
| Small | 0.63 | 0.37 | 0.45 | 0.86 | 0.63 | 0.70 |
| Medium | 1.56 | 0.71 | 1.02 | 1.58 | 1.21 | 1.35 |
| Large | 1.38 | 1.08 | 1.38 | 1.57 | 1.69 | 1.63 |
| Overall | 1.25 | 0.59 | 0.86 | 1.31 | 0.99 | 1.10 |

## 5.3 Area Under Cotton

The cultivation of Bt cotton was found under both irrigated and unirrigated conditions among the sample households except in Gujarat where it was grown only under irrigated condition. But the Bt cotton area under irrigated conditions was more among the sample farmer households (Table 5.5). The average area under cotton among the sample households on a per farm basis was highest in Gujarat followed by Maharashtra and the least was in Andhra Pradesh.

## 5.4 Cotton Varieties Used by the Sample Households

Many Bt and non Bt varieties were reported. The use of non-genuine Bt seeds was reported by both Gujarat and Andhra Pradesh sample households (Table 5.6). This accounted for 56 % of the Bt sample in Gujarat and 20 % in Andhra Pradesh. The Bt seeds that were used by the sample farmer households were of RCH and Mahyco varieties in Andhra Pradesh and Gujarat besides the use of non-confirmed Bt seeds. While in Maharashtra both Mahyco and Rasi Bt seeds were used by the sample

**Table 5.6** Different varieties of cotton used by the sample households

| State | Cotton type | Varieties |
|---|---|---|
| Andhra Pradesh | BT | Rasi (RCH), MECH, Other Non-confirmed |
| | Non-BT | Bunny, Super Bunny, Brahma, Satya, Attara, JK, Tagore, Bindu, others |
| Gujarat | BT | Rasi (RCH), Mahyco, other non-confirmed |
| | Non-BT | Sankar, Vikram, Navbharat Deshi, Other deshi |
| Maharashtra | BT | MECH 184, MECH 12, MECH 162, Rasi 2, MECH + Rasi |
| | Non-BT | Ankur, Bunny Ajit, others |
| Tamil Nadu | BT | RCH-2 BT |
| | Non-BT | RCH-2 non-BT |

**Table 5.7** Brand/variety-wise use of cotton seeds by sample farmers: Andhra Pradesh

| | Brand/variety | Number | Percentage |
|---|---|---|---|
| Bt | Rasi | 53 | 58.89 |
| | MCH | 19 | 21.11 |
| | Non-confirmed | 18 | 20.00 |
| | Total Bt | 90 | 100 |
| Non-Bt | Bunny | 34 | 37.78 |
| | Super bunny | 20 | 22.22 |
| | Brahma | 5 | 5.56 |
| | JK | 6 | 6.67 |
| | Satya | 4 | 4.44 |
| | Attara | 3 | 3.33 |
| | Others | 18 | 20.00 |
| | Total non-Bt | 90 | 100 |

households, in Tamil Nadu only Rasi Bt seeds were used. A large number of non-Bt hybrid seeds were also reported among the sample households.

The use of non-confirmed Bt cotton was found in 20 % of the Bt cotton sample households in Andhra Pradesh (Table 5.7). Among the genuine Bt seeds, Rasi seeds were more popular among the sample farmers (58.9 %). A large number of non-Bt varieties have been used by the sample farmers in the state.

In Gujarat the use of non-confirmed Bt cotton seeds was very common accounting for over 57 % of the sample farmers. The two most popular varieties among the Bt sample farmers were Navbharat and Dhanlaxmi (Table 5.8). The two confirmed Bt varieties used were RCH and Mahyco. Under non-Bt seeds, a large number of hybrids and non-hybrids were in use during the survey year.

In Maharashtra two Bt varieties have been used by the sample farmers namely MECH and Rasi (Table 5.9) and among them MECH was more popular among the farmers. Among the non-Bt farmers, the prominent varieties were Ankur and Ajit.

In Tamil Nadu the Bt farmers used only RCH-2 Bt and the non-Bt sample farmers used only RCH-2 Non. Bt (Table 5.10).

**Table 5.8** Brand/variety-wise use of cotton seeds by sample farmers: Gujarat

| | Brand/variety | Number | Percentage |
|---|---|---|---|
| Bt-G | RCH | 20 | 22.22 |
| | Mahyco | 19 | 21.11 |
| Bt-NC | Navbharat | 32 | 35.56 |
| | Dhanlaxmi | 19 | 21.11 |
| | Total Bt | 90 | 100 |
| Non-Bt | Sankar(H) | 15 | 16.67 |
| | Vikram(H) | 33 | 36.67 |
| | Navbharat Deshi(non-H) | 25 | 27.78 |
| | Other deshi(non-H) | 17 | 18.89 |
| | Total non-Bt | 90 | 100 |

**Table 5.9** Brand/variety-wise use of cotton seeds by sample farmers: Maharashtra

| | Brand/variety | Number | Percentage |
|---|---|---|---|
| Bt cotton | MECH 184 – 1 | 25 | 29.41 |
| | MECH 12 – 2 | 19 | 22.35 |
| | MECH 162 – 3 | 6 | 7.06 |
| | Rasi 2 – 4 | 12 | 14.12 |
| | 2 MECH – 5 | 20 | 23.53 |
| | MECH + Rasi – 6 | 3 | 3.53 |
| | All varieties | 85 | 100 |
| Non-Bt cotton | Ankur | 14 | 20.29 |
| | Bunny | 11 | 15.94 |
| | Ajit | 19 | 27.54 |
| | Others | 25 | 36.23 |
| | All varieties | 69 | 100 |

**Table 5.10** Brand/variety-wise use of cotton seeds by sample farmers: Tamil Nadu

| | Brand/variety | Number | Percentage |
|---|---|---|---|
| Bt | RCH-2 | 90 | 100 |
| Non-Bt | RCH-2 | 90 | 100 |

## 5.5 Average Cost of Seeds Used

The average cost of Bt Cotton seeds used per hectare by the sample households varied substantially across the states. The cost of Bt seeds was the least in Gujarat at Rs 3079 per hectare followed by Andhra Pradesh with Rs. 3313 (Table 5.11). This could be because the use of non-confirmed Bt seeds was very common in these states and the cost of such seeds was reported to be lower than the confirmed Bt seeds. On an average the cost of Bt seeds per hectare was more than double than the non-Bt seeds in all the four states (Fig. 5.1).

**Table 5.11** Average cost of seed used by the sample households (Rs/hectare)

| Particulars | Andhra Pradesh | Gujarat | Maharashtra | Tamil Nadu |
|---|---|---|---|---|
| Bt- cotton | 3313 | 3079 | 3857 | 3977 |
| Non-Bt cotton | 1213 | 1324 | 1319 | 1180 |
| Bt-cotton over non-Bt cotton | 2100 | 1756 | 2538 | 2797 |

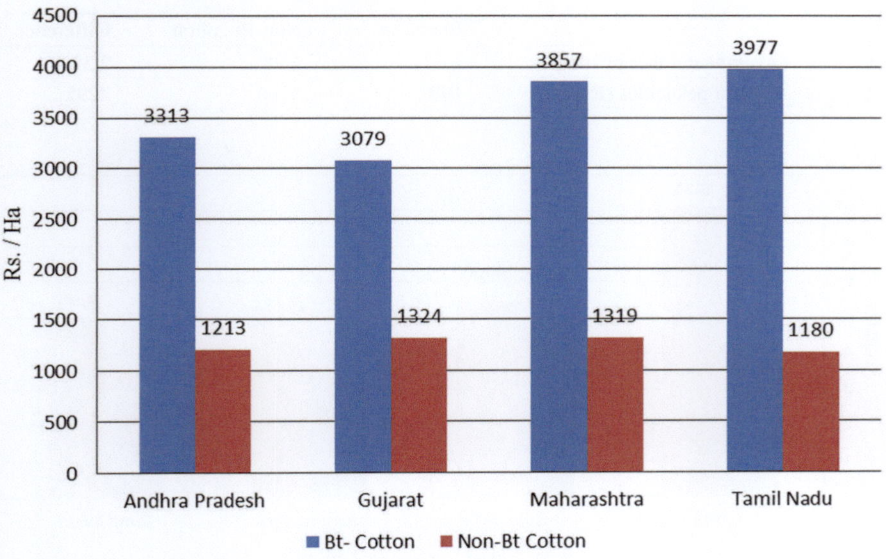


**Fig. 5.1** Average cost of cotton seed used by sample farmers

**Table 5.12** Practice followed by sample farmers in pesticides spraying: AP

| | | Bt cotton | Non-Bt cotton | Difference |
|---|---|---|---|---|
| 1 | Average number of sprays | 4.27 | 8.11 | 3.84 |
| 2 | Total quantity sprayed (litre/ha) | 5.22 | 6.93 | 1.71 |
| 3 | Cost of pesticides (Rs/ha) | 7926 | 10675 | 2749 |

**Table 5.13** Practice followed by sample farmers in pesticides spraying: Gujarat

| | | Bt cotton | Non-Bt cotton | Difference |
|---|---|---|---|---|
| 1 | Average number of sprays | 4.92 | 5.44 | 0.52 |
| 2 | Total quantity sprayed (litre/ha) | 2.53 | 3.90 | 1.37 |
| 3 | Cost of pesticides (Rs/ha) | 2732 | 3168 | 436 |

## 5.6 Application of Pesticides

The average number of pesticides sprays as well as its cost per hectare was relatively higher on non-Bt cotton in all the selected states (Tables 5.12, 5.13, 5.14, and 5.15). But the difference across states was very high. For example, average number of spray

**Table 5.14** Practice followed by sample farmers in pesticides spraying: Maharashtra

| | | Bt cotton | Non-Bt cotton | Difference |
|---|---|---|---|---|
| 1 | Average number of sprays | 3.37 | 5.28 | 1.91 |
| 2 | Cost of pesticides (Rs/ha) | 3242 | 4120 | 878 |

**Table 5.15** Practice followed by sample farmers in pesticides spraying: Tamil Nadu

| | | Bt cotton | Non-Bt cotton | Difference |
|---|---|---|---|---|
| 1 | Average number of sprays | 2.67 | 4.98 | 2.31 |
| 2 | Cost of pesticides (Rs/ha) | 1901 | 4196 | 2295 |

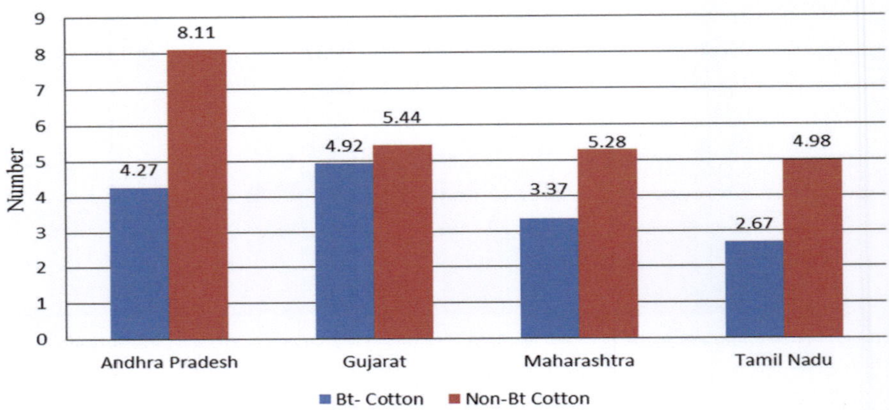


**Fig. 5.2** Average number of insecticides spray among sample households

on non-Bt Cotton in Gujarat was 0.52, but it was as high as 3.84 in Andhra Pradesh. Similarly the cost of pesticides was higher on non-Bt cotton by Rs. 436 per hectare in Gujarat, and Rs. 2749 in Andhra Pradesh. Thus the average number of pesticides spray as well as its cost was higher on non-Bt cotton in all the selected states (Fig. 5.2).

# Chapter 6
# Economics of Bt Cotton vs. Non-Bt Cotton Across the States

## 6.1 Introduction

In this chapter we examine the cost of cultivation, various input costs, the value of output, and the profitability of Bt and Non-Bt cotton under irrigated and unirrigated conditions. These aspects have also been examined for different farm sizes and these results are given at the end of the chapter.

## 6.2 Average Bt and Non-Bt Cotton Yields

The average yields observed of Bt and non-Bt cotton under irrigated and unirrigated conditions among different sizes of farming households are given in Table 6.1 and Figs. 6.1, 6.2, and 6.3. In Andhra Pradesh the Bt yields were higher but the difference in the yield levels under irrigated and unirrigated was not consistent. The yield of Bt cotton irrigated was significantly higher in Maharashtra and Tamil Nadu. There was no consistent yield difference across farm sizes. The average yield of Bt cotton over non-Bt cotton was higher in all the states in the study and it was higher by 18.2 % in Andhra Pradesh, 28.4 % in Gujarat, 46.4 % in Maharashtra and 28.5 % in Tamil Nadu.

## 6.3 Cost of Production, Value of Output and Profit

The total cost of cultivation under irrigated condition was in general higher than under unirrigated conditions both for Bt and non-Bt cotton. Comparing the input costs and value of output of Bt and non-Bt cotton overall – without making distinction between irrigated and unirrigated, the following major observations can be


V.P. Gandhi, D. Jain, *Introduction of Biotechnology in India's Agriculture*, DOI 10.1007/978-981-10-1091-0_6

**Table 6.1** Average cotton yield among sample households (Quintal/Ha)

| Farm size | Bt cotton | | | Non-Bt cotton | | | Difference between Bt cotton and non-Bt cotton (percentage) | | |
|---|---|---|---|---|---|---|---|---|---|
| | I | UI | Total | I | UI | Total | I | UI | Total |
| Andhra Pradesh | | | | | | | | | |
| Small | 30.16 | 25.00 | 27.96 | 26.40 | 23.74 | 24.81 | 14.24 | 5.29 | 12.51 |
| Medium | 31.20 | 32.34 | 31.83 | 26.27 | 20.77 | 23.39 | 18.76 | 55.66 | 36.84 |
| Large | 27.48 | 29.50 | 28.60 | 30.73 | 23.30 | 26.74 | −10.59 | 26.57 | 6.96 |
| Overall | 29.73 | 29.51 | 29.44 | 28.06 | 22.34 | 24.90 | 4.68 | 32.03 | 18.19 |
| Gujarat | | | | | | | | | |
| Small | 29.24 | | 29.24 | 21.07 | | 21.07 | 38.78 | | 38.78 |
| Medium | 34.34 | | 34.34 | 25.61 | | 25.61 | 34.09 | | 34.09 |
| Large | 32.13 | | 32.13 | 26.95 | | 26.95 | 19.22 | | 19.22 |
| Overall | 32.20 | | 32.20 | 25.07 | | 25.07 | 28.44 | | 28.44 |
| Maharashtra | | | | | | | | | |
| Small | 24.65 | 21.46 | 23.0 | 18.66 | 16.09 | 17.10 | 32.10 | 33.37 | 34.31 |
| Medium | 26.81 | 24.49 | 25.90 | 17.42 | 17.39 | 17.40 | 53.90 | 40.83 | 48.88 |
| Large | 28.85 | 25.17 | 27.30 | 19.24 | 18.44 | 18.64 | 49.95 | 36.50 | 46.44 |
| Overall | 27.56 | 24.10 | 26.05 | 18.56 | 17.47 | 17.80 | 48.49 | 37.95 | 46.36 |
| Tamil Nadu | | | | | | | | | |
| Small | 24.24 | 12.85 | 17.47 | 10.32 | 18.99 | 29.31 | 134.88 | −32.33 | −20.01 |
| Medium | 23.84 | 13.47 | 19.26 | 16.78 | 13.15 | 14.71 | 42.07 | 2.43 | 30.93 |
| Large | 23.40 | 13.76 | 19.64 | 16.94 | 13.12 | 14.96 | 38.13 | 4.88 | 31.28 |
| Overall | 23.75 | 13.35 | 18.93 | 16.97 | 12.10 | 14.73 | 39.95 | 10.33 | 28.51 |

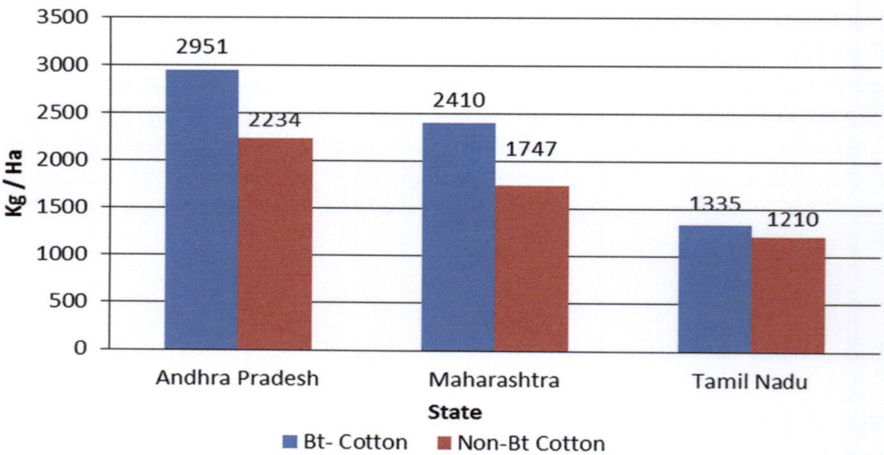


**Fig. 6.1** Average yield per hectare of Bt and Non-Bt cotton under irrigated condition (Kg/Ha)

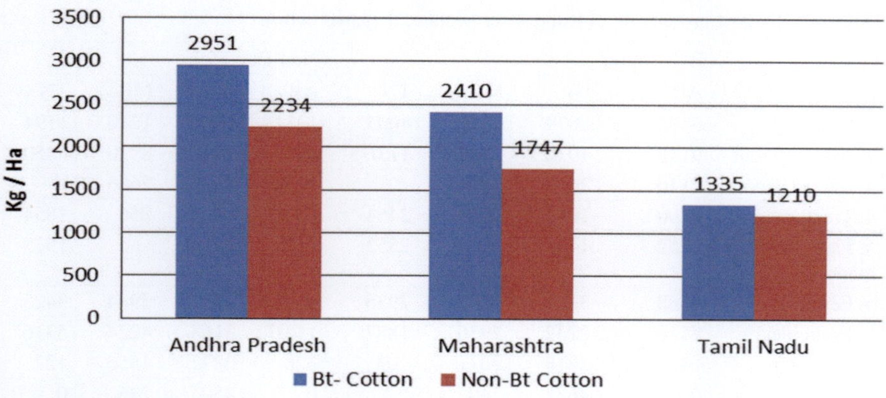


**Fig. 6.2** Average yield per hectare of Bt and Non-Bt cotton under unirrigated condition (Kg/Ha)

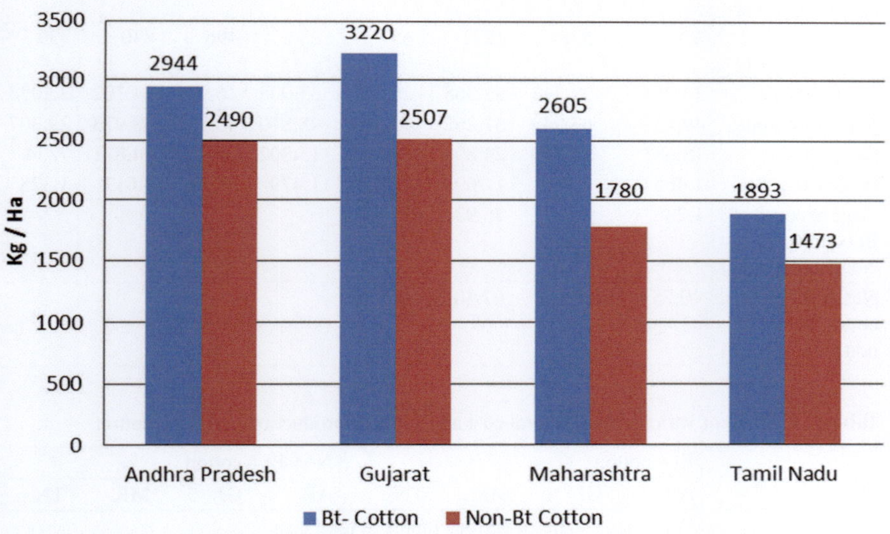


**Fig. 6.3** Average yield per hectare of Bt and Non-Bt cotton (Kg/Ha)

made (Tables 6.2, 6.3, 6.4, 6.5, 6.6, 6.7, and 6.8 and Figs. 6.4, 6.5, 6.6, 6.7, 6.8, 6.9, 6.10, and 6.11)

- The cost of cultivation per hectare of Bt cotton exceeded that of non-Bt cotton in all the states and the difference was the highest in Maharashtra exceeding Rs. 8250.
- The cost of irrigation for both Bt and non-Bt was relatively much higher in Maharashtra and Gujarat. This may be because of the dependence on tube-well irrigation as against canal irrigation in Andhra Pradesh and Tamil Nadu.

**Table 6.2** Cost of production in Rupees per hectare: irrigated cotton

| | Bt – cotton | | | | Non-Bt cotton | | | |
|---|---|---|---|---|---|---|---|---|
| Item | AP | GJ | MR. | TN | AP | GJ | MR. | TN |
| 1. Seed | 3495 | 3079 | 3776 | 4005 | 1234 | 1324 | 1240 | 1194 |
| 2. Human labour | 9140 | 10,827 | 11,801 | 11,073 | 8786 | 9587 | 8920 | 8250 |
| 3. Bullock labour | 2146 | 2468 | 2004 | | 2056 | 2527 | 2606 | |
| 4. Tractor | 1800 | 1042 | 1037 | 2464 | 1685 | 848 | 856 | 1854 |
| 5. Farm yard manure | 2252 | 1488 | | 2553 | 2109 | 1429 | | 1575 |
| 6. Fertilizer | 5888 | 3247 | 7193 | 2936 | 4605 | 3292 | 3984 | 3423 |
| 7. Pesticides | 8116 | 2732 | 2936 | 1887 | 11,972 | 3168 | 4224 | 5316 |
| 8. Irrigation | 642 | 3818 | 2011 | 103 | 328 | 3872 | 1556 | 122 |
| 9. Other operational costs | 0 | 467 | 340 | 0 | 0 | 450 | 235 | 0 |
| 10. Total operational cost | 33,478 | 29,167 | 31,098 | 25,021 | 32,775 | 26,497 | 23,621 | 21,734 |
| 11. Total marketing cost | 276 | 576 | 1271 | 651 | 233 | 496 | 830 | 339 |
| 12. Total cost | 33,754 | 29,743 | 32,368 | 25,672 | 33,008 | 26,993 | 24,102 | 22,073 |
| Value of output | 49,437 | 63,942 | 57,262 | 45,599 | 48,810 | 48,873 | 38,973 | 29,307 |
| Net profit | 15,683 | 34,199 | 24,894 | 19,927 | 15,802 | 21,880 | 14,871 | 7234 |
| Benefit-cost ratio | 1.465 | 2.150 | 1.769 | 1.776 | 1.479 | 1.811 | 1.617 | 1.328 |
| Value of output of Bt cotton over Non-Bt cotton (%) | 1.28 | 30.83 | 46.93 | 55.59 | | | | |
| Net profit of Bt cotton over non-Bt cotton (%) | –0.75 | 56.07 | 67.40 | 175.46 | | | | |

**Table 6.3** Share of various costs in total cost and value of production: irrigated cotton

| | Bt – cotton | | | | Non-Bt – cotton | | | |
|---|---|---|---|---|---|---|---|---|
| Item | AP | GJ | MR. | TN | AP | GJ | MR. | TN |
| | Percentage share of various inputs in total cost | | | | | | | |
| 1. Seed | 10.35 | 10.41 | 11.67 | 15.60 | 3.74 | 5.00 | 5.07 | 5.41 |
| 2. Farm power | 38.77 | 48.23 | 45.85 | 52.73 | 37.95 | 48.02 | 50.64 | 45.78 |
| 3. Manure | 24.11 | 15.56 | 22.22 | 21.38 | 20.34 | 16.88 | 16.29 | 22.64 |
| 4. Pesticides | 24.04 | 8.66 | 9.07 | 7.35 | 36.27 | 11.99 | 17.28 | 24.08 |
| 5. Other inputs | 2.72 | 17.15 | 11.19 | 2.94 | 1.70 | 18.10 | 10.72 | 2.09 |
| 6. Total cost | 100 | 100 | 100 | 100 | 100 | 100 | 100 | 100 |
| | Costs as percentage of value of output | | | | | | | |
| 1. Seed | 7.07 | 5.02 | 6.59 | 8.78 | 2.53 | 2.95 | 3.18 | 4.07 |
| 2. Farm power | 26.47 | 23.26 | 25.92 | 29.69 | 25.66 | 28.34 | 31.77 | 34.48 |
| 3. Manure | 16.47 | 7.51 | 12.56 | 12.04 | 13.76 | 9.97 | 10.22 | 17.05 |
| 4. Pesticides | 16.42 | 4.17 | 5.13 | 4.14 | 24.53 | 7.08 | 10.84 | 18.14 |
| 5. Other costs | 1.86 | 8.27 | 6.33 | 1.65 | 1.15 | 10.69 | 6.73 | 1.57 |
| 6. Total cost | 68.28 | 48.23 | 56.53 | 56.30 | 67.63 | 59.03 | 62.74 | 75.32 |
| 7. Value of output | 100 | 100 | 100 | 100 | 100 | 100 | 100 | 100 |

**Table 6.4** Cost of production in Rupees per hectare: unirrigated cotton

| | Bt – cotton | | | | Non-Bt – cotton | | | |
|---|---|---|---|---|---|---|---|---|
| Item | AP | GJ | MR. | TN | AP | GJ | MR. | TN |
| 1. Seed | 3137 | | 3963 | 3945 | 1215 | | 1354 | 1167 |
| 2. Human labour | 10,378 | | 11,694 | 6791 | 7846 | | 9251 | 7177 |
| 3. Bullock labour | 1982 | | 1794 | | 2029 | | 1915 | |
| 4. Tractor | 1611 | | 989 | 2267 | 1642 | | 701 | 1613 |
| 5. Farm yard manure | 1955 | | | 1852 | 1928 | | | 1075 |
| 6. Fertilizer | 3752 | | 7017 | 2905 | 3631 | | 4131 | 4057 |
| 7. Pesticides | 7509 | | 3638 | 1934 | 10,004 | | 4074 | 3074 |
| 8. Irrigation | 0 | | 0 | | 0 | | 0 | |
| 9. Other operational costs | 0 | | 321 | 0 | 0 | | 72 | 0 |
| 10. Total operational cost | 30,323 | | 29,416 | 19,694 | 28,295 | | 21,498 | 12,163 |
| 11. Total marketing cost | 139 | | 1369 | 297 | 167 | | 1334 | 291 |
| 12. Total cost | 30,463 | | 30,785 | 19,991 | 28,462 | | 22,815 | 18,454 |
| Value of output | 52,847 | | 50,487 | 29,797 | 28,372 | | 36,890 | 23,632 |
| Net profit | 22,384 | | 19,702 | 9806 | –90 | | 14,075 | 5178 |
| Benefit-cost ratio | 1.735 | | 1.640 | 1.491 | 0.997 | | 1.617 | 1.281 |
| Value of output of Bt cotton over non-Bt cotton (%) | 86.26 | | 36.86 | 26.09 | | | | |
| Net profit of Bt cotton over non-Bt cotton (%) | | | 39.98 | 89.38 | | | | |

**Table 6.5** Share of various costs in total cost and value of production: unirrigated cotton

| | Bt – cotton | | | | Non-Bt – cotton | | | |
|---|---|---|---|---|---|---|---|---|
| Item | AP | GJ | MR. | TN | AP | GJ | MR. | TN |
| | Percentage share of various inputs in total cost | | | | | | | |
| 1. Seed | 10.30 | | 12.87 | 19.73 | 4.27 | | 5.93 | 6.32 |
| 2. Farm power | 45.86 | | 47.03 | 45.31 | 40.46 | | 51.98 | 47.63 |
| 3. Manure | 18.73 | | 22.79 | 23.80 | 19.53 | | 18.09 | 27.81 |
| 4. Pesticides | 24.65 | | 11.82 | 9.67 | 35.15 | | 17.84 | 16.66 |
| 5. Other inputs | 0.46 | | 5.49 | 1.49 | 0.59 | | 6.16 | 1.58 |
| 6. Total cost | 100 | | 100 | 100 | 100 | | 100 | 100 |
| | Costs as percentage of value of output | | | | | | | |
| 1. Seed | 5.94 | | 7.85 | 13.24 | 4.28 | | 3.67 | 4.94 |
| 2. Farm power | 26.44 | | 28.67 | 30.40 | 40.59 | | 32.17 | 37.20 |
| 3. Manure | 10.80 | | 13.90 | 15.96 | 19.59 | | 11.20 | 21.72 |
| 4. Pesticides | 14.21 | | 7.21 | 6.49 | 35.26 | | 11.04 | 13.01 |
| 5. Other costs | 0.26 | | 3.35 | 1.00 | 0.59 | | 3.81 | 1.23 |
| 6. Total cost | 57.64 | | 60.98 | 67.09 | 100.32 | | 61.89 | 78.09 |
| 7. Value of output | 100 | | 100 | 100 | 100 | | 100 | 100 |

**Table 6.6** Cost of production in Rupees per hectare: overall

| | Bt – cotton | | | | Non-Bt – cotton | | | |
|---|---|---|---|---|---|---|---|---|
| Item | AP | GJ | MR. | TN | AP | GJ | MR. | TN |
| 1. Seed | 3313 | 3079 | 3857 | 3977 | 1213 | 1324 | 1319 | 1180 |
| 2. Human labour | 9818 | 10,827 | 11,754 | 9089 | 8249 | 9587 | 9150 | 7714 |
| 3. Bullock labour | 2062 | 2468 | 1913 | | 2024 | 2527 | 2125 | |
| 4. Tractor | 1705 | 1042 | 1016 | 2373 | 1648 | 848 | 748 | 1734 |
| 5. Farm yard manure | 2103 | 1488 | 0 | 2228 | 2000 | 1429 | 0 | 1325 |
| 6. Fertilizer | 4804 | 3247 | 7116 | 2922 | 4078 | 3292 | 4086 | 3740 |
| 7. Pesticides | 7806 | 2732 | 3242 | 1909 | 10,878 | 3168 | 4120 | 4195 |
| 8. Irrigation | 319 | 3818 | 1136 | 55 | 163 | 3872 | 474 | 60 |
| 9. Other operational costs | 0 | 467 | 332 | 0 | 0 | 450 | 122 | 0 |

**Table 6.7** Share of various costs in total cost and value of production: overall

| | Bt – cotton | | | | Non-Bt – cotton | | | |
|---|---|---|---|---|---|---|---|---|
| Item | AP | GJ | MR. | TN | AP | GJ | MR. | TN |
| | Percentage share of various inputs in total cost | | | | | | | |
| 1. Seed | 10.31 | 10.56 | 12.17 | 17.26 | 3.98 | 5.00 | 5.65 | 5.82 |
| 2. Farm power | 42.27 | 49.15 | 46.35 | 49.75 | 39.16 | 48.92 | 51.55 | 46.63 |
| 3. Manure | 21.49 | 16.23 | 22.46 | 22.35 | 19.96 | 17.82 | 17.52 | 25.00 |
| 4. Pesticides | 24.29 | 9.37 | 10.23 | 8.29 | 35.73 | 11.96 | 17.66 | 20.71 |
| 5. Other inputs | 1.65 | 14.69 | 8.78 | 2.35 | 1.17 | 16.31 | 7.62 | 1.84 |
| 6. Total cost | 100 | 100 | 100 | 100 | 100 | 100 | 100 | 100 |
| | Costs as percentage of value of output | | | | | | | |
| 1. Seed | 6.50 | 4.82 | 7.10 | 10.39 | 3.38 | 2.71 | 3.52 | 4.53 |
| 2. Farm power | 26.65 | 22.42 | 27.03 | 29.94 | 33.23 | 26.52 | 32.04 | 36.29 |
| 3. Manure | 13.55 | 7.41 | 13.10 | 13.45 | 16.94 | 9.66 | 10.89 | 19.46 |
| 4. Pesticides | 15.31 | 4.27 | 5.97 | 4.99 | 30.33 | 6.48 | 10.98 | 16.11 |
| 5. Other costs | 1.04 | 6.70 | 5.12 | 1.42 | 0.99 | 8.89 | 4.74 | 1.43 |
| 6. Total cost | 63.06 | 46.61 | 58.33 | 60.18 | 84.88 | 54.22 | 62.16 | 77.83 |
| 7. Value of output | 100 | 100 | 100 | 100 | 100 | 100 | 100 | 100 |

- The average per hectare cost of seed for Bt cotton was higher compared to non-Bt cotton and the difference ranged from Rs. 1600 per ha in Gujarat to over Rs. 2700 per ha in Tamil Nadu. The relatively low cost of Bt cotton seeds in Gujarat and Andhra Pradesh as opposed to Maharashtra and Tamil Nadu could be due to wide availability and use of non-genuine Bt seeds whose price (as well as quality) was comparatively lower than genuine Bt cotton seeds.

**Table 6.8** Average yield, value of output and net profit of Bt and non-Bt cotton among sample households

| | Bt cotton | | | | Non-Bt cotton | | | | Bt over non-Bt cotton | | | |
|---|---|---|---|---|---|---|---|---|---|---|---|---|
| | AP | GUJ | MH | TN | AP | GUJ | MH | TN | AP | GUJ | MH | TN |
| Yield (Kg/ha) | 2944 | 3220 | 2605 | 1893 | 2490 | 2507 | 1780 | 1473 | 18.2 | 28.4 | 46.4 | 28.5 |
| Value of output (Rs/ha) | 50,970 | 63,942 | 54,313 | 38,282 | 35,870 | 48,873 | 37,524 | 26,032 | 42.1 | 30.8 | 44.7 | 47.1 |
| Net profit (Rs/ha) | 18,831 | 34,199 | 22,634 | 15,242 | 5426 | 21,880 | 14,317 | 5772 | 217.1 | 53.1 | 58.1 | 164.1 |

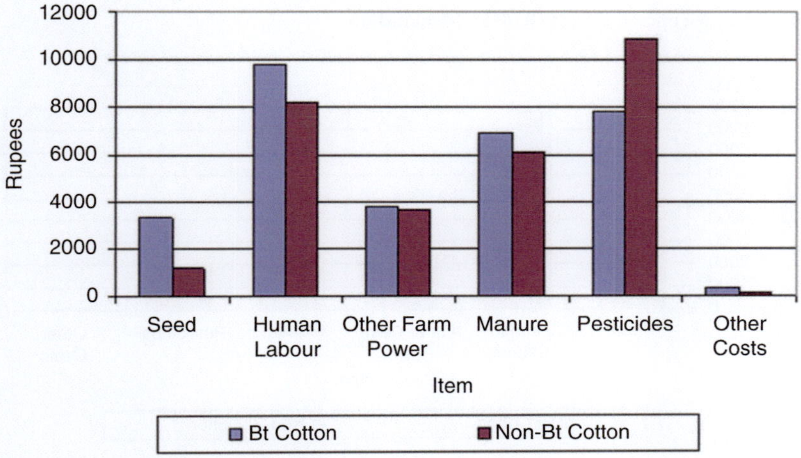


**Fig. 6.4** Average operating cost (Rs/Ha): Andhra Pradesh

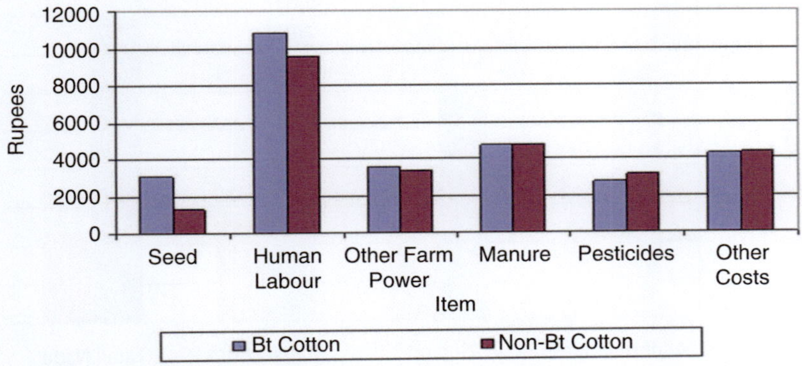


**Fig. 6.5** Average operating cost (Rs/Ha): Gujarat

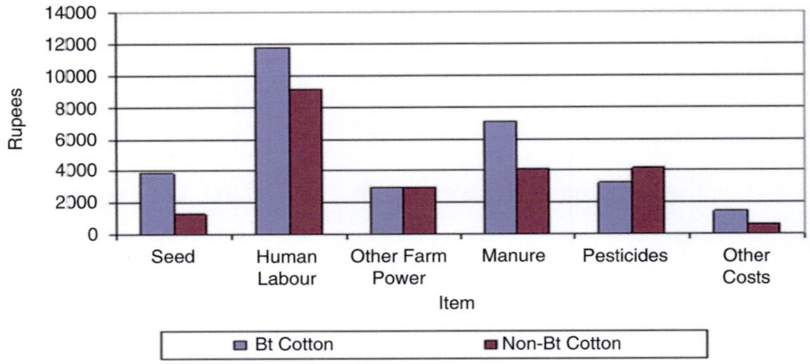


**Fig. 6.6** Average operating cost (Rs/Ha): Maharashtra

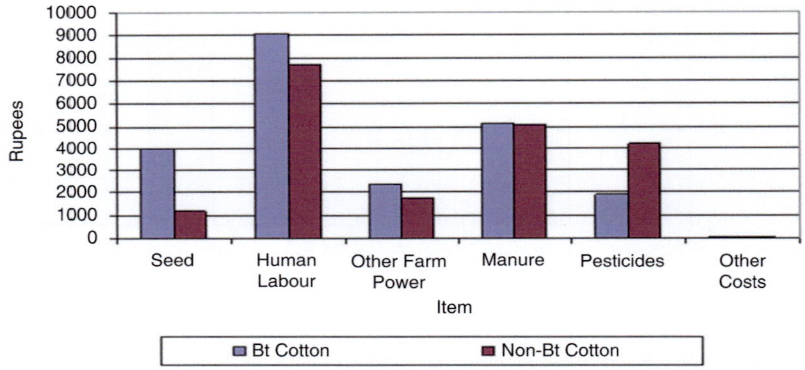


**Fig. 6.7** Average operating cost (Rs/Ha): Tamil Nadu

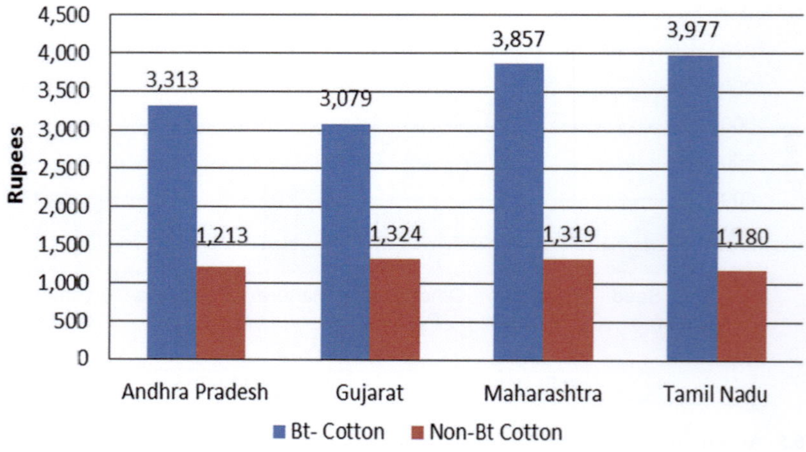


**Fig. 6.8** Average cost of seed per hectare under Bt and Non-Bt Cotton

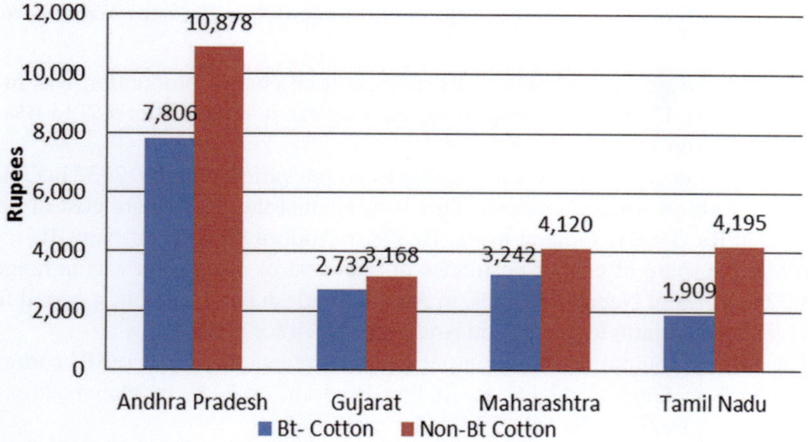


**Fig. 6.9** Average cost of pesticides used per hectare under Bt and Non-Bt cotton

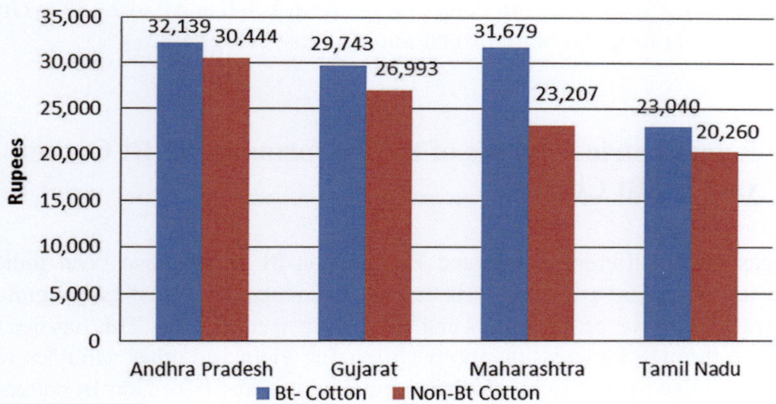


**Fig. 6.10** Average cost of cultivation per hectare for Bt and Non-Bt cotton

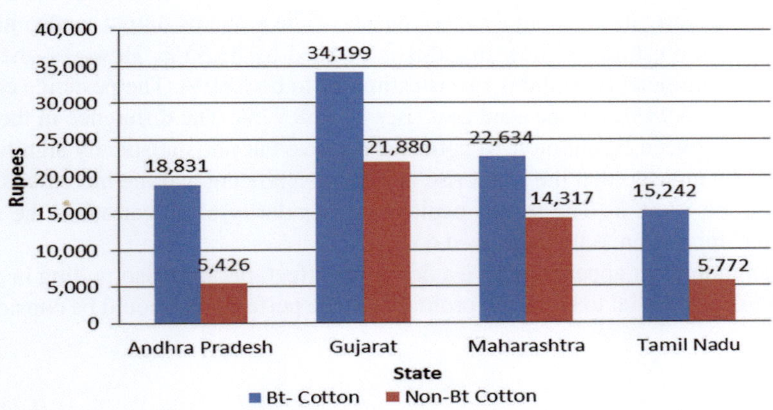


**Fig. 6.11** Average net revenue per hectare under Bt and Non-Bt cotton

- While the share of seed cost of Bt seeds in total cost of production was in the range of 10–17 % in the selected states, whereas it varied from 4 % to 6 % for non-Bt cotton seeds.
- The per hectare cost of pesticides under Bt cotton varied from Rs. 2732 in Gujarat to Rs. 7806 in Andhra Pradesh. This is as against the per hectare cost of pesticides at Rs. 3168 in Gujarat to Rs. 10,878 in Andhra Pradesh for Non-Bt
- While the share of cost of pesticides in total cost of cultivation was in range of 8.29 % in Tamil Nadu to 24.29 % in Andhra Pradesh for Bt cotton, it varied from 11.96 % in Gujarat to 35.73 % in Andhra Pradesh for non-Bt
- The value of output of Bt cotton per hectare exceeded that of non-Bt cotton by 30.843 % in Gujarat, 42.1 % in Andhra Pradesh, 44.74 % in Maharashtra and 47.06 % in Tamil Nadu.
- The net profit per hectare from Bt cotton ranged from Rs. 15,242 in Tamil Nadu to Rs. 34,199 in Gujarat as opposed to Rs. 5772 in Tamil Nadu to Rs. 21,880 in Gujarat for non-Bt cotton.
- The Benefit-Cost Ratio of Bt cotton ranges from 1.586 in AP to 2.150 in Gujarat and is higher than that of non-Bt cotton in almost every case.

## 6.4 Econometric Analysis of the Performance of Bt Cotton vs. Non-Bt Cotton

Whereas many differences between Bt and non-Bt cotton have been indicated above, it is important to examine whether the differences are statistically significant, and for this purpose, econometrics analysis has been carried out. This has been first examined through a regression approach relating yield and other variables with a dummy variable of Bt cotton, which is 1 for Bt cotton and 0 for Non-Bt cotton. The results of this would be identical to that of Analysis of Variance (ANOVA). The results in Table 6.9 below indicate that Bt cotton clearly has a statistically significant impact on the yield, significant at the 99 % level. The estimates indicate that Bt cotton yields are 30.71 % higher. The impact of the value of output is also highly significant and estimates show that this is boosted by 33.35 %. However, the cost also rises significantly, and this rise is estimated to be 6.69 %. The pesticide cost is reduced by 23.98 %, but the seed cost rises by 168.77 %. The difference in the output price between Bt and non-Bt cotton is positive but not statistically significant. The results indicate that the profit rise is highly significant and the increase is estimated to be 87.58 %. The results explain the popularity of Bt cotton, at the same time, the opposition to the high seed cost.

Even though Bt appears to have a dominant effect, perhaps also pulling in other inputs to its potential to boost the profitability, the performance could be considered

**Table 6.9** Regression results: impact of Bt cotton

| | | Independent variables | | N=515 |
|---|---|---|---|---|
| Dependent variable | | Constant | Bt | Percent impact of Bt |
| Yield | Coefficient | 2212.25 | 679.45 | 30.71 |
| | t-stat | 47.05 | 10.37 | |
| | Signifi. | *** | *** | |
| Value of output | Coefficient | 41,861 | 13,960 | 33.35 |
| | t-stat | 45.2 | 10.81 | |
| | Signifi. | *** | *** | |
| Total cost | Coefficient | 28,066 | 1878.56 | 6.69 |
| | t-stat | 71.5 | 3.43 | |
| | Signifi. | *** | *** | |
| Pesticide cost | Coefficient | 7387.95 | −1771.47 | −23.98 |
| | t-Stat | 33.01 | −5.68 | |
| | Signifi. | *** | *** | |
| Seed cost | Coefficient | 1296.12 | 2187.41 | 168.77 |
| | t-stat | 28.71 | 34.76 | |
| | Signifi. | *** | *** | |
| Price | Coefficient | 19.04 | 0.28679 | 1.51 |
| | t-stat | 140.45 | 1.52 | |
| | Signifi. | *** | NS | |
| Profit | Coefficient | 13,795 | 12,081 | 87.58 |
| | t-stat | 16.1 | 10.11 | |
| | Signifi. | *** | *** | |

*NS* not significant
*** = significant at 99 %
** = significant at 95 %
* = significant at 90 %

a function of other inputs as well. The model below relates the performance of the dependent variables to various inputs/factors including Bt, pesticide, seed, fertilizer, irrigation, and state of location, together. These results would be affected to some extent by the multicollinearity across the explanatory variables – since the inputs are associated and complementary. The results indicate that Bt alone is still statistically highly significant as a determinant of the yield, value of output and profitability. The impact on yield is estimated to be about 22 % and the impact on profitability about 35 %. Profit is negatively related to pesticide cost and positively related to seed cost (reflecting use of Bt seeds) and irrigation. The adverse relationship with fertilizer cost is perhaps a result of multicollinearity. The state dummies indicate that whereas profits are significantly higher in Maharashtra as compared to Gujarat, there is no statistically significant difference in the profitability between Gujarat and Andhra Pradesh (Table 6.10).

**Table 6.10** Regression results: impact of Bt cotton and other determinants

| | | Independent variables (N=515) | | | | | | | |
|---|---|---|---|---|---|---|---|---|---|
| Dependent variable | | Constant | Bt | Pesticide cost | Seed cost | Fertilizer cost | Irrigation status | MH dummy | AP dummy |
| Yield | Coefficient | 1912.78 | 428.03 | 0.0318 | 0.1469 | −0.0819 | 475.07 | −333.41 | −37.70 |
| | t-stat | 16.36 | 3.78 | 2.27 | 3.44 | −4.43 | 6.22 | −2.94 | −0.3 |
| | Signifi. | *** | *** | ** | *** | *** | *** | ** | NS |
| Value of output | Coefficient | 35,854 | 7568.29 | 0.6136 | 3.3843 | −1.2864 | 8440.21 | −2875.29 | −6810.50 |
| | t-stat | 15.08 | 3.28 | 2.16 | 3.9 | −3.42 | 5.43 | −1.25 | −2.67 |
| | Signifi. | *** | *** | ** | *** | *** | *** | NS | *** |
| Total cost | Coefficient | 15,637 | 392.30 | 1.2683 | 1.6172 | 1.2548 | 1117.76 | −9105.81 | −6689.49 |
| | t-stat | 20.7 | 0.54 | 14.03 | 5.86 | 10.49 | 2.26 | −12.43 | −8.25 |
| | Signifi. | *** | NS | *** | *** | *** | ** | *** | *** |
| Profit | Coefficient | 20,217 | 7175.99 | −0.6547 | 1.7671 | −2.5412 | 7322.45 | 6230.52 | −121.014 |
| | t-stat | 9.85 | 3.61 | −2.67 | 2.36 | −7.82 | 5.46 | 3.13 | −0.05 |
| | Signifi. | *** | *** | *** | ** | *** | *** | *** | NS |

*NS* not significant
*** significant at 99 %
** significant at 95 %
* significant at 90 %

# Chapter 7
# Farmers' Perceptions on Various Features of Bt Cotton

This chapter is devoted to examining the general observations and perceptions of the sample farmer households growing Bt cotton on various aspects such as pest incidence, economic benefits, availability of Bt seeds, extension and information sources, environmental impact, and other features. The Bt cotton farmers were asked questions on various aspects such as input use, seed availability and quality, various farm operations, marketing of cotton fiber, and other related aspects. Since the exact format with which the opinions of the farmer households were collected was not uniform across the states, these depend on the availability of such information, and have been presented separately for each state.

## 7.1 Observations of Farmer Households in Andhra Pradesh

As expressed by the sample farmer households in Andhra Pradesh, the major advantages of Bt cotton over non-Bt cotton varieties were lesser need for pesticides, better yields and profits (Table 7.1). No major differences were expressed in terms of the use of inputs other than pesticides, and price of cotton. Two major disadvantages expressed by the sample households were the availability of Bt seeds and their high price.

About 57.6 % of the sample households reported that the seed rate for Bt cotton was reduced by about 25 %, because the seeds are very expensive. More than three-fourth of the households reported that the Bt cotton flowers earlier by 1–10 days than non-Bt cotton. Other major features observed were shorter plant size of Bt cotton, bigger boll size, and lesser number of pickings as compared to non-Bt cotton (Table 7.2). Ninety-six percent of the sample households reported that Bt cotton has more resistance to boll worm attack. Over 90 % of the sample households expressed their willingness to continue with Bt seeds considering better yields and profitability. About ten per of the sample households expressed their unwillingness to continue with Bt cultivation due to high cost of seeds and its unsuitability to their land.



V.P. Gandhi, D. Jain, *Introduction of Biotechnology in India's Agriculture*, DOI 10.1007/978-981-10-1091-0_7

**Table 7.1** Advantages or disadvantages of Bt cotton-G vis-à-vis non-Bt cotton reported by Bt growers in Andhra Pradesh: percentages

| Sr. No | Particulars | Strong advantage | Advantage | No difference | Disadvantage | Strong disadvantage |
|---|---|---|---|---|---|---|
| 1 | Availability of seeds | 0.0 | 17.8 | 8.8 | 71.4 | 2.0 |
| 2 | Seed cost/price | 3.4 | 16.6 | 12.5 | 40.3 | 28.0 |
| 3 | Quality of avail. seeds | 15.4 | 75.8 | 8.8 | 0.0 | 0.0 |
| 4 | Pest incidence/ problem | 19.0 | 56.7 | 21.3 | 3.5 | 0.0 |
| 5 | Pesticide need/ cost | 37.0 | 54.4 | 7.6 | 1.0 | 0.0 |
| 6 | Fertilizer need/ cost | 10.0 | 32.2 | 58.1 | 0.0 | 0.0 |
| 7 | Labour need/cost | 7.8 | 48.6 | 43.6 | 0.0 | 0.0 |
| 8 | Machine need/ cost | 2.2 | 19.8 | 78.0 | 0.0 | 0.0 |
| 9 | Irrigation need/ cost | 1.0 | 45.5 | 53.5 | 0.0 | 0.0 |
| 10 | Harvesting cost | 4.2 | 49.1 | 46.7 | 0.0 | 0.0 |
| 11 | Cotton quality | 38.0 | 52.2 | 9.8 | 0.0 | 0.0 |
| 12 | Market preference | 12.0 | 23.3 | 64.2 | 0.0 | 0.0 |
| 13 | Staple length | 19.8 | 74.2 | 4.5 | 1.0 | 0.0 |
| 14 | Fiber colour | 25.5 | 67.9 | 5.6 | 1.0 | 0.0 |
| 15 | Cotton price | 5.4 | 23.3 | 70.8 | 0.0 | 0.0 |
| 16 | Easy marketing | 2.2 | 27.6 | 70.2 | 0.0 | 0.0 |
| 17 | By-product output | 1.2 | 18.8 | 80.0 | 0.0 | 0.0 |
| 18 | Yield | 19.1 | 68.7 | 5.6 | 0.0 | 0.0 |
| 19 | Profit | 29.0 | 57.8 | 4.1 | 3.5 | 0.0 |
| 20 | Livestock feeding | 0.0 | 22.1 | 77.9 | 0.0 | 0.0 |
| 21 | Water saving | 2.2 | 27.8 | 70.0 | 0.0 | 0.0 |
| 22 | Suitable for early sowing | 2.2 | 23.3 | 74.5 | 0.0 | 0.0 |
| 23 | Suitable for late sowing | 2.2 | 21.2 | 76.6 | 0.0 | 0.0 |

## 7.2 Observations of Farmer Households in Gujarat

The sample farmer households in Gujarat expressed that the major advantages of Bt cotton over non-Bt cotton varieties were better yields and profits, less pest incidence and pesticides cost, and suitability for early sowing (Table 7.3). No major differences were expressed in terms of the use of inputs other than pesticides, and price

**Table 7.2** Sample households opinion of seed rate, flowering date, plant size, boll size, date of picking etc. in Andhra Pradesh (percent response)

| | |
|---|---|
| **Sample households opinion on seed rate** | |
| Seed rate reduced 25 % | 57.60 |
| **Sample households opinion on flowering** | |
| 1–10 days early flowering | 75.54 |
| No change | 24.46 |
| **Sample households opinion on plant size** | |
| Plant size shorter | 74.90 |
| Plant size same | 25.10 |
| **Sample households opinion on boll size** | |
| Bigger boll size | 74.80 |
| Smaller boll size | 11.00 |
| Same boll size | 14.20 |
| **Sample households opinion on picking**: | |
| No. of picking more | 0.00 |
| No. of picking less | 88.70 |
| No of picking same | 11.30 |
| **Resistance of Bt cotton to pests** | |
| Percentage of households reporting more resistance to boll worm | 96.80 |
| **Resistance level reported** | |
| Resistance below 25 % | 61.20 |
| Resistance 25–50 % | 38.80 |
| **Willingness to continue Bt cotton cultivations**: | |
| Percentage of households reporting willingness to continue Bt cultivation | 90.00 |
| **Why to continue Bt cultivation**? | |
| Resistance to boll worm and better yield | 90.00 |
| **Why not to continue Bt cultivation**? | |
| High seed cost | 10.00 |
| Poor performance | 6.00 |
| Bt not suitable to the area | 6.00 |

of cotton. Two major disadvantages expressed by the sample households were high seed price and the availability of Bt seeds.

## 7.3 Observations of Farmer Households in Maharashtra

The most important reason for the adoption of Bt cotton is its resistance to pests, particularly boll worms, which can be a devastating problem for cotton. Observations on the pest incidence on the crops were available from Maharashtra farmers. Table 7.4 below reports this for Maharashtra for Bt growers. For boll worms, including American, pink and spotted boll worms, no infestation is indicated by over 70 %

**Table 7.3** Advantages/disadvantages of Bt cotton vis-à-vis non Bt cotton in Gujarat: percentage of Bt cotton growers reporting

| Particulars | Strong advantage | Advantage | No difference | Disadvantage | Strong disadvantage |
|---|---|---|---|---|---|
| 1. Seed availability | 16.67 | 24.44 | 38.89 | 20.00 | 0.0 |
| 2. Seed price | 0.0 | 0.0 | 4.44 | 40.00 | 55.56 |
| 3. Pest incidence | 4.44 | 86.67 | 8.89 | 0.0 | 0.0 |
| 4. Pesticide cost | 5.56 | 88.89 | 5.56 | 0.0 | 0.0 |
| 5. Fertilizer cost | 1.11 | 12.22 | 80.00 | 5.56 | 1.11 |
| 6. Irrigation cost | 0.0 | 4.44 | 91.12 | 3.33 | 1.11 |
| 7. Labour cost | 0.0 | 5.56 | 84.44 | 10.00 | 0.0 |
| 8. Harvesting cost | 0.0 | 0.0 | 68.89 | 0.0 | 31.11 |
| 9. Fibre quality | 13.33 | 52.22 | 34.44 | 0.0 | 0.0 |
| 10. Cotton price | 15.56 | 24.44 | 58.89 | 1.11 | 0.0 |
| 11. Yield | 54.44 | 38.89 | 6.67 | 0.0 | 0.0 |
| 12. Profit | 52.22 | 42.22 | 5.56 | 0.0 | 0.0 |
| 13. Suitability of early sowing | 21.11 | 68.89 | 10.00 | 0.0 | 0.0 |
| 14. Market preference | 17.78 | 18.89 | 63.33 | 0.0 | 0.0 |
| 15. Improvement in village economy | 8.88 | 88.89 | 2.22 | 0.0 | 0.0 |

on Bt cotton, whereas no infestation is reported by only 2–30 % on non-Bt cotton. Only about 4–6 % report of moderate to heavy infestation on Bt, whereas this number is as high as 20–60 % on non-Bt. Surprisingly, there is also a difference in the infestation by sucking and foliage feeding pests, for which the incidence is none to light in the case of Bt, whereas it is moderate to heavy in the case of non-Bt. Thus Bt cotton seems to provide considerable resistance to boll worms, and even shows resistance to other pests. A small number of farmers indicate incidence of boll worms on Bt, particularly other kinds of boll worms.

Other responses indicate that Bt cotton growers were not approached or pressured by private sales agents for the promotion of Bt cotton (Table 7.5). As regards the plant size and boll size, the respondents did not indicate any major difference compared to non-Bt cotton. About 96.5 % of the sample farmer households indicated that the numbers of pickings under Bt and non-Bt cotton are same. No government agencies had approached them for the inspection of Bt cotton, and none of the sample households felt specific problems with respect to the marketing of Bt cotton fibre. None of the Bt growers had observed any adverse environmental impact as a result of the cultivation of Bt cotton. They also did not indicate any increase in the pest attack on other crops as a result of the cultivation of Bt cotton. All the sample farmers indicated that they need to buy Bt seed every year for cultivation. Almost 98 % of the sample farmers did not face any difficulty in getting quality Bt seeds in time. As high as 94.1 % of the sample farmers were positive on continuing with Bt cultivation in the future.

**Table 7.4** Pest/insect attack on cotton: response of Bt cotton growers (percent)

| Pest/insect | Bt | | | | | Non-Bt | | | | |
|---|---|---|---|---|---|---|---|---|---|---|
| | Percent reporting | Infestation reported | | | | Percent reporting | Infestation reported | | | |
| | | None | Light | Moderate | Heavy | | None | Light | Moderate | Heavy |
| A. Boll worm | | | | | | | | | | |
| 1. American boll worm | 96.47 | 74.39 | 21.95 | 3.66 | 0.00 | 11.76 | 30.00 | 50.00 | 20.00 | 0.00 |
| 2. Pink boll worm | 87.06 | 75.68 | 20.27 | 4.05 | 0.00 | 61.18 | 3.85 | 36.54 | 30.77 | 28.85 |
| 3. Spotted boll worm | 90.59 | 72.73 | 20.78 | 6.49 | 0.00 | 56.47 | 2.08 | 41.67 | 39.58 | 16.67 |
| 4. Others | 18.82 | 50.00 | 6.25 | 18.75 | 25.00 | 56.47 | 4.17 | 41.67 | 35.42 | 18.75 |
| B. Sucking pests | | | | | | | | | | |
| 1. Thrips | 96.47 | 4.88 | 56.10 | 35.37 | 3.66 | 11.76 | 30.00 | 0.00 | 20.00 | 50.00 |
| 2. Leafhopper | 95.29 | 3.70 | 58.02 | 30.86 | 7.41 | 58.82 | 0.00 | 18.00 | 56.00 | 26.00 |
| 3. Whitefly | 95.29 | 6.17 | 58.02 | 33.33 | 2.47 | 57.65 | 0.00 | 20.41 | 55.10 | 24.49 |
| 4. Others | 3.53 | 33.33 | 0.00 | 66.67 | 0.00 | 60.00 | 1.96 | 19.61 | 47.06 | 31.37 |
| C. Foliage feeding pests | | | | | | | | | | |
| 1. Leaf roller | 94.12 | 27.50 | 45.00 | 26.25 | 1.25 | 42.35 | 2.78 | 38.89 | 50.00 | 8.33 |
| 2. Caterpillar | 89.41 | 27.63 | 40.79 | 28.95 | 2.63 | 40.00 | 2.94 | 29.41 | 61.76 | 5.88 |
| 3. Others | 7.06 | 50.00 | 33.33 | 16.67 | 0.00 | 5.88 | 20.00 | 40.00 | 40.00 | 0.00 |
| D. Soil pests | | | | | | | | | | |
| 1. Termite | 88.24 | 34.67 | 17.33 | 36.00 | 12.00 | 40.00 | 2.94 | 14.71 | 55.88 | 26.47 |

**Table 7.5** Response of Bt cotton sample farmer households on various aspects

| Particulars | Yes | No | No opinion |
|---|---|---|---|
| 1. Did any private sales agent approach you for promoting Bt cotton? | 0 | 100 | 0 |
| 2. Is Bt cotton plant shorter? | 0 | 100 | 0 |
| 3. Does Bt cotton have smaller bolls? | 0 | 100 | 0 |
| 4. Does Bt cotton give lesser number of cotton pickings? | 0 | 96.5 | 3.5 |
| 5. Did any Government agency approach you for inspecting the cotton variety you have sown? | 0 | 100 | 0 |
| 6. Do you face any problem in marketing Bt cotton? | 0 | 100 | 0 |
| 7. In your opinion is Bt cotton is more pest resistant than non-Bt? | 82.4 | 14.1 | 3.5 |
| 8. Do you need to buy Bt cotton seed every year? | 100 | 0 | 0 |
| 9. Is Bt cotton seed easily available | 97.6 | 2.4 | 0 |
| 10. Will you continue with Bt cotton cultivation? | 94.1 | 2.4 | 2.4 |
| 11. Do you feel that the pest/insect attack on other crop is higher or lower, when Bt Cotton is cultivated?? | 0 | 0 | 100 |
| 12. Have you observed any adverse effect on the environment due to Bt Cotton cultivation?? | 0 | 100 | 0 |

**Table 7.6** Bt cotton farmer households' response to "Who Recommended" in growing Bt cotton (percentage)

| | |
|---|---|
| Percent of farmers responding | 95.30 |
| 1. Extension worker | 2.47 |
| 2. Fellow farmer | 50.62 |
| 3. Village leader | 7.41 |
| 4. Village cooperative | 7.41 |
| 5. Seed company | 20.99 |
| 6. Seed dealer | 11.11 |

**Table 7.7** Advantages of Bt seed as opposed to traditional cotton conveyed by agents referred in above table (percentage)

| | |
|---|---|
| Percent of farmers responding | 83.53 |
| 1. More profit | 73.24 |
| 2. Less pesticides spraying | 66.20 |
| 3. No boll shedding | 15.49 |
| 4. Comparatively more bolls | 59.15 |

Over 50.6 % of the sample households adopted Bt seeds with the recommendation of fellow farmers, and another major sources of information was seed company agents/ dealers (Table 7.6). Some farmers came to know about the Bt technology from the village cooperatives and the village leaders. But government extension agencies did not play much role in creating farmer awareness of Bt cotton. The main communication from the agents was about its superiority in terms of better profits and lesser amount of pesticides sprays, more bolls per plant, no bolls shedding etc. (Table 7.7).

**Table 7.8** Seed rate used in Bt compared to non-Bt

| | |
|---|---|
| Percent of farmers responding | 100 |
| Percentage of farmers reporting use of lower seed rate | 100 |
| 25 % | 57.65 |
| 25–50 % | 42.45 |

**Table 7.9** Farmers response on early flowering of Bt cotton

| | |
|---|---|
| Percent of farmers responding early flowering | 100 |
| 1–10 days | 50.59 |
| 10–20 days | 36.47 |
| Above 20 days | 12.94 |

**Table 7.10** Response on comparison of Bt vis-à-vis traditional varieties by sample farmer households

| | |
|---|---|
| Percent of farmers responding | 74.12 |
| 1. More bolls | 66.23 |
| 2. Better yield | 71.56 |
| 3. Less pest attack | 30.00 |
| 4. High cost of seeds | 82.56 |

Regarding the seed rate for Bt cotton as compared to non-Bt cotton, all of sample farmer households indicated that the seed rate used for Bt was lower compared to non-Bt About 57.65 % of them reported the seed rate used was lower by up to 25 %, and the rest 42.35 % reported 25–50 % lower seed rate compared to non-Bt cotton (Table 7.8). This indicates an adjustment to the high price of seeds and their more efficient use.

All the sample farmer households reported early flowering of Bt cotton compared to non-Bt cotton. About 50.6 % reported that Bt cotton flowers 1–10 days early, 36.5 % reported it at 10–20 days earlier, and 12.9% reported that Bt cotton flowers more than 20 days earlier than non-Bt cotton (Table 7.9).

The sample farmers were asked about the reasons behind their preference and comparison of Bt cotton over non-Bt cotton. Over 74 % of the households responded to this question (Table 7.10). Better yield, more bolls, and less pest attack were the major reasons expressed by them. On the other hand, 83 % of them expressed that the cost of Bt cotton seed was very high.

The sample farmers were asked about their suggestion to improve the Bt technology and 74.12 % of the sample households responded to it (Table 7.11). The most important suggestion given was to reduce the cost of Bt cotton seed. Other suggestions in order of their importance are: arranging field demonstrations, seed packages with smaller quantities, and assurance of seed quality.

**Table 7.11** Suggestion by Bt cotton farmer households for improving the acceptance of Bt cotton

| Percent of farmers responding | 74.12 |
|---|---|
| 1. Reduce seed cost | 53.97 |
| 2. Seed packages with less quantity seeds | 14.29 |
| 3. Field demonstration | 33.33 |
| 4. Assurance of seed quality | 6.35 |

**Table 7.12** Advantages or disadvantages of Bt cotton-G vis-à-vis non-Bt cotton reported by Bt cotton-growers: percentage – Maharashtra

| | Strong advantage | Advantage | No difference | Disadvantage | Strong disadvantage | |
|---|---|---|---|---|---|---|
| 1 | Availability of seeds | 0.0 | 11.0 | 80.5 | 8.5 | 0.0 |
| 2 | Seed cost/price | 0.0 | 3.7 | 2.5 | 35.8 | 58.0 |
| 3 | Quality of avail. seeds | 4.8 | 46.4 | 44.0 | 4.8 | 0.0 |
| 4 | Pest incidence/problem | 3.6 | 58.3 | 23.8 | 14.3 | 0.0 |
| 5 | Pesticide need/cost | 8.3 | 60.7 | 23.8 | 6.0 | 1.2 |
| 6 | Fertilizer need/cost | 1.2 | 15.9 | 47.6 | 35.4 | 0.0 |
| 7 | Labour need/cost | 2.4 | 8.4 | 73.5 | 15.7 | 0.0 |
| 8 | Machine need/cost | 2.4 | 6.0 | 89.3 | 1.2 | 1.2 |
| 9 | Irrigation need/cost | 4.8 | 12.0 | 45.8 | 37.3 | 0.0 |
| 10 | Harvesting cost | 2.4 | 6.0 | 56.0 | 34.5 | 1.2 |
| 11 | Cotton quality | 4.8 | 51.8 | 41.0 | 2.4 | 0.0 |
| 12 | Market preference | 2.4 | 16.7 | 73.8 | 6.0 | 1.2 |
| 13 | Staple length | 2.4 | 34.9 | 50.6 | 12.0 | 0.0 |
| 14 | Fiber colour | 2.4 | 16.7 | 73.8 | 6.0 | 1.2 |
| 15 | Cotton price | 2.4 | 2.4 | 88.0 | 7.2 | 0.0 |
| 16 | Easy marketing | 2.4 | 15.7 | 74.7 | 7.2 | 0.0 |
| 17 | By-product output | 2.4 | 9.6 | 88.0 | 0.0 | 0.0 |
| 18 | Yield | 14.5 | 81.9 | 2.4 | 1.2 | 0.0 |
| 19 | Profit | 12.0 | 77.1 | 9.6 | 1.2 | 0.0 |
| 20 | Livestock feeding | 2.4 | 7.2 | 90.4 | 0.0 | 0.0 |
| 21 | Water saving | 8.5 | 20.7 | 53.7 | 17.1 | 0.0 |
| 22 | Suitable for early sowing | 14.6 | 58.5 | 26.8 | 0.0 | 0.0 |
| 23 | Suitable for late sowing | 2.4 | 2.4 | 79.3 | 15.9 | 0.0 |

The Bt cotton sample farmer households were specifically asked about their opinion on advantages and disadvantages of Bt cotton. The major advantages of Bt cotton that are expressed by majority of the sample farmer households are yield superiority, more profit, lesser need of pesticides, better quality, and its suitability for early sowing (Table 7.12).

On the other hand a common disadvantage expressed was the high cost of seed. Other issues expressed include higher fertilizer and irrigation cost, and higher harvest cost. No major differences were observed in other matters.

## 7.4 Observations of Sample Farmer Households from Tamil Nadu

Majority of the sample farmer households from Tamil Nadu said that the plant size of Bt cotton is shorter than non-Bt cotton but the boll size of Bt cotton is bigger (Table 7.13). Ninety-two percent of the sample farmers indicated that the number of pickings is same for Bt and non-Bt cotton.

The sample farmer households were unanimous in terms of the reasons behind the adoption of Bt cotton: less boll worm attack and the crop being more remunerative compared to other crops (Table 7.14).

Majority of the sample households did not observe any major difference in terms of the flowering time between Bt and non-Bt cotton either under irrigated or unirrigated conditions (Table 7.15). As of now, none of the sample framers had observed any adverse impact on the environment caused by the cultivation of Bt cotton (Table 7.16).

**Table 7.13** Attributes of Bt cotton compared to non-Bt cotton: response of sample households (%): Tamil Nadu

| | |
|---|---|
| **Plant size** | |
| Shorter | 93.4 |
| Taller | 2.2 |
| Same | 4.4 |
| **Boll size** | |
| Bigger | 87.8 |
| Smaller | 10.0 |
| Same | 2.2 |
| **Number of picking** | |
| More | 3.3 |
| Less | 4.4 |
| Same | 92.3 |

**Table 7.14** Reason for cultivation Bt cotton (%): Tamil Nadu

| | |
|---|---|
| Less boll worm and less pesticides use | 100 |
| More remunerative compared to other crops | 100 |

**Table 7.15** Flowering of Bt cotton as compared to non-Bt cotton: farmers' response (%): Tamil Nadu

| | Irrigated | Unirrigated | Overall |
|---|---|---|---|
| Early | 6.3 | 8.6 | 7.8 |
| No difference | 93.7 | 91.4 | 92.2 |

**Table 7.16** Impact of Bt cotton on environment: Tamil Nadu

| | | Bt | Non-Bt |
|---|---|---|---|
| Adverse impact on environment observed (%) | Yes | 0 | 0 |
| | No | 100 | 100 |

**Table 7.17** Measures suggested by farmers to help Bt cotton cultivation: Tamil Nadu

| | Percentage |
|---|---|
| Reduce the price of seed | 100 |
| Field visit and guidance by extension agencies | 88 |
| Traders should not sell spurious seeds | 100 |

Major measures suggested by the sample households to help Bt cultivation are reduction in the seed price, guidance from extension agencies, and prevention of the sale of spurious Bt cotton seeds (Table 7.17).

# Chapter 8
# Consolidated Summary and Conclusions

The study has examined the performance and returns to Bt cotton vs. Non-Bt cotton in the state of Andhra Pradesh, Gujarat, Maharashtra and Tamil Nadu. The study has been undertaken at the request of Ministry of Agriculture, Government of India. It is part of a coordinated project undertaken to objectively examine performance of Bt cotton across the states of Gujarat, Maharashtra, Andhra Pradesh and Tamil Nadu, in light of various conflicting voices and opinions.

## 8.1 Bt Cotton Technology

There have been major advances in biotechnology in the recent years and this has made it possible to directly identify genes, isolate them, know their functions, sequence them and transfer them from one organism to another. These developments have spanned the entire biological sciences. The development of Bt cotton is one outcome of this. Between 1996 and 2003 the global area under transgenic crops has increased 25-fold from 1.7 million hectare to 68 million hectare. In the year 2010, which marked the 15th anniversary of commercialization of transgenic or biotech crops, the global area was estimated to be over 1 billion hectares in 29 countries, 19 developing and 10 industrialized countries. This is tenfolds increase over the area coverage in year 2005, when the area under cultivation was around 90 million hectare spreading across 21 countries, 11 developing and 10 industrialized countries. In all, over 59 countries have approved biotech crops, 29 are cultivating biotech crops while remaining 30 countries have approved import of biotech crops. Bt Cotton was developed by Monsanto and it is now one of the most widely grown transgenic crops currently grown in many countries including United States, China, India, Australia, Argentina, South Africa and Indonesia. The adoption of Bt cotton has been rapid, from an estimated 0.8 million hectares in 1996 to over 23 million hectares in 2010 and over 23 million hectares globally.


V.P. Gandhi, D. Jain, *Introduction of Biotechnology in India's Agriculture*, DOI 10.1007/978-981-10-1091-0_8

The reported advantages of Bt cotton include agronomic, economic and environmental benefits. The major agronomic advantage of Bt cotton over the conventional cotton is the resistance to the bollworm pest. The major economic benefits are reduced need for pesticides, and yield superiority through the resistance over non-Bt cotton varieties. Even though there are some potential environmental risks, the major environmental benefits include reduction in number of pesticides sprays, less exposure to pesticides for human beings and animals, and less pesticides in the water and soil.

Many countries have reported positive experiences with Bt cotton. This includes USA, China and Australia. Bt cotton has spread rapidly in China. Great demand for it is reported from the farmers since it reduces the cost of pesticide applications and provides effective yield superiority. India entered late after much hesitation. The Government of India allowed the growing of three genetically modified Bt cotton hybrids initially for 3 years from April 2002 to March 2005. The Indian trial data over several years demonstrated the superiority of Bt technology in terms reduced pesticides application and increase in effective yield. Even though the performance of Bt cotton has been projected to be satisfactory, there is great discontent in some quarters. Those in favour indicate reduction in the use of insecticides, better yield per unit of input use, equal or better quality, and lesser residue of pesticides in the fiber. Those against indicate concerns such as: the gene may spread and have adverse impact in the eco-system, higher cost of Bt cotton seeds as compared to non-Bt seeds, inadequate pest resistance thus requiring the farmers to still use insecticides, and other issues. It is in this context that this study has been undertaken to examine the advantages/disadvantages, the economics Bt cotton vs. non-B cotton at the farmer level, and other aspects such as pest incidence, impact on environment etc.

## 8.2 Cotton in India

Though India ranks first in area cultivated of cotton in the world, it occupies the second position in production after China and US because of low ranking in yields. About 65 % of the cotton cultivation in India is unirrigated and therefore less productive and subject to vagaries of monsoon. Cotton fiber accounts for about 73 % of the total raw material mix of the textile industry. The cotton crop is highly susceptible to insects/pests and about 166 different species of insect pests are reported to attack cotton at various stages of its growth. It is estimated that the pests and diseases cause over 50 % damage to cotton in India, compared to 24.5 % world over. Of about 96,000 metric tons of technical grade pesticides produced in the country, about 54 % is estimated to be used on cotton.

Area under cotton in India was about 9 million hectares in 2004–2005 which is about 5 % of the total cropped area in the country. It has further risen to about 12 million hectare in 2010–2011 and about 85 % of total cotton area. Large variation in the area under cotton is observed over the years due to the vagaries of rainfall, as well as fluctuating prices and profitability of cotton. The cotton yield in India is one of the lowest in the world and it stagnated or declined during the 1990s. However, there is significant growth after 2002–2003 in the wake of Bt cotton. The estimated production of cotton in 2004–2005 was a record in the history of cotton cultivation in the country at 21.3 million bales (1 bale = 170 kg), and this has further risen to 33.0 million bales by 2010–2011. The annual rate of growth of cotton production during the period 1990–1991 to 2001–2002 was 2.68 % but taken during 2002–2003 and 2010–2011, it is more than three times at 8.63 % per annum. There was significant increase in yield after the introduction of Bt cotton technology and The average cotton yield during 1990–2091 to 2001–2002 was about 300 kg per hectare and it increased substantially to over 400 kg per hectare during 2002–2003 to 2010–2011.

The Bt cotton was approved for commercial cultivation in India in 2002. In March 2002 the Genetic Engineering Approval Committee (GEAC), the regulatory authority for transgenic crops in India, approved the commercial cultivation of three Bt cotton varieties viz., Bt Mech 12, Bt Mech 162 and Bt Mech 184. This remained and only after several years in 2005, the GEAC approved large scale field trials and seed production of 12 more varieties of Bt hybrids. Gujarat and Maharastra were the early adopters of Bt cotton on a large scale that commenced from 2002, followed by Andhra Pradesh, Karnataka, Tamil Nadu and Madhya Pradesh.

## 8.3 Cotton in the Study States

Based on cotton production in India during the recent triennium ending 2007–2008, Gujarat ranks top with a share of 36 %, followed by Maharashtra with 17.8 % and Andhra Pradesh with 13.2 %. Tamil Nadu has a share of only 1.86 % in the national production. Together, Gujarat, Maharashtra, Andhra Pradesh and Tamil Nadu accounted for 69 % of the cotton production. In terms of area under cotton, Maharashtra occupies the top position with a share of 33.2 % in the 9.2 million hectares of area under cotton cultivation in the country, followed by Gujarat with 25.36 % and Andhra Pradesh with 11.3 % during triennium ending 2007–2008. However, the average yield of cotton is among the lowest in Maharashtra at 273 kg per hectare as against 514 kg per hectare for the country as a whole.

## 8.4 Data and Sampling

This study is based on the primary data collected from four states namely Andhra Pradesh, Gujarat, Maharashtra and Tamil Nadu. Effort was made to adopt similar methodology, content and survey instruments in all these states, as far as possible. Effort was made to have nearly an equal number of Bt and Non-Bt farmers in the sample. Effort was also made through stratification to cover both irrigated and unirrigated farms under Bt and non-Bt cotton, as well as small, medium and large farmers. The primary data collected pertains to the agricultural year 2004–2005.

The target sample size for each state was 180 cotton farmers with 90 Bt and 90 Non-Bt farmers. Maharashtra was an exception where the coverage was 85 Bt and 69 Non-Bt farmers. The Gujarat sample did not have unirrigated cotton as a sufficient number of such farmers were not available in the sample districts. The number of sample farmer households under unirrigated Bt cotton was relatively less in Andhra Pradesh and Maharashtra, but were relatively more in Tamil Nadu. The study had an overall sample size of 694 farm households.

## 8.5 Varieties Grown

Many Bt and non Bt varieties were reported. The use of non-genuine Bt seeds was reported by both Gujarat and Andhra Pradesh sample households. This accounted for 56 % of the Bt sample in Gujarat and 20 % in Andhra Pradesh. The genuine Bt seeds that were used by the sample farmer households were of RCH and Mahyco varieties – MECH 184, MECH 12, MECH 162, and Rasi (RCH) 2, in Andhra Pradesh and Gujarat besides the use of non-confirmed Bt seeds. While in Maharashtra both Mahyco and Rasi Bt seeds were used by the sample households, in Tamil Nadu only Rasi Bt seeds were used. A large number of non-Bt hybrid seeds were also reported among the sample households such as Bunny, Super Bunny, Brahma, Satya, Attara, JK, Tagore, Bindu, Sankar, Vikram, Navbharat-Deshi Ankur, Bunny, and Ajit.

## 8.6 Cost of Seeds

The average cost per hectare of Bt Cotton seeds used by the sample households varied substantially across the states. It was the least in Gujarat at Rs 3079 per hectare followed by Andhra Pradesh with Rs. 3313. On an average the cost of Bt seeds per hectare was more than double than the non-Bt seeds in all the four states.

## 8.7 Application of Pesticides

The average number of pesticides sprays as well as its cost per hectare was higher for non-Bt cotton in all the study states. The difference across states was very high. The average number of spray on non-Bt Cotton in Gujarat was 0.52, but it was as high as 3.84 in Andhra Pradesh. Similarly the cost of pesticides was higher on non-Bt cotton by Rs. 436 per hectare in Gujarat, and Rs. 2749 in Andhra Pradesh.

## 8.8 Yields of Bt and Non-Bt Cotton

In Andhra Pradesh the Bt yields were higher but the difference in the yield levels under irrigated and unirrigated was not consistent. The yield of Bt cotton irrigated was significantly higher in Maharashtra and Tamil Nadu. There was no consistent yield difference across farm sizes. The average yield of Bt cotton over non-Bt cotton was higher in all the states in the study and it was higher by 18.2 % in Andhra Pradesh, 28.4 % in Gujarat, 46.4 % in Maharashtra and 28.5 % in Tamil Nadu.

## 8.9 Cost of Production, Value of Output and Profit

The total cost of cultivation under irrigated condition was in general higher than under unirrigated conditions both for Bt and Non-Bt cotton. The cost of cultivation per hectare of Bt cotton exceeded that of non-Bt cotton in all the states and the difference was the highest in Maharashtra exceeding Rs. 8250.

The average per hectare seed cost for Bt cotton was higher compared to non-Bt cotton and the difference ranged from Rs. 1600 per ha in Gujarat to over Rs. 2700 per ha in Tamil Nadu. The lower cost of Bt cotton seeds in Gujarat and Andhra Pradesh as opposed to Maharashtra and Tamil Nadu could be due to use of non-genuine Bt seeds in Gujarat and Andhra Pradesh. The share of seed cost of Bt seeds in total cost of production was about 10–17 % in the selected states, whereas it varied from 4 % to 6 % for non-Bt cotton seeds.

The per hectare cost of pesticides under Bt cotton varied from Rs. 2732 in Gujarat to Rs. 7806 in Andhra Pradesh. This is as against Rs. 3168 in Gujarat to Rs. 10,878 in Andhra Pradesh for Non-Bt While the share of pesticide cost in total cost of cultivation ranged from 8.29 % in Tamil Nadu to 24.29 % in Andhra Pradesh for Bt cotton, it varied from 11.96 % in Gujarat to 35.73 % in Andhra Pradesh for non-Bt.

The value of output of Bt cotton per hectare exceeded that of non-Bt cotton by 30.84 % in Gujarat, 42.1 % in Andhra Pradesh, 44.74 % in Maharashtra and 47.06 % in Tamil Nadu. The net profit per hectare from Bt cotton ranged from Rs. 15,242 in Tamil Nadu to Rs. 34,199 in Gujarat as opposed to Rs. 5772 in Tamil Nadu to Rs. 21,880 in Gujarat for non-Bt cotton. The Benefit-Cost Ratios of Bt cotton is higher than that of non-Bt cotton.

## 8.10 Results of Econometric Analysis

The results of statistical/econometric analysis of the whole sample indicate that the positive impact of Bt cotton on the yields has strong statistically significance. It is significant at the 99 % level and the estimates indicate that Bt cotton yields are 30.71 % higher. The impact of the value of output is also highly significant and estimates show that this is boosted by 33.35 %. However, the cost also rises significantly, and this rise is estimated to be 6.69 %. The pesticide cost is reduced by 23.98 %, but the seed cost rises by 168.77 %. The difference in the output price between Bt and Non-Bt cotton is positive but not statistically significant. The results indicate that the profit rise is also highly significant and the increase is estimated to be 87.58 %. The results explain the popularity of Bt cotton, at the same time, the opposition to the high seed cost.

The performance varies from state to state. The results for Gujarat indicate that the positive impact on yield and value of output is greater than the combined results, but the cost increase is also greater. The reduction in the pesticide cost is somewhat lower, but the increase in the seed cost is also lower. The price increase is statistically significant but small, and the profit increase is 73.81 %. In the case of Maharashtra indicate that the impact on the yield and the value of output at 42.67 % and 42.79 % respectively are the highest among the three states, and the impact on the total cost is relatively low at 5.18 %. Pesticide cost is reduced by 22.38 %, and the profit increase is the highest at 120.08 %.

In the case of Andhra Pradesh, the impact on the yields as well as the value of outputs is the lowest at about 21.33 %, but the rise in total cost is also lower. The fall in the pesticide cost is the highest in Andhra Pradesh at −28.17 %, but the rise in the seed cost is also the highest at 192.53 %. This is supports the great opposition to the seed prices in Andhra Pradesh. The rise in the profits is statistically highly significant and amounts to a 78.18 %, which is in between Gujarat and Maharashtra. The absolute level of profitability of cotton in Andhra Pradesh is lowest amongst the three states.

The model relating the performance to all the covered inputs/factors including Bt, pesticide, seed, fertilizer, irrigation, and state of location together, though affected by multicollinearity, indicates that Bt alone is still statistically highly significant as a determinant of the yield, value of output and profitability. Its exclusive impact on yield is estimated to be about 22 % and the impact on profitability about 35 %. Profit is negatively related to pesticide cost and positively related to seed cost (reflecting use of Bt seeds) and irrigation. The results also indicate that whereas profits are significantly higher in Maharashtra as compared to Gujarat, there is no statistically significant difference in the profitability between Gujarat and Andhra Pradesh.

## 8.11 General Observations from Farmers

The sample farmer households of Andhra Pradesh expressed that the major advantages of Bt cotton over non-Bt cotton varieties were lesser need for pesticides, better yields, and profits. No major differences were expressed in terms of the use of inputs other than pesticides, and price of cotton. Two major disadvantages expressed by the sample households were the availability of Bt seeds and its high price. About 57.6 % of the sample households reported that the seed rate for Bt cotton was reduced by about 25 %, because the seeds are very expensive. Other major features observed were shorter plant size of Bt cotton, bigger boll size, and lesser number of pickings as compared to non-Bt cotton. 96 % of the sample households reported that Bt cotton has more resistance to boll worm attack. Over 90 % of the sample households expressed their willingness to continue with Bt seeds considering better yields and profitability. About 10 per of the sample households expressed their unwillingness to continue with Bt cultivation due to high cost of seeds and its unsuitability to their land.

The sample farmer households in Gujarat expressed that the major advantages of Bt cotton over non-Bt cotton varieties were better yields and profits, less pest incidence and pesticides cost, and suitability for early sowing. No major differences were expressed in terms of the use of inputs other than pesticides, and price of cotton. Two major disadvantages expressed by the sample households were high seed price and the availability of Bt seeds.

Observations on the pest incidence are available from the Maharashtra Bt growers. They indicate that for boll worms, including American, pink and spotted boll worms, no infestation on Bt cotton is indicated by over 70 % of growers, whereas no infestation is reported by only 2–30 % on non-Bt cotton. Only about 4–6 % report moderate to heavy infestation on Bt, whereas this number is as high as 20–60 % on non-Bt Surprisingly, there is also a difference in the infestation by sucking and foliage feeding pests, for which the incidence is none to light in the case of Bt, whereas it is moderate to heavy in the case of non-Bt Thus, Bt cotton appears to provide considerable resistance to boll worms, and even shows resistance for other pests.

Over 50.6 % of the Maharashtra sample households adopted Bt seeds with the recommendation of fellow farmers, and the next major sources of information was seed company agents/dealers. Some farmers came to know about the Bt technology from the village cooperatives and the village leaders. But government extension agencies did not play much role. The main communication from the agents was about its superiority in terms of better profits and lesser amount of pesticides sprays, more bolls per plant, no bolls shedding etc.

Other responses from Maharashtra Bt cotton growers indicate that no government agencies had approached them for the inspection of Bt cotton, and none of the sample households felt specific problems with respect to the marketing of Bt cotton. None of the Bt growers indicate having observed any adverse environmental impact as a result of the cultivation of BT cotton. They also do not indicate any increase in the pest attack on other crops as a result of the cultivation of Bt cotton. Almost 98 % of the sample farmers did not face any difficulty in getting quality Bt seeds in time. As high as 94.1 % of the sample farmers, responded affirmatively on continuing with Bt cultivation in the future.

The major advantages of Bt cotton that are expressed by majority of the Maharashtra sample farmer households were yield superiority, more profit, lesser need of pesticides, better quality, and its suitability for early sowing. On the other hand a common disadvantage expressed was the high cost of seed. Other differences expressed include higher fertilizer and irrigation cost, and higher harvest cost. No major differences were observed in other matters.

To improve the use of Bt technology the most frequent suggestion given by the Maharashtra farmers was to reduce the cost of Bt cotton seed. Other suggestions were: arranging field demonstrations, seed packages with smaller quantities, and assurance of seed quality.

Majority of the sample farmer households from Tamil Nadu reported that the plant size of Bt cotton is shorter than non-Bt cotton but the boll size of Bt cotton is bigger. Ninety-two percent of the sample farmers indicated that the number of pickings is same for Bt and non-Bt cotton. Majority of the sample households did not observe any major difference in terms of the flowering time between Bt and non-Bt cotton either under irrigated or unirrigated conditions. None of the sample framers had observed any adverse impact on the environment caused by the cultivation of Bt cotton. Major measures suggested by the Tamil Nadu sample households to help Bt cultivation are reduction in the seed price, guidance from extension agencies, and prevention of the sale of spurious Bt cotton seeds.

# Part III
# State Study 1: Andhra Pradesh

The part is based on research study done by N. Ramgopal, Agro-Economic Research Centre Andhra University, Visakhapatnam, under the guidance of Vasant P. Gandhi CMA-IIMA, and the Head of the Agro-Economic Research Centre.

# Chapter 9
# Cotton Cultivation in Andhra Pradesh

## 9.1 Importance of Cotton in the Cropping Pattern of the State

Soils and climate in Andhra Pradesh are conducive for growing a variety of crops. The prominent crops grown in the state are Rice, Jowar, Bajra, Maize, Ragi, Bengal gram, Red gram, Green gram, Horse gram, Black gram, Groundnut, Castor, Sesamum, Sunflower, Coconut, Cotton, Sugarcane, Tobacco, Chillies, Fruits and Vegetables. Among these, the dominant crops are Rice, Groundnut, cotton, Jowar, Maize, Green gram, Black gram and Red gram. The cotton crop stands in the third place in terms of area. Over the past two and half decades there had been a significant increase in the area under this crop in the state. It increased from 404,191 ha (3.22 % of GCA) in TE 1980–1981 to 644,359 ha (4.88 % of GCA) in TE 1990–1991 and further increased to 916,006 ha (7.5 % of GCA) in TE 2003–2004 (Table 9.1).

## 9.2 Area, Production and Productivity Trends in Cotton in the State

Since cotton is the third largest crop in the state, economic analysis of this crop assumes importance. In this connection the trends in area, production and productivity are analysed in terms of trend growth rates. For computing the trend growth rates, linear and exponential trend curves are estimated and the best fitted form is chosen on the basis of the maximum value of R2. The growth rate based on the best filled form is used in the analysis. The time period covered in this analysis is 1970–1971 to 2003–2004. It is divided into sub-periods 1970–1971 to 1980–1981, 1980–1981 to 1990–1991 and 1990–1991 to 2003–2004 and analysis is carried out separately.


V.P. Gandhi, D. Jain, *Introduction of Biotechnology in India's Agriculture*, DOI 10.1007/978-981-10-1091-0_9

**Table 9.1** Changes in the cropping pattern in A.P triennium ending 1980–1981, 1990–1991, 2000–2001, 2003–2004 (% in gross cropped area)

| Crops | 1981 | 1991 | 2001 | 2004 |
|---|---|---|---|---|
| Rice | 29.32 | 31.46 | 31.28 | 26.23 |
| Jowar | 18.04 | 9.63 | 5.39 | 5.21 |
| Bajra | 4.07 | 3.15 | 0.96 | 0.86 |
| Wheat | 0.15 | 0.08 | 0.10 | 0.10 |
| Maize | 2.46 | 2.28 | 3.43 | 4.57 |
| Ragi | 2.17 | 1.27 | 0.74 | 0.63 |
| Small millets | 4.80 | 1.55 | 0.46 | 0.43 |
| Total cereals and millets | 61.00 | 49.43 | 42.37 | 38.03 |
| Bengal Gram | 0.42 | 0.50 | 1.27 | 2.99 |
| Red Gram | 1.66 | 2.60 | 3.33 | 3.84 |
| Green Gram | 4.17 | 3.82 | 3.61 | 4.49 |
| Horse Gram | 2.72 | 1.06 | 0.72 | 0.66 |
| Black Gram | 1.87 | 3.68 | 3.59 | 4.84 |
| Other pulses | 0.40 | 0.10 | 0.20 | 0.09 |
| Total pulses | 11.23 | 11.77 | 12.72 | 16.91 |
| Total food grains | 72.24 | 61.20 | 55.09 | 54.94 |
| Groundnut | 10.42 | 17.65 | 14.08 | 12.69 |
| Castor | 1.94 | 2.57 | 20.16 | 2.20 |
| Linseed | 0.10 | 0.05 | 0.04 | 0.04 |
| Sesamum | 1.40 | 1.22 | 1.28 | 1.11 |
| Sunflower | 0.27 | 0.17 | 0.13 | 0.15 |
| Rap and Mustard | 0.01 | 0.03 | 0.03 | 0.02 |
| Coconut | 0.33 | 0.46 | 0.75 | 0.85 |
| Niger | 0.09 | 0.13 | 0.14 | 0.14 |
| Other Oil Seeds | 0.04 | 1.00 | 2.26 | 0.27 |
| Total oilseeds | 14.60 | 23.27 | 20.86 | 20.87 |
| Cotton | 3.22 | 4.88 | 8.33 | 7.49 |
| Sugarcane | 1.25 | 1.38 | 1.70 | 1.80 |
| Tobacco | 1.43 | 1.22 | 1.04 | 1.03 |
| Chillies | 1.27 | 1.67 | 1.81 | 1.90 |
| Other crops | 5.99 | 6.38 | 11.17 | 11.97 |
| Gross cropped area | 100.00 | 100.00 | 100.00 | 100.00 |

The estimates of the trend equations presented in Table 9.2 show that in a large number of cases, exponential form is relatively a better fit. Therefore exponential growth rates are used in the analysis.

After 1970, cotton cultivation in the state experienced remarkable growth. The area under the crop increased from 316,215 ht in 1970–1971 to 836,892 ht in 2003–2004 at an exponential rate of 4.17 % per annum, while the yield increased from 0.24 to 2.25 tonnes/ha at an exponential rate of 3.50 % per annum and the total production increased from 1.6 to 3.18 at an exponential rate of 7.67 %. But this rate

**Table 9.2** Estimated values of the parameters of trend equations and growth of area, production and Productivity of cotton in Andhra Pradesh: 1970–1971 to 2003–2004

| Item | Functional form | 1970–1971 to 1980–1981 | | | | | | 1980–1981 to 1990–1991 | | | | | |
|---|---|---|---|---|---|---|---|---|---|---|---|---|---|
| | | A | B | t | $R^2$ | F | Gr | a | b | t | $R^2$ | F | Gr |
| Area | Linear | 3.03 | 0.0905 | 1.772 | 0.2586 | 3.14 | 2.53 | 4.01 | 0.2266* | 3.912 | 0.6279 | 15.30 | 4.22 |
| | Exponential | 1.11 | 0.0249 | 1.631 | 0.2281 | 2.66 | 2.49 | 1.41 | 0.0420* | 3.672 | 0.5997 | 13.48 | 4.20 |
| Production | Linear | 172.63 | 52.9134* | 3.238 | 0.5381 | 10.48 | 0.00 | 1032.61 | 41.1017 | 1.180 | 0.1339 | 1.39 | 3.21 |
| | Exponential | 5.25 | 0.1341* | 3.521 | 0.5794 | 12.40 | 13.41 | 6.93 | 0.0309 | 1.147 | 0.1276 | 1.32 | 3.09 |
| Productivity | Linear | 63.21 | 11.7570* | 3.176 | 0.5285 | 10.09 | 8.79 | 253.74 | –2.2663 | 0.366 | 0.0146 | 0.13 | –0.94 |
| | Exponential | 4.14 | 0.1093* | 3.188 | 0.5303 | 10.16 | 10.93 | 5.52 | –0.0112 | 0.429 | 0.0200 | 0.18 | –1.12 |
| **Item** | **Functional form** | **1990–1991 to 2003–2004** | | | | | | **1970–1971 to 2003–2004** | | | | | |
| | | A | b | t | $R^2$ | F | Gr | a | b | t | $R^2$ | F | Gr |
| Area | Linear | 7.54 | 0.2147** | 2.025 | 0.2547 | 4.10 | 2.35 | 1.91 | 0.2534* | 12.301 | 0.8254 | 151.28 | 3.99 |
| | Exponential | 2.01 | 0.0250** | 2.224 | 0.2918 | 4.94 | 2.50 | 1.02 | 0.0417* | 14.947 | 0.8747 | 223.39 | 4.17 |
| Production | Linear | 2057.77 | 61.6894** | 2.222 | 0.2915 | 4.94 | 2.45 | –36.22 | 89.5140* | 13.788 | 0.8559 | 190.07 | 5.85 |
| | Exponential | 7.64 | 0.0238** | 2.104 | 0.2694 | 4.42 | 2.38 | 5.72 | 0.0767* | 12.203 | 0.8231 | 148.89 | 7.67 |
| Productivity | Linear | 277.21 | 0.2067 | 0.066 | 0.0004 | 0.00 | 0.07 | 117.07 | 5.9642* | 5.902 | 0.5212 | 34.83 | 2.69 |
| | Exponential | 5.63 | –0.0012 | 0.105 | 0.0009 | 0.01 | –0.12 | 4.69 | 0.0350* | 5.839 | 0.5158 | 34.09 | 3.50 |

Note: *** = significant at 99 percent, ** = significant at 95 percent, * = significant at 90 percent.

of increase is not uniform throughout the period 1970–1971 to 2003–2004. Area, production and productivity exhibited different trends in the 1970s, 1980s and after.

In the 1970s, there was no significant increase in the area under the crop but yield increased tremendously and consequently the production. During this period, yields recorded a growth rate of 10.81 % per annum and production recorded a growth rate of 12.40 % per annum.

The trends in area, production and productivity observed in the 1970s were reversed in the 1980s. In the 1980s area increased significantly at the rate of 4.20 % per annum. The rates of growth of both yield and production are not statistically significant.

The increasing trend in the area has been continued after 1990. It increased at a rate of 2.50 % per annum during the period 1990–1991 to 2003–2004. Production also increased at the rate of 2.38 % per annum during this period. But Yield recorded a statistically not significant negative rate of growth.

Thus, in Andhra Pradesh after 1980, there has been a continuous increase in the area under cotton with no significant improvement in productivity. In fact, there has been a decline in yields. However due to the area effect, production has been increasing over time.

## 9.3 District Wise Pattern of Cotton Cultivation

Of the 23 districts in the state, 21 districts have cotton cultivation now. Among these, the prominent are, Adilabad, Guntur, Warangal, Khammam, Nalgonda, Karimnagar, Kurnool, Prakasam, Krishna and Mahbubbnagar. All these districts together accounted for 88.64 % of the total area (836,892 ha) in 2003–2004. In total production of 1,890,394 these districts accounted for 90.67 % (Table 9.3).

Also in the districts of Guntur, Adilabad, Karimnagar, Warangal, Khammam and Nalgonda cropping pattern is dominated by cotton crop after 2000. The proportion of area under cotton in GCA varied from 10 % to 30 %. The domination of cotton is observed even earlier in the districts of Guntur and Adilabad (Table 9.4).

Cotton is grown on both irrigated and unirrigated (rain fed) lands in the state, but mostly on unirrigated lands. Out of 837,000 ha area under cotton 674,000 ha (80.53) is under unirrigated cultivation. Cotton under ensured irrigation is found mostly in the districts Warangal, Karimnagar, Nellore, Kadapa and Prakasam (Table 9.5).

The district wise data on yields of cotton (lint) on irrigated and unirrigated lands furnished by Directorate of Economics and Statistics, Hyderabad are given in Table 9.6 at the state level and in all the prominent cotton grown districts the yields on irrigated lands are much higher than the yields on the unirrigated lands. However, the yields on unirrigated lands in some districts are much higher than the average yields on the irrigated lands at the district level. This may be due to the lower area under cultivating districts.

**Table 9.3** District wise area, production and yield of cotton in AP 2009–2010

| District | Area (hectares) | % of area in total area | Production (bales) | % of production in total production | Yield (kg/ha) |
|---|---|---|---|---|---|
| Adilabad | 280.50 | 19.12 | 552.80 | 17.13 | 335.00 |
| Anantapur | 2.10 | 0.14 | 1.20 | 0.04 | 101.00 |
| Chittoor | 0.10 | 0.01 | 0.10 | 0.00 | 242.10 |
| East Godavari | 6.80 | 0.46 | 19.60 | 0.61 | 488.00 |
| Guntur | 151.50 | 10.33 | 524.90 | 16.27 | 589.00 |
| Karimnagar | 199.60 | 13.61 | 461.30 | 14.30 | 393.00 |
| Khammam | 131.90 | 8.99 | 300.30 | 9.31 | 387.00 |
| Krishna | 46.90 | 3.20 | 140.70 | 4.36 | 510.00 |
| Kurnool | 36.10 | 2.46 | 52.60 | 1.63 | 248.00 |
| Mahbubnagar | 118.70 | 8.09 | 150.10 | 4.65 | 215.00 |
| Medak | 66.00 | 4.50 | 54.00 | 1.67 | 139.00 |
| Nalgonda | 136.40 | 9.30 | 216.60 | 6.71 | 270.00 |
| Nellore | 4.60 | 0.31 | 14.70 | 0.46 | 546.00 |
| Nizamabad | 15.10 | 1.03 | 38.70 | 1.20 | 435.00 |
| Prakasam | 33.00 | 2.25 | 93.60 | 2.90 | 483.00 |
| Rangareddi | 26.60 | 1.81 | 35.70 | 1.11 | 228.00 |
| Srikakulam | 1.20 | 0.08 | 4.00 | 0.12 | 546.00 |
| Visakhapatnam | 0.70 | 0.05 | 2.20 | 0.07 | 546.10 |
| Vizianagaram | 8.80 | 0.60 | 15.10 | 0.47 | 292.00 |
| Warangal | 185.30 | 12.63 | 527.70 | 16.35 | 484.00 |
| West Godavari | 2.50 | 0.17 | 8.10 | 0.25 | 546.00 |
| Y.S.R. | 12.60 | 0.86 | 17.90 | 0.55 | 241.00 |
| Andhra Pradesh | 1467.00 | 100.00 | 3227.00 | 100.00 | 374.00 |

Many improved and hybrid varieties of cotton are grown in the state. They are broadly classified into five major categories viz., (1) Superior long staple, (2) long staple, (3) Superior Medium Staple, (4) Medium staple and (5) short staple. Among the varieties grown, Suvin, Seas island, MCU-5, and Varalaxmi come under superior long staple, H-4 and Nandium, come under long staple, N-14 Coconda's ordinary Kumpta, Munagari and Gowri come under medium staple and Chinapathi, Hyderabad Oomaras and Munagari come under short staple. The district wise data on the area covered under these varieties are presented in Table 9.7. However, the area particulars relating to area under the latest varieties like Bt Cotton, Bunny, Brahma etc., are not included in the data furnished by the Directorate of Economics and Statistics. As per the data available, the most widely grown varieties at the state as well as at the district level are MCU-5 and H-4. The other varieties like Varalaxmi, Nandium, Chinapathi, and Mungari are also grown to a considerable extent.

**Table 9.4** District wise area under cotton in Andhra Pradesh – Triennium ending in 1980, 1990, 2000 and 2004 (area in hectares)

| District | AREA TE 1980 | % of Area in total area in the State | % of area in GCA-district wise | AREA TE 1990 | % of area in total area in the State | % of Area in GCA-District wise | AREA TE 2000 | % of area in total area in the State | % of Area in GCA-District wise | AREA TE 2004 | % of Area in total area in the State | % of area in GCA-district wise |
|---|---|---|---|---|---|---|---|---|---|---|---|---|
| Srikakulam | 9 | 0.00 | 0.00 | 0 | 0.00 | 0.00 | 1513 | 0.14 | 0.37 | 1652 | 0.18 | 0.43 |
| Vizianagaram | 28 | 0.01 | 0.01 | 82 | 0.01 | 0.02 | 9294 | 0.86 | 2.02 | 15,836 | 1.73 | 3.88 |
| Visakhapatnam | 50 | 0.01 | 0.01 | 25 | 0.00 | 0.01 | 1328 | 0.12 | 0.31 | 1080 | 0.12 | 0.29 |
| East godavari | 1357 | 0.34 | 0.24 | 5425 | 0.89 | 0.80 | 7374 | 0.68 | 0.98 | 6604 | 0.72 | 0.94 |
| West godavari | 109 | 0.03 | 0.02 | 1648 | 0.27 | 0.25 | 2690 | 0.25 | 0.39 | 2656 | 0.29 | 0.42 |
| Krishna | 1798 | 0.45 | 0.26 | 13,524 | 2.21 | 1.85 | 33,121 | 3.07 | 4.61 | 34,059 | 3.72 | 5.32 |
| Guntur | 69,480 | 17.25 | 9.40 | 170,679 | 27.90 | 21.15 | 165,104 | 15.32 | 19.19 | 121,279 | 13.24 | 15.60 |
| Prakasam | 37,593 | 9.33 | 5.81 | 57,821 | 9.45 | 9.02 | 48,086 | 4.46 | 8.52 | 32,601 | 3.56 | 5.70 |
| Nellore | 2732 | 0.68 | 0.71 | 2730 | 0.45 | 0.78 | 6798 | 0.63 | 1.92 | 7448 | 0.81 | 2.13 |
| Chittor | 139 | 0.03 | 0.02 | 2 | 0.00 | 0.00 | 14 | 0.00 | 0.00 | 2 | 0.00 | 0.00 |
| Cuddapah | 12,165 | 3.02 | 2.86 | 1654 | 0.27 | 0.41 | 21,077 | 1.96 | 4.94 | 16,288 | 1.78 | 3.79 |
| Anantapur | 25,275 | 6.28 | 2.68 | 9994 | 1.63 | 1.03 | 10,248 | 0.95 | 1.01 | 6801 | 0.74 | 0.65 |
| Kurnool | 127,415 | 31.64 | 12.59 | 55,927 | 9.14 | 5.82 | 116,826 | 10.84 | 12.17 | 59,060 | 6.45 | 6.36 |
| Mahabubnagar | 8262 | 2.05 | 0.83 | 24,178 | 3.95 | 2.65 | 68,856 | 6.39 | 8.49 | 47,222 | 5.16 | 5.63 |
| Ranga reddy | 161 | 0.04 | 0.05 | 11,261 | 1.84 | 3.52 | 22,243 | 2.06 | 7.04 | 16,320 | 1.78 | 5.68 |
| Hyderabad | 91 | 0.02 | – | – | – | – | – | – | – | – | – | – |
| Medak | 393 | 0.10 | 0.08 | 2764 | 0.45 | 0.59 | 16,424 | 1.52 | 3.23 | 9707 | 1.06 | 1.80 |
| Nizamabad | 4316 | 1.07 | 1.21 | 7301 | 1.19 | 2.15 | 17,607 | 1.63 | 4.82 | 9422 | 1.03 | 2.90 |
| Adilabad | 107,431 | 26.67 | 17.81 | 146,412 | 23.93 | 24.12 | 157,097 | 14.58 | 27.91 | 160,952 | 17.57 | 28.79 |
| Karimnagar | 189 | 0.05 | 0.04 | 21,998 | 3.60 | 4.09 | 73,169 | 6.79 | 13.74 | 58,504 | 6.39 | 11.72 |
| Warangal | 2840 | 0.71 | 0.53 | 47,467 | 7.76 | 8.29 | 133,892 | 12.42 | 23.54 | 141,835 | 15.48 | 25.09 |
| Khammam | 195 | 0.05 | 0.05 | 21,886 | 3.58 | 4.60 | 79,983 | 7.42 | 16.85 | 81,839 | 8.93 | 19.58 |
| Nalgonda | 828 | 0.21 | 0.11 | 9116 | 1.49 | 1.28 | 84,900 | 7.88 | 13.29 | 84,841 | 9.26 | 15.87 |
| Andhra Pradesh | 402,752 | 100.00 | 3.18 | 611,835 | 100.00 | 3.21 | 1,077,646 | 100.00 | 8.34 | 916,006 | 100.00 | 7.49 |

**Table 9.5** Irrigated and unirrigated area under cotton (area in 000' hectares)

| District | Irrigated | Unirrigated | Total |
|---|---|---|---|
| 1. Srikakulam | – | 2 | 2 |
| | – | (100.00) | (100.00) |
| 2. Vizianagaram | – | 19 | 19 |
| | – | (100.00) | (100.00) |
| 3. Visakhapatnam | – | 1 | 1 |
| | – | (100.00) | (100.00) |
| 4. East Godavari | – | 9 | 9 |
| | – | (100.00) | (100.00) |
| 5. West Godavari | 1 | 3 | 4 |
| | (25.00) | (75.00) | (100.00) |
| 6. Krishna | Negl | 35 | 35 |
| | – | (100.00) | (100.00) |
| 7. Guntur | 9 | 130 | 139 |
| | (6.47) | (93.53) | (100.00) |
| 8. Prakasam | 15 | 21 | 36 |
| | (41.67) | (58.33) | (100.00) |
| 9. Nellore | 8 | 2 | 10 |
| | (80.00) | (20.00) | (100.00) |
| 10. Chittor | – | – | – |
| | – | – | – |
| 11. Cuddapah | 8 | 8 | 16 |
| | (50.00) | (50.00) | (100.00) |
| 12. Anantapur | 2 | 2 | 4 |
| | (50.00) | (50.00) | (100.00) |
| 13. Kurnool | 7 | 32 | 39 |
| | (17.95) | (82.05) | (100.00) |
| 14. Mahabubnagar | 5 | 30 | 35 |
| | (14.29) | (85.71) | (100.00) |
| 15. Rangareddy | – | 15 | 15 |
| | – | (100.00) | (100.00) |
| 16. Hyderabad | – | – | – |
| | – | – | – |
| 17. Medak | – | 6 | 6 |
| | – | (100.00) | (100.00) |
| 18. Nizamabad | – | – | 9 |
| | – | (100.00) | (100.00) |
| 19. Adilabad | 4 | 141 | 145 |
| | (2.76) | (97.24) | (100.00) |
| 20. Karimnagar | 17 | 32 | 49 |
| | (34.69) | (65.31) | (100.00) |
| 21. Warangal | 78 | 46 | 124 |
| | (62.90) | (37.10) | (100.00) |

(continued)

**Table 9.5** (continued)

| District | Irrigated | Unirrigated | Total |
|---|---|---|---|
| 22. Khammam | 9 | 67 | 76 |
| | (11.84) | (88.16) | (100.00) |
| 23. Nalgonda | Negl | 64 | 64 |
| | – | (100.00) | (100.00) |
| Andhra Pradesh | 163 | 674 | 873 |
| | (19.47) | (80.53) | (100.00) |

**Table 9.6** Yield of cotton: district wise 2003–2004. Lint Kgs/Ha

| District | Irrigated | Unirrigated |
|---|---|---|
| 1. Srikakulam | 0 | 506 |
| 2. Vizianagaram | 0 | 234 |
| 3. Visakhapatnam | 0 | 506 |
| 4. East Godavari | 0 | 506 |
| 5. West Godavari | 433 | 506 |
| 6. Krishna | 503 | 503 |
| 7. Guntur | 695 | 585 |
| 8. Prakasam | 254 | 348 |
| 9. Nellore | 506 | 595 |
| 10. Chittor | 0 | 0 |
| 11. Cuddapah | 83 | 212 |
| 12. Anantapur | 216 | 172 |
| 13. Kurnool | 355 | 142 |
| 14. Mahabubnagar | 335 | 269 |
| 15. Rangareddy | 0 | 399 |
| 16. Hyderabad | 0 | 0 |
| 17. Medak | 0 | 393 |
| 18. Nizamabad | 0 | 127 |
| 19. Adilabad | 338 | 339 |
| 20. Karimnagar | 429 | 345 |
| 21. Warangal | 457 | 254 |
| 22. Khammam | 549 | 373 |
| 23. Nalgonda | 300 | 300 |
| *Andhra Pradesh* | *418* | *376* |

## 9.4 Introduction of Bt Cotton to the State

Bt cotton is introduced to the state in 2002. The GEAC had permitted Mahyco limited for commercial release of three Bt cotton hybrids, namely MECH-12 BT, MECH-162BT and MECH-184 BT But Mahyco has marketed only the first two varieties in Andhra Pradesh during Kharif 2002.

**Table 9.7** Area under different varieties of cotton (Ha)

| District | Superior long staple | | | | | Long staple | | | Superior medium staple | | | | | | | |
|---|---|---|---|---|---|---|---|---|---|---|---|---|---|---|---|---|
| | Suvin | Sea island | MCU-5 | Varalakshmi | Total | H – 4 | Nandium | Total | Lakshmi | Parbhani American | L-147 | Buri 1007 | Jaya Dhar | Humpi | Krishna | Total |
| Srikakulam | 0 | 180 | 1263 | 13 | 1456 | 0 | 0 | 0 | 34 | 0 | 0 | 0 | 0 | 0 | 0 | 34 |
| % | – | 10.52 | 73.82 | 0.76 | 85.1 | – | – | – | 1.99 | – | – | – | – | – | – | 1.99 |
| Vizianagaram | 0 | 0 | 0 | 0 | 0 | 0 | 0 | 0 | 0 | 0 | 0 | 0 | 0 | 0 | 0 | 0 |
| % | – | – | – | – | – | – | – | – | – | – | – | – | – | – | – | – |
| Visakhapatnam | 0 | 0 | 127 | 1076 | 1203 | 0 | 0 | 0 | 0 | 0 | 0 | 0 | 0 | 0 | 0 | 0 |
| % | – | – | 9.34 | 79.18 | 88.52 | – | – | – | – | – | – | – | – | – | – | – |
| East Godavari | 0 | 0 | 3904 | 2601 | 6505 | 0 | 0 | 0 | 1244 | 0 | 0 | 0 | 0 | 0 | 0 | 1244 |
| % | – | – | 43.53 | 29 | 72.53 | – | – | – | 13.87 | – | – | – | – | – | – | 13.87 |
| West Godavari | 0 | 0 | 3077 | 543 | 3620 | 0 | 0 | 0 | 83 | 0 | 0 | 0 | 0 | 0 | 0 | 83 |
| % | – | – | 81.85 | 14.45 | 96.3 | – | – | – | 2.21 | – | – | – | – | – | – | 2.21 |
| Krishna | 0 | 0 | 4103 | 413 | 4516 | 0 | 0 | 0 | 2550 | 21,097 | 2571 | 3683 | 34 | 81 | 85 | 30,101 |
| % | – | – | 11.58 | 1.16 | 12.74 | – | – | – | 7.2 | 59.54 | 7.25 | 10.39 | 0.1 | 0.23 | 0.24 | 84.95 |
| Guntur | 0 | 840 | 18,277 | 5344 | 24,461 | 92,762 | 5532 | 98,294 | 1930 | 0 | 1729 | 3911 | 0 | 0 | 1944 | 9514 |
| % | – | 0.6 | 13.16 | 3.85 | 17.61 | 66.77 | 3.98 | 70.75 | 1.39 | – | 1.24 | 2.82 | – | – | 1.4 | 6.85 |
| Prakasam | 0 | 0 | 14,932 | 1548 | 16,480 | 2322 | 0 | 2322 | 0 | 0 | 0 | 0 | 0 | 0 | 3620 | 3620 |
| % | – | – | 41.69 | 4.32 | 46.01 | 6.48 | – | 6.48 | – | – | – | – | – | – | 10.11 | 10.11 |
| Nellore | 0 | 0 | 1158 | 155 | 1313 | 603 | 0 | 603 | 21 | 0 | 0 | 0 | 0 | 225 | 5840 | 6086 |
| % | – | – | 11.59 | 1.55 | 13.14 | 6.04 | – | 6.04 | 0.21 | – | – | – | – | 2.25 | 58.44 | 60.9 |
| Chittoor | 0 | 0 | 0 | 0 | 0 | 0 | 0 | 0 | 0 | 0 | 0 | 0 | 0 | 0 | 0 | 0 |
| % | – | – | – | – | – | – | – | – | – | – | – | – | – | – | – | – |
| Cuddapah | 0 | 294 | 63 | 0 | 357 | 5177 | 0 | 5177 | 260 | 0 | 0 | 0 | 0 | 0 | 0 | 260 |
| % | – | 1.86 | 0.4 | – | 2.26 | 32.8 | – | 32.8 | 1.65 | – | – | – | – | – | – | 1.65 |
| Ananthapur | 0 | 0 | 181 | 10 | 191 | 941 | 0 | 941 | 0 | 0 | 0 | 0 | 0 | 0 | 0 | 0 |

(continued)

**Table 9.7** (continued)

| District | Superior long staple | | | | | Long staple | | | Superior medium staple | | | | | | | |
|---|---|---|---|---|---|---|---|---|---|---|---|---|---|---|---|---|
| | Suvin | Sea island | MCU-5 | Varalakshmi | Total | H – 4 | Nandium | Total | Lakshmi | Parbhani American | L-147 | Buri 1007 | Jaya Dhar | Humpi | Krishna | Total |
| % | – | – | 4.3 | 0.24 | 4.54 | 22.34 | – | 22.34 | – | – | – | – | – | – | – | – |
| Kurnool | 31 | 21 | 2282 | 37 | 2371 | 13,591 | 42 | 13,633 | 19 | 0 | 0 | 0 | 9667 | 0 | 0 | 9686 |
| % | 0.08 | 0.05 | 5.81 | 0.09 | 6.04 | 34.61 | 0.11 | 34.71 | 0.05 | – | – | – | 24.61 | – | – | 24.66 |
| Mahbubnagar | 566 | 338 | 9001 | 1957 | 11,862 | 14,567 | 4280 | 18,847 | 760 | 0 | 778 | 0 | 0 | 0 | 0 | 1538 |
| % | 1.62 | 0.97 | 25.73 | 5.59 | 33.91 | 41.64 | 12.2 | 53.88 | 2.17 | – | 2.22 | – | – | – | – | 4.4 |
| Rangareddy | 0 | 0 | 10,851 | 18 | 10,869 | 3063 | 0 | 3063 | 0 | 0 | 1383 | 0 | 0 | 0 | 0 | 1383 |
| % | – | – | 70.61 | 0.62 | 70.73 | 19.93 | – | 19.93 | – | – | 9 | – | – | – | – | 9 |
| Hyderabad | 0 | 0 | 0 | 0 | 0 | 0 | 0 | 0 | 0 | 0 | 0 | 0 | 0 | 0 | 0 | 0 |
| % | – | – | – | – | – | – | – | – | – | – | – | – | – | – | – | – |

As against the area of 803,274 ha of cotton crop in the state during 2002–2003, 3315 ha were occupied by Bt cotton (0.41 %). By 2003 Kharif the area under the crop has gone up to 4859 ha. The incidence of pest attack is also less in this season. During 2004, in Andhra Pradesh only two companies i.e. Mahyco seeds and Rasi seeds have distributed the approved Bt cotton varieties. They are MECH-12 Bt, MECH-184 Bt and RCH-2 Bt, RCH-20 Bt Based on the estimates of the seed companies 72,874 ha were under licensed Bt seed in Kharif 2004.

Evaluation studies were conducted on the performance of Bt cotton by Acharya N.G. Ranga Agricultural University, Rajendranagar at Regional Agricultural Research station, Lam, Guntur district during the year 1999–2000 to 2001–2002. All Bt cotton hybrids showed good vigour and growth with significantly higher boll number and yield as compared to non-Bt.

Genomic investigations and its results revealed that increased application of Nitrogen fertilizer than the recommended dose showed marginal improvement in boll number and yield. The Bt hybrids are found to be high yielding and early maturing by 15–30 days due to early produced bolls.

According to APAU study in Kurnool district the Bt cotton hybrid has resistance to boll worms. But MECH-162 is not having good yield potential when compared with hybrids like Bunny, Savitha and NH-44.

## 9.5 Pest Incidence

Based on the average performance of Bt cotton hybrids over four seasons of testing it is noticed that the incidence of Helicoverpa larva is less i.e., population/ten plants ranging from 0.002 to 3.85 against 0.60–16.00/10 plants in check hybrids, such as 'Savitha' and NHH-44.

The Bt cotton hybrids have exhibited more susceptibility to three major diseases namely bacterial blight, Alternaria and grey mildew.

Farmers in Warangal district incurred losses by cultivating Monsanto Mahyco varieties in the year 2003–2004. Under pressure from farmers' organisations and voluntary groups government ordered an enquiry. Agricultural officials confirmed that the losses were indeed a fact. They have estimated 16,000 farmers have incurred losses averaging Rs 1496 per acre. There were allegations that the officials have under-reported the loss. After some more deliberations the government ordered that Monsanto should pay Rs 3500 per acre as compensation and its seed sales were banned until it complies with the order. Monsanto pleaded for review of the order and conveyed its willingness to pay Rs 800 per acre. Government rejected the proposal.

Another major issue that is bothering the Bt farmers is the high price of licensed seed. Monsanto is charging Rs 1600 for 450 g of seed packet. Out of this Rs 281 is paid to the seed dealers as commission. An all-party meet was arranged by the state government. All of them have shown solidarity with the government in support of measures to pressurise the seed companies to reduce the seed cost. The state government filed a case before the Monopolies and Restrictive Trade Practices Commission (MRTPC) against Monsanto. The government has complained that while the seed growers get less than Rs 250 for 750 g of seed, the farmer was asked to pay Rs 1850 for 450 g including a royalty or technology licensing fee of Rs 1250. The strategy worked to some extent as Monsanto offered to reduce the cost by Rs 300 per packet. The other seed companies like Nuzivid, Ankur, and Rasi have agreed to reduce by Rs 82, Rs 72 and Rs 80 respectively.

# Chapter 10
# Sampling and Methodology

This study is mainly based on the primary data collected from a sample of cotton cultivators. Therefore the validity of the results of the study depends on the type of sample design used for the selection of the farmers, and the method of collection and analysis of the data. All these aspects are described in this section.

## 10.1 Sample Design

In this study a four-stage sample design has been used. The first stage sampling units are the districts, the second stage sampling units are the mandals with in the selected districts and the third stage sampling units are the villages with in the selected mandals and the fourth stage sampling units are the cotton cultivators with in the selected villages.

### *10.1.1 Selection of Sample Districts/Mandals/Villages*

Keeping in view extensive cotton cultivation and sufficiently large area under Bt cotton cultivation, two districts, one with larger cotton area under irrigated conditions and the other with rain fed conditions, are selected. From each of the selected districts one or more mandals is selected based on the same criteria. Again three villages from each of the selected mandals are selected using the same criteria with the help of agricultural officials. As the maximum area under cotton cultivation and sufficiently large area under Bt cotton cultivation are the criteria of selection, the selected villages include both Bt cotton cultivators and non-Bt cotton cultivators. The selected districts, mandals and villages are shown below (Table 10.1).




V.P. Gandhi, D. Jain, *Introduction of Biotechnology in India's Agriculture*, DOI 10.1007/978-981-10-1091-0_10

**Table 10.1** Selected sample districts, mandals and villages

| District | Mandals | Villages |
|---|---|---|
| Guntur (Unirrigated) | Pattipadu | Gottipadu |
| | Tadikonda | Tadikonda |
| | Amaravathi | Dharanikota |
| Warangal (Irrigated) | Geesukonda | Machapur |
| | | Manukonda |
| | | Vanchanagiri |

**Table 10.2** Sample size

| Farm size (acre) | Distribution of sample farmers | | | | | | | | | | | |
|---|---|---|---|---|---|---|---|---|---|---|---|---|
| | Warangal (with high level of Irrigation) | | | | | | Guntur (with low level of Irrigation) | | | | | |
| | Village – 1 | | Village – 2 | | Village – 3 | | Village – 1 | | Village – 2 | | Village – 3 | |
| Bt cotton | | | | | | | | | | | | |
| | I | UI | I | UI | I | UI | I | UI | I | UI | I | UI |
| Below 5 | 4 | 3 | 4 | 3 | 4 | 3 | 4 | 3 | 4 | 3 | 4 | |
| 5–10 | 3 | 2 | 3 | 2 | 3 | 2 | 3 | 2 | 3 | 2 | 3 | 2 |
| Above 5 | 2 | 1 | 2 | 1 | 2 | 1 | 2 | 1 | 2 | 1 | 2 | 1 |
| Total | 9 | 6 | 9 | 6 | 9 | 6 | 9 | 6 | 9 | 6 | 9 | 6 |
| Non-Bt cotton | | | | | | | | | | | | |
| | I | UI | I | UI | I | UI | I | UI | I | UI | I | UI |
| Below 5 | 4 | 3 | 4 | 3 | 4 | 3 | 4 | 3 | 4 | 3 | 4 | 3 |
| 5–10 | 3 | 2 | 3 | 2 | 3 | 2 | 3 | 2 | 3 | 2 | 3 | 2 |
| Above 5 | 2 | 1 | 2 | 1 | 2 | 1 | 2 | 1 | 2 | 1 | 2 | 1 |
| Total | 9 | 6 | 9 | 6 | 9 | 6 | 9 | 6 | 9 | 6 | 9 | 6 |

*I* Irrigated, *UI* Unirrigated

### *10.1.2 Selection of Sample Farmer Households*

For selecting the sample farmer households, cotton cultivators in each village are classified into two groups viz., Bt cotton cultivators and non-Bt cotton cultivators. Again farmers in each group are classified in to three-landholding groups viz., below 5 acres, 5–10 acres and above 10 acres. Farmers in each landholding group are further classified into two categories viz., farmers cultivating cotton on irrigated land and farmers cultivating cotton on un- irrigated land. From each category representing a particular irrigation status, land holding group and type of cotton cultivation (Bt or Non Bt) farmer households are selected at random. The number of farmer households selected under each category is presented in Table 10.2. The total sample size chosen is 180.

**Table 10.3** Distribution of sample households by type of seed used, irrigation status and size of land holding (Guntur)

| Farm size in hectare | Number of farming households | | | | | | Percentage distribution | | | | | |
|---|---|---|---|---|---|---|---|---|---|---|---|---|
| | Bt cotton-G | | Bt cotton-NC | | Non-Bt cotton | | Bt cotton-G | | Bt cotton-NC | | Non-Bt cotton | |
| | I | UI | I | UI | I | UI | I | UI | I | UI | I | UI |
| Small (below 2) | 10 | 6 | 2 | 3 | 12 | 9 | 62.50 | 54.55 | 18.18 | 42.86 | 44.44 | 50.00 |
| Medium (2–4) | 5 | 3 | 4 | 3 | 9 | 6 | 31.25 | 27.27 | 36.36 | 42.86 | 33.33 | 33.33 |
| Large (above 4) | 1 | 2 | 5 | 1 | 6 | 3 | 6.25 | 18.18 | 45.45 | 14.28 | 22.23 | 16.67 |
| Total | 16 | 11 | 11 | 7 | 27 | 18 | 100.00 | 100.00 | 100.00 | 100.00 | 100.00 | 100.00 |

*Bt Cotton-G* Genuine, *Bt Cotton* Bt cotton, *NC* Non-confirmed Bt Cotton

### 10.1.3 Method of Data Collection and Reference Year

The data has been collected from each of the selected households by canvassing a pre- designed schedule at a single point of time, during the period March-April 2005. The reference year chosen for the study is 2004–2005.

### 10.1.4 Method of Analysis

As the focus of this study is on the economic analysis of Bt cotton Vs. non-Bt cotton cultivation, a comparative study is made between Bt Cotton and non-Bt cotton cultivation in respect of all-important parameters. The performance of both Bt cotton and non-Bt cotton cultivation are also analysed separately across the land holding groups under both irrigated and unirrigated conditions, since the size of landholding and irrigation status are assumed to have a significant bearing on the performance.

Some farmers are using non-confirmed Bt seed and its performance may be different from that of confirmed Bt seed (genuine Bt. seed). The distribution of Bt sample farmers by type of seed used (confirmed Bt and non-confirmed Bt), irrigation status and size of land holding is presented in Table 10.3.

It shows that in Guntur district the Bt farmers raising the crop under irrigated conditions are going for genuine seed and that is more marked in small holding group. Larger farmers seem to take risk and using non-confirmed Bt cotton more. In unirrigated conditions this observation is not so marked. Therefore, a comparative analysis is also made between farmers using genuine Bt seed and farmers using non-confirmed Bt seed in this study.

**Table 10.4** Distribution of Sample Households by Type of Seed Used, Irrigation Status and Size of Land Holding (Warangal)

| Farm size in hectare | Number of farming households | | | | | | Percentage distribution | | | | | |
|---|---|---|---|---|---|---|---|---|---|---|---|---|
| | Bt cotton-G | | Bt cotton-NC | | Non-Bt cotton | | Bt cotton-G | | Bt cotton-NC | | Non-Bt cotton | |
| | I | UI | I | UI | I | UI | I | UI | I | UI | I | UI |
| Small (below 2) | 12 | 9 | 0 | 0 | 12 | 9 | 44.44 | 50.00 | 0.00 | 0.00 | 44.44 | 50.00 |
| Medium (2–4) | 9 | 6 | 0 | 0 | 9 | 6 | 33.33 | 33.33 | 0.00 | 0.00 | 33.33 | 33.33 |
| Large (above 4) | 6 | 3 | 0 | 0 | 6 | 3 | 22.22 | 16.67 | 0.00 | 0.00 | 22.22 | 16.67 |
| Total | 27 | 18 | 0 | 0 | 27 | 18 | 100 | 100 | 100 | 100 | 100 | 100 |

*Bt Cotton-G* Genuine Bt Cotton, *Bt Cotton-NC* Non-Confirmed Bt Cotton

In Warangal district all the farmers are using licensed Bt seed and there are no reports of illegal Bt seeds in the sample area (Table 10.4). The entire analysis is carried out in terms of percentages and averages.

# Chapter 11
# Nature and Performance of Bt Cotton vs. Non-Bt Cotton in Andhra Pradesh

## 11.1 Introduction

As a backdrop for the study, this chapter is aimed at examining the area under Bt cotton and non-Bt cotton, cotton varieties used by the sample households, average seed cost, pesticides used in Bt and Non-Bt cotton and their cropping pattern basing on the sample data.

## 11.2 Cropping Pattern

All the three categories of farmers viz., small, medium and large in Guntur grow Bt- cotton, Non-Bt Cotton, Paddy, Maize and Chillies under irrigated and unirrigated conditions during the Kharif season. The other crops like Turmeric, Groundnut, Green gram, Black gram, Red gram, Tomato, Jute etc., are grown by a few farmers in a very limited extent of land.

Tables 11.1 and 11.2 gives the cropping pattern and types of cotton varieties grown in Guntur and Warangal districts of Andhra Pradesh. In percentage terms, Bt-NC leads Bt-G. Similarly, area of cotton under unirrigated cultivation is more than the area under irrigated cultivation. It is also observed from Table 11.1 that in Guntur district, Bunny is the most preferred variety among non-Bt varieties. Chilli is the next commercial crop raised in 8.84 % of irrigated area and 7.86 in rain fed conditions. Paddy is the main food crop grown exclusively in irrigated area.

Unlike in Guntur district, few cultivators are able to raise crops like Maize and Vegetables in summer season under well irrigation in Warangal district (see Table 11.2). Bt cotton farmers are procuring their seed only from licensed seed dealers until now. Hence, no farmer has reported any non-confirmed Bt crop. A wide variety of non-Bt crops are raised in the district. Among these Bunny and Super Bunny are popular.


V.P. Gandhi, D. Jain, *Introduction of Biotechnology in India's Agriculture*, DOI 10.1007/978-981-10-1091-0_11

**Table 11.1** Cropping pattern (Guntur)

| | Average area sown per farm | | | | | | | |
|---|---|---|---|---|---|---|---|---|
| | Small | | Medium | | Large | | Overall | |
| Season/crcp ha. | I | UI | I | UI | I | UI | I | UI |
| Summer, 2004 | | | | | | | | |
| Crop 1 | – | – | – | – | – | – | – | – |
| Crop 2 | – | – | – | – | – | – | – | – |
| Crop 3 | – | – | – | – | – | – | – | – |
| Kharif, 2004 | | | | | | | | |
| Bt Cotton-G | 0.14 | 0.13 | 011 | 0.43 | – | 0.49 | 0.10 | 0.31 |
| Bt Cotton-NC | 0.05 | 0.11 | 0.20 | 0.28 | 0.56 | 1.12 | 0.20 | 0.37 |
| Non-Bt cotton (by variety) | | | | | | | | |
| Bunny | 0.08 | 0.13 | 0.23 | 0.09 | 0.83 | 0.47 | 0.28 | 0.18 |
| Satya | 0.03 | 0.02 | 0.03 | 0.04 | – | 0.11 | 0.02 | 0.04 |
| Super bunny | 0.02 | 0.11 | 0.15 | 0.22 | – | – | 0.06 | 0.12 |
| Bindu | 0.04 | – | – | – | – | 0.16 | 0.02 | 0.03 |
| Durga | – | – | 0.07 | 0.03 | – | – | 0.02 | 0.01 |
| H-8 | – | – | 0.09 | – | – | – | 0.03 | – |
| Saraswathi | – | – | 0.01 | 0.05 | – | – | 0.005 | 0.02 |
| Tulasi | – | – | 0.03 | – | – | – | 0.01 | – |
| Gemini | – | – | – | 0.09 | – | – | – | 0.03 |
| Sindur | – | – | – | – | 0.04 | 0.18 | 0.01 | 0.04 |
| JK | – | – | – | – | – | 0.11 | – | 0.02 |
| Chetna | – | 0.03 | – | – | – | – | – | 0.01 |
| Lakshmi | – | – | – | – | 0.11 | – | 0.02 | – |
| Amog | – | – | – | – | – | 0.11 | – | 0.02 |
| Other crops | | | | | | | | |
| 1. Paddy | 0.22 | 0.00 | 0.20 | 0.00 | 0.23 | 0.00 | 0.22 | 0.00 |
| 2. Maize | 0.00 | 0.12 | 0.03 | 0.13 | 0.00 | 0.70 | 0.01 | 0.24 |
| 3. Chillies | 0.08 | 0.13 | 0.19 | 0.24 | 0.88 | 0.27 | 0.27 | 0.19 |
| 4. Turmeric | 0.00 | 0.00 | 0.00 | 0.00 | 0.07 | 0.00 | 0.01 | 0.00 |
| 5. Others | 0.00 | 0.01 | 0.02 | 0.17 | 0.17 | 0.22 | 0.04 | 0.10 |
| Total: | 0.66 | 0.79 | 1.36 | 1.77 | 2.89 | 3.94 | 1.33 | 1.73 |
| Rabi, 2004–2005 | | | | | | | | |
| Crop 1 | Nil | | | | | | | |
| Crop 2 | | | | | | | | |
| Crop 3 | | | | | | | | |

Paddy and Maize are the main food crops raised in the kharif season. Other commercial crops are chillies and Turmeric. In rabi season Paddy and Vegetables are cultivated by small farmers in a limited extent under well irrigation

**Table 11.2** Cropping pattern (Warangal)

| | Average area sown per farm | | | | | | | |
|---|---|---|---|---|---|---|---|---|
| | Small | | Medium | | Large | | Overall | |
| Season/crop | I | UI | I | UI | I | UI | I | UI |
| Summer, 2004 | | | | | | | | |
| Maize | 0.00 | 0.00 | 0.00 | 0.00 | 0.04 | 0.00 | 0.01 | 0.00 |
| Groundnut | 0.00 | 0.00 | 0.00 | 0.00 | 0.01 | 0.00 | 0.00 | 0.00 |
| Turmeric | 0.00 | 0.00 | 0.00 | 0.00 | 0.02 | 0.00 | 0.00 | 0.00 |
| Chillies | 0.00 | 0.00 | 0.00 | 0.00 | 0.01 | 0.00 | 0.00 | 0.00 |
| Tomato | 0.00 | 0.00 | 0.00 | 0.00 | 0.01 | 0.00 | 0.00 | 0.00 |
| Vegetables | 0.01 | 0.00 | 0.00 | 0.00 | 0.01 | 0.00 | 0.01 | 0.00 |
| Kharif, 2004 | | | | | | | | |
| Bt cotton-G | 0.30 | 0.13 | 0.36 | 0.12 | 0.97 | 0.31 | 0.45 | 0.16 |
| Bt cotton-NC | 0.00 | 0.00 | 0.00 | 0.00 | 0.00 | 0.00 | 0.00 | 0.00 |
| Non-Bt cotton (by variety) | | | | | | | | |
| Bunny | 0.06 | 0.03 | 0.05 | 0.22 | 0.02 | 0.16 | 0.05 | 0.12 |
| Brahma | 0.01 | 0.00 | 0.01 | 0.03 | 0.02 | 0.05 | 0.01 | 0.02 |
| Durga | 0.01 | 0.01 | 0.00 | 0.00 | 0.02 | 0.00 | 0.01 | 0.00 |
| J.K. | 0.00 | 0.03 | 0.01 | 0.08 | 0.07 | 0.09 | 0.02 | 0.06 |
| Super bunny | 0.03 | 0.00 | 0.07 | 0.12 | 0.07 | 0.00 | 0.05 | 0.04 |
| Tagore | 0.00 | 0.01 | 0.00 | 0.00 | 0.00 | 0.00 | 0.00 | 0.00 |
| Amoga | 0.01 | 0.00 | 0.00 | 0.00 | 0.00 | 0.00 | 0.00 | 0.00 |
| Bhavani | 0.01 | 0.00 | 0.00 | 0.00 | 0.00 | 0.00 | 0.00 | 0.00 |
| Rudhra | 0.00 | 0.00 | 0.00 | 0.00 | 0.05 | 0.00 | 0.01 | 0.00 |
| Dasara | 0.00 | 0.00 | 0.03 | 0.00 | 0.00 | 0.05 | 0.01 | 0.01 |
| Sumo | 0.00 | 0.00 | 0.01 | 0.00 | 0.00 | 0.00 | 0.00 | 0.00 |
| Attar | 0.00 | 0.00 | 0.05 | 0.05 | 0.02 | 0.05 | 0.02 | 0.03 |
| Satya | 0.00 | 0.00 | 0.01 | 0.00 | 0.00 | 0.00 | 0.00 | 0.00 |
| Prasanna | 0.00 | 0.00 | 0.03 | 0.00 | 0.00 | 0.00 | 0.01 | 0.00 |
| Akshyasad | 0.00 | 0.00 | 0.00 | 0.00 | 0.05 | 0.00 | 0.01 | 0.00 |
| Ankur | 0.00 | 0.00 | 0.00 | 0.00 | 0.05 | 0.00 | 0.01 | 0.00 |
| Assess-II | 0.00 | 0.00 | 0.00 | 0.00 | 0.05 | 0.00 | 0.01 | 0.00 |
| Suguna | 0.00 | 0.02 | 0.00 | 0.00 | 0.00 | 0.00 | 0.00 | 0.01 |
| Pradam | 0.00 | 0.03 | 0.00 | 0.00 | 0.00 | 0.00 | 0.00 | 0.01 |
| Chitra | 0.00 | 0.00 | 0.00 | 0.00 | 0.00 | 0.00 | 0.00 | 0.00 |
| Siddardha | 0.00 | 0.00 | 0.00 | 0.00 | 0.00 | 0.00 | 0.00 | 0.00 |
| Sigma | 0.00 | 0.00 | 0.00 | 0.03 | 0.00 | 0.00 | 0.00 | 0.01 |
| Dhan | 0.00 | 0.00 | 0.00 | 0.01 | 0.00 | 0.00 | 0.00 | 0.00 |
| Maruthi | 0.00 | 0.00 | 0.00 | 0.01 | 0.00 | 0.00 | 0.00 | 0.00 |
| *X*2-35 | 0.00 | 0.00 | 0.00 | 0.00 | 0.00 | 0.05 | 0.00 | 0.01 |
| Augora | 0.00 | 0.00 | 0.00 | 0.00 | 0.00 | 0.05 | 0.00 | 0.01 |
| Paddy | 0.23 | 0.02 | 0.24 | 0.04 | 0.39 | 0.07 | 0.27 | 0.04 |
| Maize | 0.03 | 0.17 | 0.11 | 0.36 | 0.52 | 0.45 | 0.16 | 0.29 |

(continued)

**Table 11.2** (continued)

| | Average area sown per farm | | | | | | | |
|---|---|---|---|---|---|---|---|---|
| | Small | | Medium | | Large | | Overall | |
| Season/crop | I | UI | I | UI | I | UI | I | UI |
| Chillies | 0.05 | 0.02 | 0.09 | 0.05 | 0.20 | 0.12 | 0.09 | 0.05 |
| Turmeric | 0.00 | 0.00 | 0.05 | 0.01 | 0.30 | 0.15 | 0.08 | 0.04 |
| Jowar | 0.00 | 0.00 | 0.00 | 0.03 | 0.00 | 0.00 | 0.00 | 0.01 |
| Others | 0.04 | 0.07 | 0.07 | 0.26 | 0.06 | 0.27 | 0.05 | 0.17 |
| Rabi, 2004–2005 | | | | | | | | |
| Paddy | 0.00 | 0.00 | 0.00 | 0.00 | 0.00 | 0.00 | 0.00 | 0.00 |
| Vegetables | 0.01 | 0.00 | 0.00 | 0.00 | 0.00 | 0.00 | 0.00 | 0.00 |

**Table 11.3** Area under cotton (Guntur)

| | Area under cotton (per farm) in ha | | | | | | Percentage share of irrigated and unirrigated area | | | | | |
|---|---|---|---|---|---|---|---|---|---|---|---|---|
| | Bt cotton-G | | Bt cotton-NC | | Non-Bt cotton | | Bt cotton-G | | Bt cotton-NC | | Non-Bt cotton | |
| Farm size | I | UI | I | UI | I | UI | I | UI | I | UI | I | UI |
| Small | 0.38 | 0.35 | 0.40 | 0.88 | 0.32 | 0.56 | 52.05 | 47.95 | 31.25 | 68.75 | 36.36 | 63.34 |
| Medium | 0.40 | 1.62 | 0.87 | 1.91 | 1.21 | 1.05 | 19.80 | 80.20 | 42.23 | 57.77 | 53.54 | 46.46 |
| Large | – | 2.97 | 1.69 | 3.37 | 1.98 | 2.29 | – | 100.00 | 33.40 | 66.60 | 46.37 | 53.63 |
| Overall | 0.34 | 1.02 | 1.01 | 1.83 | 0.95 | 1.07 | 25.00 | 75.00 | 35.56 | 64.44 | 47.03 | 52.97 |

## 11.3 Area Under Cotton

Bt cotton cultivation, which started in the state in a big way in 2001–2002, moved in a down ward direction during the years 2002–2003 and 2003–2004 and again picked up well during 2004–2005. This pattern of cultivation of Bt cotton as a whole is more or less the same in all the land holding groups indicating the fact that adoption of Bt seed is not influenced by size of land holding. However, this pattern differs in the use of Bt-G seed, Bt-NC seed on the irrigated and unirrigated lands. On the irrigated lands, there has been a continuous increase in the cultivation of Bt (genuine and non-confirmed), while; it is on a declining trend on the un irrigated lands. Regarding the farmers adoption of Bt -G seed Vs. Bt-NC seed, the sample data reveal that the use of Bt-NC seed is more popular in the state, and it is gaining momentum over time.

The Tables 11.3, and 11.4 presents the land holding group wise and irrigation status wise estimates of per holding area under Bt cotton-G, Bt cotton-NC and Non Bt cotton in Guntur and Warangal districts during the reference year. Within the Bt cotton cultivation, relatively more area is covered by non-confirmed Bt seed in Guntur. The cultivation of these three varieties of cotton (Bt-G, Bt-NC and non-Bt) is relatively more on unirrigated lands. However, between the Bt and Non Bt cotton

**Table 11.4** Area under Cotton (Warangal)

| | Area under cotton (per farm) in ha | | | | | | Percentage share of irrigated and unirrigated area | | | | | |
|---|---|---|---|---|---|---|---|---|---|---|---|---|
| | Bt cotton-G | | Bt cotton-NC | | Non-Bt cotton | | Bt cotton-G | | Bt cotton-NC | | Non-Bt cotton | |
| Farm size | I | UI | I | UI | I | UI | I | UI | I | UI | I | UI |
| Small | 0.44 | 0.17 | 0.00 | 0.00 | 0.15 | 0.21 | 72.13 | 27.87 | 0.00 | 0.00 | 41.67 | 58.33 |
| Medium | 0.65 | 0.19 | 0.00 | 0.00 | 0.35 | 0.59 | 77.38 | 22.62 | 0.00 | 0.00 | 37.23 | 62.77 |
| Large | 1.28 | 0.18 | 0.00 | 0.00 | 0.45 | 0.54 | 87.67 | 12.33 | 0.00 | 0.00 | 45.45 | 54.55 |
| Overall | 0.68 | 0.18 | 0.00 | 0.00 | 0.28 | 0.40 | 79.07 | 20.93 | 0.00 | 0.00 | 41.18 | 58.82 |

varieties, cultivation of Bt cotton is more on irrigated lands. All these observations are more or less the same in the three land holding groups indicating that size of land holding is not a factor influencing the spread of Bt cotton cultivation.

In Warangal district area under Bt cotton is slightly higher than the area under Non-Bt in the reference year. Bt cotton-G farmers have larger area, 73.98 %, under irrigation while non-Bt farmers have only 41.23 % of cotton area under irrigation.

Thus the spread of Bt cotton has been taking place on a limited scale in the state. It appears that farmers are going for Bt cotton cultivation on trial and error basis, without knowing fully, its advantages and limitations.

## 11.4 Cotton Varieties Used by the Sample Households

A number of brands of Bt-G, Hybrid, and non-Hybrid seeds are available in the state and they are being used by the farmers (see Table 11.5).

Genuine Bt cotton varieties are being grown since 2 years. They are Rasi and Monsanto-Mahyco seeds. In the first year of the crop they were favoured by equal number of farmers. But by second year, i.e., the reference year, Rasi seeds are grown by 81 % of the farmers whereas Mahyco could not increase its popularity and remained at 19 %.

Farmers in Bt-NC group could not name the seed company as they are Illegal and a few of them could identify the source, i.e. Kurnool district. In this group only 67 % of the farmers had raised Bt-NC variety in 2003–2004. So, 33 % of the farmers raised this crop for the first time in the reference year (see Table 11.5).

Among the non-Bt groups most popular brands of seeds are Bunny, Super Bunny, JK, Satya, Durga ad Sri Tulasi.

In Warangal district 69 % of the farmers have used Rasi seeds and 31 % farmers have raised the Mahyco variety in Bt category. The other hybrids that are very popular are Bunny, Super Bunny and Brahma. The farmers felt that Rasi variety is more suitable for their region where unirrigated conditions prevail most of the time (Table 11.6).

**Table 11.5** Different brands of cotton seeds bought by sample farmers in Guntur

| Year/brand name | Bt Cotton-G | | Bt cotton-NC | | Hybrid | | Non-hybrid | |
|---|---|---|---|---|---|---|---|---|
| | number | % | Number | % | Number | % | Number | % |
| 2004–2005 | | | | | | | | |
| Rasi | 22 | 81.48 | 0 | 0.00 | 0 | 0.00 | 0 | 0.00 |
| Mch | 5 | 18.52 | 0 | 0.00 | 0 | 0.00 | 0 | 0.00 |
| Bt | 0 | 0.00 | 13 | 72.22 | 0 | 0.00 | 0 | 0.00 |
| Knl.Bt | 0 | 0.00 | 5 | 27.78 | 0 | 0.00 | 0 | 0.00 |
| Bunny | 0 | 0.00 | 0 | 0.00 | 23 | 51.11 | 0 | 0.00 |
| Super bunny | 0 | 0.00 | 0 | 0.00 | 11 | 24.44 | 0 | 0.00 |
| Satya | 0 | 0.00 | 0 | 0.00 | 4 | 8.89 | 0 | 0.00 |
| Bindu | 0 | 0.00 | 0 | 0.00 | 2 | 4.44 | 0 | 0.00 |
| Ratna | 0 | 0.00 | 0 | 0.00 | 1 | 2.22 | 0 | 0.00 |
| Durga | 0 | 0.00 | 0 | 0.00 | 1 | 2.22 | 0 | 0.00 |
| Gemini | 0 | 0.00 | 0 | 0.00 | 1 | 2.22 | 0 | 0.00 |
| 118 | 0 | 0.00 | 0 | 0.00 | 1 | 2.22 | 0 | 0.00 |
| Sindur | 0 | 0.00 | 0 | 0.00 | 1 | 2.22 | 0 | 0.00 |
| | 27 | 100 | 18 | 100 | 45 | 100 | 0 | 0.00 |
| 2003–2004 | | | | | | | | |
| Rasi | 5 | 18.52 | 0 | 0.00 | 0 | 0.00 | 0 | 0.00 |
| Mch | 5 | 18.52 | 0 | 0.00 | 0 | 0.00 | 0 | 0.00 |
| Bt | 0 | 0.00 | 8 | 4.44 | 0 | 0.00 | 0 | 0.00 |
| Knl.Bt | 0 | 0.00 | 4 | 22.22 | 0 | 0.00 | 0 | 0.00 |
| Bunny | 6 | 22.22 | 4 | 22.22 | 15 | 33.33 | 0 | 0.00 |
| Super bunny | 8 | 29.63 | 1 | 5.56 | 13 | 28.89 | 0 | 0.00 |
| Gemini | 1 | 3.70 | 0 | 0.00 | 1 | 22.22 | 0 | 0.00 |
| Sritulasi | 1 | 3.70 | 0 | 0.00 | 0 | 0.00 | 0 | 0.00 |
| Durga | 0 | 0.00 | 0 | 0.00 | 2 | 4.44 | 0 | 0.00 |
| Hyma | 0 | 0.00 | 0 | 0.00 | 1 | 2.22 | 0 | 0.00 |
| Bharath | 0 | 0.00 | 0 | 0.00 | 1 | 2.22 | 0 | 0.00 |
| ITC 160 | 0 | 0.00 | 0 | 0.00 | 1 | 2.22 | 0 | 0.00 |
| Jk | 1 | 3.70 | 0 | 0.00 | 8 | 17.78 | 0 | 0.00 |
| Bindu | 0 | 0.00 | 1 | 5.56 | 1 | 22.22 | 0 | 0.00 |
| Satya | 0 | 0.00 | 0 | 0.00 | 1 | 22.22 | 0 | 0.00 |
| Ratna | 0 | 0.00 | 0 | 0.00 | 1 | 2.22 | 0 | 0.00 |
| | 27 | 100 | 18 | 100 | 45 | 100 | 0 | 0.00 |
| 2002–2003 | | | | | | | | |
| Bunny | 6 | 22.22 | 7 | 38.89 | 15 | 33.33 | 0 | 0.00 |
| Super bunny | 11 | 40.74 | 4 | 22.22 | 6 | 13.33 | 0 | 0.00 |
| Sritulasi | 1 | 3.70 | 0 | 0.00 | 1 | 2.22 | 0 | 0.00 |
| Attara | 1 | 3.70 | 0 | 0.00 | 1 | 2.22 | 0 | 0.00 |
| Durga | 1 | 3.70 | 0 | 0.00 | 4 | 8.89 | 0 | 0.00 |
| Janaki | 1 | 3.70 | 0 | 0.00 | 2 | 4.44 | 0 | 0.00 |

(continued)

**Table 11.5** (continued)

| | Bt Cotton-G | | Bt cotton-NC | | Hybrid | | Non-hybrid | |
|---|---|---|---|---|---|---|---|---|
| Year/brand name | number | % | Number | % | Number | % | Number | % |
| Brahma | 0 | 0.00 | 1 | 5.56 | 0 | 0.00 | 0 | 0.00 |
| Amrutha | 0 | 0.00 | 1 | 5.56 | 0 | 0.00 | 0 | 0.00 |
| Jk | 5 | 18.52 | 4 | 22.22 | 5 | 11.11 | 0 | 0.00 |
| Ankur | 1 | 3.70 | 0 | 0.00 | 1 | 2.22 | 0 | 0.00 |
| Sandeep | 0 | 0.00 | 1 | 5.56 | 0 | 0.00 | 0 | 0.00 |
| Ratna | 0 | 0.00 | 0 | 0.00 | 3 | 6.67 | 0 | 0.00 |
| Satya | 0 | 0.00 | 0 | 0.00 | 2 | 4.44 | 0 | 0.00 |
| Saraswathi | 0 | 0.00 | 0 | 0.00 | 1 | 2.22 | 0 | 0.00 |
| Bindu | 0 | 0.00 | 0 | 0.00 | 1 | 2.22 | 0 | 0.00 |
| | 27 | 100 | 18 | 100 | 45 | 100 | 0 | 0.00 |
| 2001–2002 | | | | | | | | |
| Bunny | 9 | 33.33 | 8 | 44.44 | 15 | 33.33 | 0 | 0.00 |
| Super Bunny | 2 | 7.41 | 3 | 16.67 | 9 | 20.00 | 0 | 0.00 |
| Rama | 1 | 3.70 | 0 | 0.00 | 0 | 0.00 | 0 | 0.00 |
| Attara | 1 | 3.70 | 0 | 0.00 | 0 | 0.00 | 0 | 0.00 |
| Dugra | 4 | 14.81 | 0 | 0.00 | 0 | 0.00 | 0 | 0.00 |
| Amog | 4 | 14.81 | 0 | 0.00 | 2 | 4.44 | 0 | 0.00 |
| Bindu | 1 | 3.70 | 0 | 0.00 | 3 | 6.67 | 0 | 0.00 |
| Alfa | 0 | 0.00 | 1 | 5.56 | 0 | 0.00 | 0 | 0.00 |
| Jk | 4 | 14.81 | 5 | 27.78 | 6 | 13.33 | 0 | 0.00 |
| Ankur | 1 | 3.70 | 1 | 5.56 | 0 | 0.00 | 0 | 0.00 |
| Satya | 0 | 0.00 | 0 | 0.00 | 1 | 2.22 | 0 | 0.00 |
| Durga | 0 | 0.00 | 0 | 0.00 | 2 | 4.44 | 0 | 0.00 |
| Sri Rama | 0 | 0.00 | 0 | 0.00 | 1 | 2.22 | 0 | 0.00 |
| Sri Rama Laxman | 0 | 0.00 | 0 | 0.00 | 1 | 2.22 | 0 | 0.00 |
| Alfa | 0 | 0.00 | 0 | 0.00 | 1 | 2.22 | 0 | 0.00 |
| Sri Tulasi | 0 | 0.00 | 0 | 0.00 | 4 | 8.89 | 0 | 0.00 |
| | 27 | 100 | 18 | 100 | 45 | 100 | 0 | 0.00 |

## 11.5 Average Cost of Seed Used

In examining the spread of Bt and Non Bt cotton cultivation, the study of related factors that assume importance are the seed rate and the cost of the seed. Tables 11.7 and 11.8 give the seed rate and cost of the seed in Guntur and Warangal. The seed rate per hectare has declined since the advent of Bt cultivation, i.e., since 2003–2004. It was around 1.30–1.40 kg per hectare for non-Bt cotton. Probably the high cost of seed restricted the seed rate of Bt-G to 1.11–1.20 kg. Bt-NC cultivation has also started simultaneously with Bt-G in the year 2003–2004. In the initial year the seed rate of Bt-NC was less than normal. This was due to false promises made to the farmers by the seed agents. In the later year the farmers raised the seed rate. There are not many variations in seed rate across different land holding groups.

**Table 11.6** Different brands of cotton seed bought (Warangal)

| Year/brand name | Bt cotton-G | | Bt cotton-NC | | Hybrid | | Non-hybrid | |
|---|---|---|---|---|---|---|---|---|
| | Number | % | Number | % | Number | % | Number | % |
| 2004–2005 | | | | | | | | |
| Rasi | 31 | 68.89 | – | – | 0 | 0.00 | – | – |
| Mch | 14 | 31.11 | – | – | 0 | 0.00 | – | – |
| Bunny | 0 | 0.00 | – | – | 11 | 24.44 | – | – |
| Super Bunny | 0 | 0.00 | – | – | 9 | 20.00 | – | – |
| Brahma | 0 | 0.00 | – | – | 5 | 11.11 | – | – |
| Attara | 0 | 0.00 | – | – | 3 | 6.67 | – | – |
| Dasara | 0 | 0.00 | – | – | 1 | 2.22 | – | – |
| Jk | 0 | 0.00 | – | – | 6 | 13.33 | – | – |
| Ankur | 0 | 0.00 | – | – | 1 | 2.22 | – | – |
| Rudra | 0 | 0.00 | – | – | 1 | 2.22 | – | – |
| Pradam | 0 | 0.00 | – | – | 1 | 2.22 | – | – |
| Siddharda | 0 | 0.00 | – | – | 1 | 2.22 | – | – |
| Tagore | 0 | 0.00 | – | – | 2 | 4.44 | – | – |
| Prasanna | 0 | 0.00 | – | – | 1 | 2.22 | – | – |
| Dhan | 0 | 0.00 | – | – | 1 | 2.22 | – | – |
| Akshyaya | 0 | 0.00 | – | – | 1 | 2.22 | – | – |
| Amoga | 0 | 0.00 | – | – | 1 | 2.22 | – | – |
| Overall | 45 | 100.00 | – | – | 45 | 100.00 | – | – |
| 2003–2004 | | | | | | | | |
| Rasi | 1 | 2.22 | – | – | 0 | 0.00 | – | – |
| Mch | 0 | 0.00 | – | – | 0 | 0.00 | – | – |
| Bunny | 22 | 48.89 | – | – | 12 | 26.67 | – | – |
| Super bunny | 8 | 17.78 | – | – | 9 | 20.00 | – | – |
| Brahma | 4 | 8.89 | – | – | 4 | 8.89 | – | – |
| Attara | 2 | 4.44 | – | – | 0 | 0.00 | – | – |
| Durga | 2 | 4.44 | – | – | 2 | 4.44 | – | – |
| Sigma | 1 | 2.22 | – | – | 1 | 2.22 | – | – |
| Arch-2 | 1 | 2.22 | – | – | 3 | 6.67 | – | – |
| Dasara | 1 | 2.22 | – | – | 1 | 2.22 | – | – |
| Jk | 0 | 0.00 | – | – | 6 | 13.33 | – | – |
| Tulasi | 1 | 2.22 | – | – | 1 | 2.22 | – | – |
| Satya | 2 | 4.44 | – | – | 0 | 0.00 | – | – |
| Rudra | 0 | 0.00 | – | – | 2 | 4.44 | – | – |
| Siddharda | 0 | 0.00 | – | – | 1 | 2.22 | – | – |
| Tagore | 0 | 0.00 | – | – | 1 | 2.22 | – | – |
| Viswanath | 0 | 0.00 | – | – | 1 | 2.22 | – | – |
| Sarala | 0 | 0.00 | – | – | 1 | 2.22 | – | – |
| Overall | 45 | 100.00 | – | – | 45 | 100.00 | – | – |

(continued)

**Table 11.6** (continued)

| Year/brand name | Bt cotton-G | | Bt cotton-NC | | Hybrid | | Non-hybrid | |
|---|---|---|---|---|---|---|---|---|
| | Number | % | Number | % | Number | % | Number | % |
| 2002–2003 | | | | | | | | |
| Rasi | 0 | 0.00 | – | – | 0 | 0.00 | – | – |
| Mch | 0 | 0.00 | – | – | 0 | 0.00 | – | – |
| Bunny | 16 | 35.56 | – | – | 14 | 31.11 | – | – |
| Super bunny | 11 | 24.44 | – | – | 7 | 15.56 | – | – |
| Brahma | 8 | 17.78 | – | – | 1 | 2.22 | – | – |
| Attara | 1 | 2.22 | – | – | 0 | 0.00 | – | – |
| Durga | 1 | 2.22 | – | – | 2 | 4.44 | – | – |
| Sigma | 1 | 2.22 | – | – | 1 | 2.22 | – | – |
| Arch-2 | 1 | 2.22 | – | – | 2 | 4.44 | – | – |
| Dasara | 1 | 2.22 | – | – | 1 | 2.22 | – | – |
| Jk | 2 | 4.44 | – | – | 6 | 13.33 | – | – |
| Ankur | 0 | 0.00 | – | – | 5 | 11.11 | – | – |
| Mico-12 | 1 | 2.22 | – | – | | 0.00 | – | – |
| Tulasi | 2 | 4.44 | – | – | 2 | 4.44 | – | – |
| Siddharda | 0 | 0.00 | – | – | 1 | 2.22 | – | – |
| Tagore | 0 | 0.00 | – | – | 3 | 6.67 | – | – |
| Overall | 45 | 100.00 | – | – | 45 | 100.00 | – | – |
| 2001–2002 | | | | | | | | |
| Rasi | 0 | 0.00 | – | – | 0 | 0.00 | – | – |
| Mch | 0 | 0.00 | – | – | 0 | 0.00 | – | – |
| Bunny | 16 | 35.56 | – | – | 11 | 24.44 | – | – |
| Super Bunny | 8 | 17.78 | – | – | 3 | 6.67 | – | – |
| Brahma | 2 | 4.44 | – | – | 2 | 4.44 | – | – |
| Attara | 1 | 2.22 | – | – | 7 | 15.56 | – | – |
| Durga | 6 | 13.33 | – | – | 5 | 11.11 | – | – |
| Sigma | 1 | 2.22 | – | – | 1 | 2.22 | – | – |
| Arch-2 | 2 | 4.44 | – | – | 1 | 2.22 | – | – |
| Dasara | 2 | 4.44 | – | – | 1 | 2.22 | – | – |
| Jk | 0 | 0.00 | – | – | 2 | 4.44 | – | – |
| Ankur | 4 | 8.89 | – | – | 2 | 4.44 | – | – |
| Sindur | 1 | 2.22 | – | – | 0 | 0.00 | – | – |
| Mico-12 | 1 | 2.22 | – | – | 0 | 0.00 | – | – |
| Tulasi | 1 | 2.22 | – | – | 4 | 8.89 | – | – |
| Siddharda | 0 | 0.00 | – | – | 1 | 2.22 | – | – |
| Tagore | 0 | 0.00 | – | – | 3 | 6.67 | – | – |
| Chilhor | 0 | 0.00 | – | – | 1 | 2.22 | – | – |
| Silver | 0 | 0.00 | – | – | 1 | 2.22 | – | – |
| Overall | 45 | 100.00 | – | – | 45 | 100.00 | – | – |

**Table 11.7** Purchase, cost and seed rate of cotton by sample farmers in Guntur

| Year/farm size | Area planted (Ha) | | Quantity of seed purchased (kgs.) | Cost per kg. | Seed rate per ha. kgs |
|---|---|---|---|---|---|
| | Irrigated | Unirrigated | | | |
| Bt-genuine | | | | | |
| Small | | | | | |
| 2004–2005 | 6.07 | 5.67 | 14.15 | 3611.99 | 1.21 |
| 2003–2004 | 4.05 | 3.64 | 8.55 | 3602.53 | 1.11 |
| 2002–2003 | 4.45 | 7.69 | 15.53 | 988.42 | 1.28 |
| 2001–2002 | 6.07 | 8.07 | 16.10 | 947.24 | 1.33 |
| Medium | | | | | |
| 2004–2005 | 0.81 | 10.52 | 12.60 | 3662.45 | 1.11 |
| 2003–2004 | 0.00 | 5.06 | 5.63 | 3647.27 | 1.11 |
| 2002–2003 | 3.24 | 0.47 | 12.83 | 978.12 | 1.32 |
| 2001–2002 | 3.24 | 7.28 | 13.73 | 959.91 | 1.30 |
| Large | | | | | |
| 2004–2005 | 0.00 | 6.88 | 7.65 | 3633.90 | 1.11 |
| 2003–2004 | 0.00 | 4.05 | 4.50 | 3607.73 | 1.11 |
| 2002–2003 | 0.00 | 5.26 | 7.18 | 969.41 | 1.36 |
| 2001–2002 | 0.00 | 5.26 | 7.18 | 950.40 | 1.36 |
| Overall | | | | | |
| 2004–2005 | 6.88 | 23.07 | 34.40 | 3636.12 | 1.15 |
| 2003–2004 | 4.05 | 12.75 | 18.68 | 3617.26 | 1.11 |
| 2002–2003 | 7.69 | 19.42 | 35.53 | 981.04 | 1.31 |
| 2001–2002 | 9.31 | 18.62 | 37.00 | 952.61 | 1.32 |
| Bt not confirmed | | | | | |
| Small | | | | | |
| 2004–2005 | 2.02 | 2.83 | 7.45 | 1997.43 | 1.53 |
| 2003–2004 | 3.04 | 2.79 | 2.93 | 1921.93 | 0.50 |
| 2002–2003 | 1.21 | 1.21 | 4.15 | 988.42 | 1.71 |
| 2001–2002 | 0.81 | 0.81 | 1.80 | 942.09 | 1.11 |
| Medium | | | | | |
| 2004–2005 | 6.07 | 7.69 | 19.95 | 1944.13 | 1.45 |
| 2003–2004 | 6.07 | 7.28 | 15.75 | 1918.43 | 1.18 |
| 2002–2003 | 3.84 | 6.47 | 12.75 | 978.73 | 1.24 |
| 2001–2002 | 3.84 | 6.68 | 12.25 | 959.91 | 1.16 |
| Large | | | | | |
| 2004–2005 | 10.12 | 8.90 | 29.48 | 1973.69 | 1.55 |
| 2003–2004 | 6.88 | 5.06 | 10.13 | 1909.83 | 0.85 |
| 2002–2003 | 4.05 | 14.77 | 37.25 | 983.11 | 1.96 |
| 2001–2002 | 0.00 | 21.25 | 30.40 | 964.89 | 1.43 |
| Overall | | | | | |
| 2004–2005 | 18.21 | 19.42 | 56.88 | 1965.95 | 1.51 |
| 2003–2004 | 15.99 | 15.14 | 28.80 | 1915.79 | 0.93 |

(continued)

**Table 11.7** (continued)

| Year/farm size | Area planted (Ha) | | Quantity of seed purchased (kgs.) | Cost per kg. | Seed rate per ha. kgs |
|---|---|---|---|---|---|
| | Irrigated | Unirrigated | | | |
| 2002–2003 | 9.11 | 22.46 | 54.15 | 982.08 | 1.72 |
| 2001–2002 | 4.65 | 28.73 | 44.45 | 962.21 | 1.33 |
| Non-Bt cotton | | | | | |
| Small | | | | | |
| 2004–2005 | 5.97 | 14.12 | 25.18 | 1062.88 | 1.25 |
| 2003–2004 | 5.67 | 11.74 | 22.53 | 1034.39 | 1.29 |
| 2002–2003 | 6.88 | 11.33 | 23.93 | 960.96 | 1.31 |
| 2001–2002 | 6.27 | 10.93 | 22.48 | 941.91 | 1.31 |
| Medium | | | | | |
| 2004–2005 | 19.02 | 20.44 | 48.83 | 1089.80 | 1.24 |
| 2003–2004 | 11.33 | 20.64 | 52.43 | 1046.29 | 1.64 |
| 2002–2003 | 11.33 | 19.42 | 50.40 | 968.91 | 1.64 |
| 2001–2002 | 11.33 | 19.42 | 46.00 | 926.64 | 1.50 |
| Large | | | | | |
| 2004–2005 | 15.78 | 38.04 | 68.78 | 1055.31 | 1.28 |
| 2003–2004 | 12.95 | 18.21 | 38.93 | 1026.93 | 1.25 |
| 2002–2003 | 5.67 | 18.21 | 31.50 | 984.23 | 1.32 |
| 2001–2002 | 4.05 | 19.02 | 30.83 | 947.24 | 1.34 |
| Overall | | | | | |
| 2004–2005 | 40.77 | 72.60 | 142.78 | 1068.65 | 1.26 |
| 2003–2004 | 29.95 | 50.59 | 113.88 | 1036.23 | 1.41 |
| 2002–2003 | 23.88 | 48.97 | 105.83 | 971.95 | 1.45 |
| 2001–2002 | 21.65 | 49.37 | 99.30 | 937.03 | 1.40 |

Bt cultivation started only in 2003–2004. Until that time cost of seed reflects only Non Bt seed

The cost of cotton seed until 2002–2003 was Rs 950–970 per kg. However there was a fourfold increase in the cost of seed for Bt-G and two times increase in the cost for Bt-NC seed from the year 2003–2004. But in the case of non-Bt seed, the increase was only marginal (Table 11.7). The abnormal cost of Bt-G seed might be acting against the spread of Bt cotton cultivation and also tempting the farmers to use Bt-NC seed in the state.

## 11.6 Application of Pesticides

As reported earlier crop damage is very high in bollworm attack. In Table 11.9 relative crop damages due to different pests reported by farmers of different groups like Bt-G, Bt-NC and non Bt are enumerated. It is observed that about half of the Bt farmers did not report any damage due to bollworms. Even for the rest of Bt farmers the attack was only light or moderate. The results are definitely encouraging when compared to non-Bt farmers.

**Table 11.8** Purchase, cost and seed rate of cotton (Warangal)

| Year/farm size | Area planted under cotton (Ha) | | Quantity of seed purchased | Cost per kg | Seed rate per ha |
|---|---|---|---|---|---|
| | Irrigated | Unirrigated | | | |
| Bt cotton-G | | | | | |
| Small | | | | | |
| 2004–2005 | 12.55 | 5.26 | 14.85 | 3559.93 | 1.11 |
| 2003–2004 | 0.40 | – | 0.45 | 3555.56 | 1.11 |
| 2003–2004 | 11.43 | 5.26 | 19.13 | 1120.00 | 1.15 |
| 2002–2003 | 9.41 | 5.26 | 19.13 | 992.94 | 1.30 |
| 2001–2002 | 9.81 | 5.26 | 20.25 | 1100.00 | 1.34 |
| Medium | | | | | |
| 2004–2005 | 10.93 | 3.64 | 17.10 | 3559.06 | 1.08 |
| 2003–2004 | 15.38 | 8.09 | 25.90 | 1141.70 | 1.10 |
| 2002–2003 | 18.62 | 4.86 | 26.80 | 1118.66 | 1.14 |
| 2001–2002 | 15.78 | 5.67 | 25.95 | 1097.88 | 1.21 |
| Large | | | | | |
| 2004–2005 | 17.60 | 5.67 | 17.33 | 3557.29 | 1.17 |
| 2003–2004 | 13.35 | 4.86 | 20.40 | 1090.49 | 1.12 |
| 2002–2003 | 13.76 | 5.67 | 20.55 | 1069.10 | 1.06 |
| 2001–2002 | 10.52 | 4.45 | 19.20 | 115.10 | 1.28 |
| Overall | | | | | |
| 2004–2005 | 41.08 | 14.57 | 49.28 | 3558.70 | 1.12 |
| 2003–2004 | 0.40 | – | 0.45 | 3555.56 | 1.13 |
| 2003–2004 | 40.17 | 18.21 | 65.43 | 1119.39 | 1.12 |
| 2002–2003 | 41.78 | 15.78 | 66.48 | 1067.17 | 1.15 |
| 2001–2002 | 36.12 | 15.38 | 65.40 | 1103.59 | 1.27 |
| Bt not confirmed | | | | | |
| Small | | | | | |
| 2004–2005 | – | – | – | – | – |
| 2003–2004 | – | – | – | – | – |
| 2002–2003 | – | – | – | – | – |
| 2001–2002 | – | – | – | – | – |
| Medium | | | | | |
| 2004–2005 | – | – | – | – | – |
| 2003–2004 | – | – | – | – | – |
| 2002–2003 | – | – | – | – | – |
| 2001–2002 | – | – | – | – | – |
| Large | | | | | |
| 2004–2005 | – | – | – | – | – |
| 2003–2004 | – | – | – | – | – |
| 2002–2003 | – | – | – | – | – |
| 2001–2002 | – | – | – | – | – |

(continued)

**Table 11.8** (continued)

| Year/farm size | Area planted under cotton (Ha) | | Quantity of seed purchased | Cost per kg | Seed rate per ha |
|---|---|---|---|---|---|
| | Irrigated | Unirrigated | | | |
| Overall | | | | | |
| 2004–2005 | – | – | – | – | – |
| 2003–2004 | – | – | – | – | – |
| 2002–2003 | – | – | – | – | – |
| 2001–2002 | – | – | – | – | – |
| Non-Bt cotton | | | | | |
| Small | | | | | |
| 2004–2005 | 5.36 | 5.26 | 21.23 | 1064.78 | 1.26 |
| 2003–2004 | 7.99 | 6.88 | 19.03 | 1033.38 | 1.28 |
| 2002–2003 | 7.99 | 6.47 | 18.89 | 1041.57 | 1.31 |
| 2001–2002 | 8.80 | 6.88 | 19.78 | 1058.41 | 1.26 |
| Medium | | | | | |
| 2004–2005 | 8.50 | 16.63 | 30.88 | 1010.85 | 1.23 |
| 2003–2004 | 8.09 | 8.90 | 20.03 | 1060.17 | 1.18 |
| 2002–2003 | 7.28 | 8.50 | 20.90 | 1090.43 | 1.32 |
| 2001–2002 | 7.69 | 9.31 | 20.45 | 1106.60 | 1.20 |
| Large | | | | | |
| 2004–2005 | 7.28 | 8.50 | 28.20 | 1071.63 | 1.09 |
| 2003–2004 | 10.12 | 7.28 | 20.00 | 1070.50 | 1.15 |
| 2002–2003 | 11.33 | 6.47 | 20.50 | 1083.90 | 1.15 |
| 2001–2002 | 11.74 | 5.67 | 19.35 | 1049.61 | 1.11 |
| Overall | | | | | |
| 2004–2005 | 21.14 | 30.39 | 80.30 | 1046.45 | 1.18 |
| 2003–2004 | 26.20 | 23.07 | 59.05 | 1055.04 | 1.20 |
| 2002–2003 | 26.61 | 21.45 | 60.29 | 1072.90 | 1.25 |
| 2001–2002 | 28.23 | 21.85 | 59.58 | 1072.09 | 1.19 |

**Table 11.9** Rating on insect or pest attack on cotton (frequency) in Guntur

| Name of pest/insect | | Bt cotton-G | | | | Bt cotton-NC | | | | Non-Bt cotton | | | |
|---|---|---|---|---|---|---|---|---|---|---|---|---|---|
| | | 4 | 3 | 2 | 1 | 4 | 3 | 2 | 1 | 4 | 3 | 2 | 1 |
| A. Boll worm | 1. American | 0 | 3 | 10 | 14 | 0 | 1 | 7 | 10 | 17 | 10 | 8 | 10 |
| | 2. Pink | 0 | 3 | 8 | 16 | 0 | 0 | 9 | 9 | 11 | 26 | 6 | 2 |
| | 3. Spotted | 0 | 0 | 10 | 17 | 0 | 1 | 3 | 14 | 12 | 18 | 9 | 6 |
| B. Sucking pests | 1. Thrips | 0 | 1 | 12 | 14 | 0 | 1 | 4 | 13 | 13 | 20 | 6 | 6 |
| | 2. Leafhopper | 0 | 3 | 7 | 17 | 0 | 1 | 3 | 14 | 12 | 18 | 8 | 7 |
| | 3. Whitefly | 0 | 4 | 8 | 15 | 1 | 3 | 7 | 7 | 13 | 22 | 7 | 3 |
| | 4. Aphid | 0 | 0 | 4 | 23 | 0 | 1 | 0 | 17 | 9 | 9 | 2 | 25 |
| | 5. Mealy bug | 0 | 3 | 0 | 24 | 1 | 2 | 1 | 14 | 2 | 6 | 5 | 32 |
| C. Foliage feeding pests | 1. Leaf roller | 0 | 0 | 6 | 21 | 0 | 1 | 2 | 15 | 15 | 5 | 1 | 24 |
| | 2. Caterpillar | 0 | 2 | 4 | 21 | 1 | 0 | 2 | 15 | 11 | 8 | 1 | 25 |
| D. Soil pests | 1. Termite | 0 | 2 | 4 | 21 | 1 | 0 | 2 | 15 | 11 | 8 | 1 | 25 |
| | 2. Blackfly | 0 | 0 | 3 | 24 | 0 | 0 | 2 | 16 | 12 | 6 | 1 | 26 |
| | 3. Grey mildew | 0 | 0 | 2 | 25 | 0 | 1 | 2 | 15 | 2 | 0 | 0 | 43 |

4 Heavy, 3 Moderate, 2 Light, 1 None

**Table 11.10** Rating on insect or pest attack on cotton (frequency) in Warangal

| Name of Pest/Insect | | Bt Cotton-G | | | | Bt Cotton-NC | | | | Non-Bt Cotton | | | |
|---|---|---|---|---|---|---|---|---|---|---|---|---|---|
| | | 4 | 3 | 2 | 1 | 4 | 3 | 2 | 1 | 4 | 3 | 2 | 1 |
| A. Boll worm | 1. American | – | 18 | 24 | 3 | – | – | – | – | – | 30 | 5 | 10 |
| | 2. Pink | – | 14 | 22 | 9 | – | – | – | – | – | 12 | 18 | 14 |
| | 3. Spotted | – | 5 | 15 | 25 | – | – | – | – | – | 12 | 22 | 11 |
| B. Sucking pests | 1. Thrips | – | 6 | 26 | 13 | – | – | – | – | 2 | 12 | 21 | 10 |
| | 2. Leafhopper | – | – | 27 | 18 | – | – | – | – | – | 12 | 22 | 11 |
| | 3. Whitefly | – | 8 | 29 | 8 | – | – | – | – | 2 | 5 | 29 | 9 |
| | 4. Aphids | – | 11 | 22 | 12 | – | – | – | – | 3 | 15 | 17 | 10 |
| C. Foliage feeding pests | 1. Leaf roller | – | 16 | 22 | 7 | – | – | – | – | 3 | 8 | 20 | 14 |
| | 2. Caterpillar | – | 8 | 24 | 12 | – | – | – | – | 5 | 7 | 25 | 8 |
| D. Soil pests | 1. Termite | – | – | 4 | 41 | – | – | – | – | – | 5 | 5 | 35 |
| | 2. Blackfly | – | – | 9 | 36 | – | – | – | – | – | 9 | 6 | 30 |

4 Heavy, 3 Moderate, 2 Light, 1 None

**Table 11.11** Practices followed by sample farmers in pesticide spraying (Guntur)

| Particulars | Bt cotton-G | Bt cotton-NC | Non-Bt cotton |
|---|---|---|---|
| 1. Average number of spray | 4.67 | 6.33 | 11.38 |
| 2. Quantity of pesticides sprayed per spray (millilitre/ha) | 1028 | 806 | 655 |
| 3. Total quantity sprayed (litre/ha) | 4.80 | 5.10 | 7.45 |
| 4. Cost of pesticides (Rs/ha) | 7319 | 7403 | 11,410 |

In Warangal district, severity of Bollworm attack is more among the non-Bt farmers when compared to Bt farmers indicating some advantage of Bt variety (Table 11.10).

The cotton farmers of Guntur districts habitually use higher quantities of pesticides on their crop. But data in Table 11.11 shows the awareness of the farmers regarding the effectiveness of Bt cotton and resultant perceptible reduction in pesticide use. Bt cotton-G farmers are applying 55 % less quantity of pesticides and incurring 55 % less costs when compared to non-Bt farmers. Bt-NC farmers have also reduced the usage by 46 %.

In Warangal district, though the number of sprays is less the average quantity per spray is more and the Bt farmers are actually using higher quantity than the farmers of Guntur. Bt-G farmers have cut down the quantity of pesticides by 16 % in comparison *to* non-Bt farmers of the same region (Table 11.12).

**Table 11.12** Practices followed by sample farmers in pesticide spraying (Warangal)

| Particulars | Bt cotton-G | Bt cotton-NC | Non-Bt cotton |
|---|---|---|---|
| 1. Average number of spray | 3.20 | – | 4.84 |
| 2. Quantity of pesticides sprayed per spray (millilitre/ha) | 1725 | – | 1322 |
| 3. Total quantity sprayed (Litre/ha) | 5.52 | – | 6.40 |
| 4. Cost of pesticides (Rs/ha) | 8499 | – | 9939 |

# Chapter 12
# Economics of Bt Cotton vs. Non-Bt Cotton in Andhra Pradesh

## 12.1 Introduction

This section deals with the pattern of input use and cost of cultivation. These two factors play a key role in accepting or rejecting a new crop or a seed variety. An attempt is made to explore the differences in harvesting practices, inter-culture operations and transport from fields to homes. Use of organic manure is widely observed among the sample farmers. There is large employment of human labour in cotton harvesting that spreads over few months. A variety of commercial crops are grown in the district and it is known for its high use of chemical pesticides. Similar observation is also seen with the sample farmers.

## 12.2 Average Bt and Non-Bt Cotton Yields

The average yield of Bt farmers is 49 % higher than the non-Bt farmers in Guntur. The small farmers in Bt category have reported higher yield than the overall average. This may be due to intensive cultivation practices followed by them. Even non-confirmed Bt category farmers have reaped higher yield of 43 % when compared to non-Bt farmers (Table 12.1).

The terrain and soils in the study area do not show much variation. Significant differences are also not seen either in input use or in cost of cultivation between irrigated and un- irrigated fields. Despite popular use of tractors bullock labour is also used for inter-culture operations and transport from fields to homes. Use of organic manures was also widely observed. There is large employment of human labour in cotton harvesting that spreads over few months. A variety of commercial crops are grown in the district and it is known for its high use of chemical pesticides. Similar observation is also seen with the sample farmers of Warangal (Table 12.2).


V.P. Gandhi, D. Jain, *Introduction of Biotechnology in India's Agriculture*, DOI 10.1007/978-981-10-1091-0_12

**Table 12.1** Area and production – irrigated, unirrigated, total (Guntur district)

| Sr no. | Particulars | Bt genuine | | | Bt not confirmed | | | Non Bt | | |
|---|---|---|---|---|---|---|---|---|---|---|
| | | Irrigated | Unirriga-ted | Total | Irriga-ted | Unirriga-ted | Total | Irrigat-ed | Unirriga-ted | Total |
| 1. | Area in Ha | 9.31 | 27.52 | 36.83 | 18.21 | 32.94 | 51.15 | 42.80 | 48.16 | 90.95 |
| 2. | Production of cotton (Kg/ha) | 3,590 | 3,357 | 3,411 | 3,300 | 3,244 | 3,271 | 3,129 | 1,819 | 2,290 |
| 3. | Value of cotton (Rs./ha) | 64,092 | 59,654 | 60,674 | 48,734 | 58,821 | 53,940 | 55,462 | 31,862 | 40,348 |
| 4. | Value of cotton and by-products Rs/ha) | 64,092 | 59,654 | 60,674 | 48,734 | 58,821 | 53,940 | 55,462 | 31,862 | 40,348 |

**Table 12.2** Area and production – irrigated, unirrigated and total (Warangal)

| Particulars | Bt genuine | | | Bt not confirmed | | | Non Bt | | |
|---|---|---|---|---|---|---|---|---|---|
| | Irrigated | Unirrigated | Total | Irrigated | Unirrigated | Total | Irrigated | Unirrigated | Total |
| 1. Area in Ha | 41.08 | 14.57 | 55.64 | – | – | – | 21.14 | 30.39 | 51.54 |
| 2. Production of cotton (kg/ha) | 2,621.38 | 1,575.17 | 2,379.58 | – | – | – | 2,113.50 | 1,271.22 | 1,622.91 |
| 3.Value of cotton (Rs./ha) | 46,426.89 | 26,482.57 | 41,817.29 | – | – | – | 35,341.58 | 22,842.11 | 27,968.55 |
| 4. Value of cotton and by-products Rs/ha) | 46,426.89 | 26,482.57 | 41,817.29 | – | – | – | 35,341.58 | 22,842.11 | 27,968.55 |

**Table 12.3** Harvesting of cotton (Guntur)

| Particulars | Bt cotton-G | | Bt cotton-NC | | Non-Bt cotton | |
|---|---|---|---|---|---|---|
| | Yield (Kg/ha) | Gap between successive picking (days) | Yield (Kg/ha) | Gap between successive picking (days) | Yield (kg/ha) | Gap between successive picking (days) |
| First picking | 360 | 118 | 310 | 128 | 290 | 135 |
| Second picking | 561 | 37 | 530 | 43 | 510 | 31 |
| Third picking | 912 | 20 | 820 | 24 | 590 | 28 |
| Fourth picking | 710 | 21 | 810 | 23 | 510 | 25 |
| Fifth picking | 470 | 22 | 517 | 27 | 490 | 26 |
| Sixth picking | 210 | 25 | 122 | 30 | 386 | 31 |
| Overall | 3,223 | | 3,109 | | 2,776 | |

**Table 12.4** Harvesting of cotton (Warangal)

| Particulars | Bt cotton-G | | Bt cotton-NC | | Non-Bt cotton | |
|---|---|---|---|---|---|---|
| | Yield (kg/ha) | Gap between successive picking (days) | Yield (kg/ha) | Gap between successive picking (days) | Yield (kg/ha) | Gap between successive picking (days) |
| Gap between sowing and first picking | 550 | 100 | | – | 325 | 115 |
| Second picking | 650 | 25 | – | – | 355 | 25 |
| Third picking | 625 | 25 | – | – | 355 | 30 |
| Fourth picking | 555 | 35 | – | – | 300 | 30 |
| Fifth picking | – | – | – | – | 295 | 30 |
| Sixth picking | – | – | – | – | – | – |
| Overall | 2,380 | | | | 1,630 | |

Bt cotton-G gives higher amount of yield even in the first picking itself due to less pest susceptibility and the colour of the cotton is also good (Tables 12.3 and 12.4). Bt cotton-NC has also yielded slightly higher quantity than the non-Bt cotton in the first picking. Thus the superiority of Bt cotton is clearly shown both in quantity and quality.

In Warangal district, the seed being genuine, Bt cotton gives early picking than the non-Bt. The yield in the early pickings is also high. Number of pickings is also limited to three or four unlike the non-Bt. The quality and colour of the bolls are also good according to the farmers.

## 12.3 Cost of Production, Value of Output and Profit

Tables 12.5 and 12.6 give the cost of cultivation per hectare in absolute and percentage terms respectively. Cost of pesticides takes the single largest share with 23 % in Bt-G category and 37 % in non-Bt category (see Table 12.6). It is 25 % in Bt-NC group. Harvesting of cotton occupies the next place with 23 %, 23 % and 17 % in Bt-G, Bt-NC and non-Bt groups respectively. A good harvest means a higher cost for the farmer, which he may not grudge. Cost of human labour involving other operations comes next with 14–15 %. Bt-G farmers feel the pinch on account of high seed cost amounting to 12 % while it is only one fourth of it, i.e. 4 % for non-Bt farmers. Costs on irrigation are insignificant as the crop's demand is low and the water supply is also very inconsistent and uncertain.

All the farmers in the sample are using licensed Bt seed. Hence the cost of the seed is high and occupies 12 % in the total costs. This is three times higher than the non-Bt seed. The costs on bullock labour and the charges on the tractor are nearly the same. Organic manure is also widely being used. The costs on chemical fertilizers are next only to costs on pesticides. Pesticides form the highest proportion of costs with 34 % in non-Bt and 25 % in Bt. On the whole costs on Bt are 14 % higher than those of non-Bt (Tables 12.5 and 12.6).

In Warangal, cost of pesticides occupy the major place with 25 % followed by fertilizers (21), seed (12), cotton picking (11) per cent in Bt category. In non-Bt category as expected, costs of pesticides are quite high with 34 % followed by cost of fertilizers, 18 %. Ten per cent of the costs are going to human labour. Cost of seeds is only 4 %.

The terrain and soils in the study area do not show much variation. Cotton crop basically does not need much water. Thus significant differences are also not seen either in input use or in cost of cultivation between irrigated and unirrigated fields. Despite popular use of tractors bullock labour is also used.

It is observed that cotton raised in irrigated fields is yielding two quintals more than the un- irrigated in Guntur. In non-Bt varieties the difference is very large. The net income derived by Bt farmers is Rs.28,493 (Table 12.7) while it is Rs.9059 for the non-Bt farmers. When compared with non-Bt farmers Bt-G farmers have gained three times more. Similarly Bt-NC farmers have gained two and half times more. Income differences among irrigated and un-irrigated category of farmers is also more marked in Bt-G group.

Yields in Warangal district are quite low when compared with those of Guntur. The soils in the sample area of the district are not well suited to cotton cultivation. The average yields of Bt cotton of Warangal are comparable with those of Non-Bt cotton of Guntur. Further, low rainfall in the reference year has contributed to lower yield. The net income of the Bt farmer is Rs. 7585. Non-Bt farmers have actually suffered negative incomes on an average of Rs. (–) 983 (see Table 12.8).

**Table 12.5** Cost of cultivation per hectare

| | | | | | | | | | (Amount in rupees) |
|---|---|---|---|---|---|---|---|---|---|
| | Average cost per hectare | | | | | | | | |
| | Bt cotton-G | | | Bt cotton-NC | | | Non-Bt cotton | | |
| Item | I | UI | Total | I | UI | Total | I | UI | Total |
| 1. Seed | 3,953.68 | 3,948.85 | 3,949.96 | 2,215.71 | 2,133.85 | 2,173.46 | 1,237.48 | 1,212.68 | 1,223.68 |
| 2. Human labour | 4,462.43 | 4,447.33 | 4,450.80 | 4,491.82 | 4,324.34 | 4,405.38 | 4,563.94 | 4,474.88 | 4,517.36 |
| 3. Bullock labour | 1,976.84 | 1,950.59 | 1,956.62 | 1,954.88 | 1,930.51 | 1,942.3 | 1,976.84 | 1,949.85 | 1,961.81 |
| 4. Tractor | 1,526.24 | 1,569.14 | 1,559.28 | 1,578.73 | 1,534.11 | 1,555.7 | 1,563.48 | 1,542.85 | 1,551.99 |
| 5. Farm yard manure | 1,976.84 | 1,950.59 | 1,956.62 | 1,957.62 | 1,925.36 | 1,940.97 | 1,953.37 | 1,924.93 | 1,937.53 |
| 6. Fertilizer | 3,459.47 | 3,381.01 | 3,399.04 | 3,486.93 | 3,294.73 | 3,387.73 | 3,403.40 | 3,332.80 | 3,364.10 |
| 7. Pesticides | 7,442.22 | 7,282.18 | 7,318.95 | 7,457.08 | 7,351.37 | 7,402.52 | 11,683.70 | 11,192.40 | 11,410.18 |
| 8. Irrigation | 0.00 | 0.00 | 0.00 | 0.00 | 0.00 | 0.00 | 0.00 | 0.00 | 0.00 |
| 9. Picking (harvesting) | 7,573.04 | 7,433.9 | 7,465.86 | 6,907.96 | 6,821.13 | 6,863.14 | 5,249.84 | 5,176.75 | 5,209.15 |
| 10. Other operational costs | 0.00 | 0.00 | 0.00 | 0.00 | 0.00 | 0.00 | 0.00 | 0.00 | 0.00 |
| **11. Total operational cost (1–10)** | **32,370.76** | **31,963.59** | **32,057.12** | **30,050.71** | **29,315.40** | **29,671.20** | **31,632.05** | **30,807.14** | **31,172.80** |

| | | | | | | | | | |
|---|---|---|---|---|---|---|---|---|---|
| 12. Transport | 266.00 | 81.92 | 124.21 | 245.46 | 82.37 | 161.28 | 176.82 | 68.63 | 116.59 |
| 13. Marketing | 0.00 | 0.00 | 0.00 | 0.00 | 0.00 | 0.00 | 0.00 | 0.00 | 0.00 |
| 14. Other marketing costs | 0.00 | 0.00 | 0.00 | 0.00 | 0.00 | 0.00 | 0.00 | 0.00 | 0.00 |
| **15. Total marketing cost (12–14)** | **266.00** | **81.92** | **124.21** | **245.46** | **82.37** | **161.28** | **176.82** | **68.63** | **116.59** |
| **16. Total cost (11 + 15)** | **32,636.76** | **32,045.51** | **32,181.33** | **30,296.17** | **29,397.77** | **29,832.48** | **31,808.87** | **30,875.77** | **31,289.39** |
| Return 1: item 5 in table 14-item 11 | 31,720.81 | 27,690.72 | 28,616.50 | 18,683.50 | 29505.40 | 24,268.48 | 23,829.69 | 1,054.72 | 9,175.28 |
| Return 2: item 5 in table 14-item 16 | 31,454.81 | 27,608.80 | 28,492.29 | 18,438.04 | 29,423.03 | 24,107.20 | 23,652.87 | 986.09 | 9,058.69 |

**Table 12.6** Cost of cultivation per hectare – percentage

| | Average cost per hectare | | | | | | | | |
|---|---|---|---|---|---|---|---|---|---|
| | Bt cotton-G | | | Bt cotton-NC | | | Non-Bt cotton | | |
| Item | I | UI | Total | I | UI | Total | I | UI | Total |
| 1. Seed | 12.11 | 12.32 | 12.27 | 7.31 | 7.26 | 7.29 | 3.89 | 3.93 | 3.91 |
| 2. Human labour | 13.67 | 13.88 | 13.83 | 14.83 | 14.71 | 14.77 | 14.35 | 14.49 | 14.43 |
| 3. Bullock labour | 6.06 | 6.09 | 6.08 | 6.45 | 6.57 | 6.51 | 6.21 | 6.32 | 6.27 |
| 4. Tractor | 4.68 | 4.90 | 4.85 | 5.21 | 5.22 | 5.21 | 4.92 | 5.00 | 4.96 |
| 5. Farm yard manure | 6.06 | 6.09 | 6.08 | 6.46 | 6.55 | 6.51 | 6.14 | 6.23 | 6.19 |
| 6. Fertilizer | 10.60 | 10.55 | 10.56 | 11.51 | 11.21 | 11.36 | 10.70 | 10.79 | 10.75 |
| 7. Pesticides | 22.80 | 22.72 | 22.74 | 24.61 | 25.01 | 24.81 | 36.73 | 36.25 | 36.47 |
| 8. Irrigation | 0.00 | 0.00 | 0.00 | 0.00 | 0.00 | 0.00 | 0.00 | 0.00 | 0.00 |
| 9. Picking (harvesting) | 23.20 | 23.20 | 23.20 | 22.80 | 23.20 | 23.01 | 16.50 | 16.77 | 16.65 |
| 10. Other operational costs | 0.00 | 0.00 | 0.00 | 0.00 | 0.00 | 0.00 | 0.00 | 0.00 | 0.00 |
| **11. Total operational cost (1–10)** | **99.18** | **99.74** | **99.61** | **99.19** | **99.72** | **99.46** | **99.44** | **99.78** | **99.63** |
| 12. Transport | 0.82 | 0.26 | 0.39 | 0.81 | 0.28 | 0.54 | 0.56 | 0.22 | 0.37 |
| 13. Marketing | 0.00 | 0.00 | 0.00 | 0.00 | 0.00 | 0.00 | 0.00 | 0.00 | 0.00 |
| 14. Other marketing costs | 0.00 | 0.00 | 0.00 | 0.00 | 0.00 | 0.00 | 0.00 | 0.00 | 0.00 |
| **15. Total marketing cost (12–14)** | **0.82** | **0.26** | **0.39** | **0.81** | **0.28** | **0.54** | **0.56** | **0.22** | **0.37** |
| **16. Total cost (11+15)** | **100.00** | **100.00** | **100.00** | **100.00** | **100.00** | **100.00** | **100.00** | **100.00** | **100.00** |

**Table 12.7** Economics of Bt cotton in Guntur

| | | | | | | | Rs. Per/ ha |
|---|---|---|---|---|---|---|---|
| | Bt-G | Bt-NC | Non Bt | | Bt-G | Bt-NC | Non Bt |
| Yield/ha kgs | 3,411 | 3,271 | 2,290 | Gross income | 10,674 | 53,940 | 40,348 |
| Value in Rs. | 60,674 | 53,940 | 40,348 | Total cost | 32,181 | 29,832 | 31,289 |
| | | | | Net income | 28,493 | 24,108 | 9,059 |

## 12.4 Econometric Analysis of the Performance of Bt Cotton vs. Non-Bt Cotton

By using dummy variable, regression analysis has been attempted to estimate impact of Bt technology on yield, profit, pesticide cost etc. and to test its statistical significance. The analysis confirms positive and significant impact of Bt cotton on yield, value of output and net profit (see Table 12.9).

**Table 12.8** Economics of Bt cotton in Warangal

| | | | | | | | Rs. Per/ ha |
|---|---|---|---|---|---|---|---|
| | Bt-G | Bt-NC | Non Bt | Per/ ha Rs. | Bt-G | Bt-NC | Non Bt |
| Yield/ha kgs | 2,380 | – | 1,623 | Gross income | 41,817 | – | 27,969 |
| Value in Rs. | 41,817 | – | 27,969 | Total cost | 34,232 | – | 28,952 |
| | | | | Net income | 7,585 | – | –983 |

**Table 12.9** Regression results: impact of Bt cotton – Andhra Pradesh

| | | Independent variables | | N = 180 |
|---|---|---|---|---|
| Dependent variable | | Constant | Bt | Percent impact of Bt |
| Yield | Coefficient | 2,377.73 | 506.89 | 21.32 |
| | t-stat | 29.75 | 4.48 | |
| | Signifi. | *** | *** | |
| Value of output | Coefficient | 41,207 | 8,789.63 | 21.33 |
| | t-stat | 30 | 4.52 | |
| | Signifi. | *** | *** | |
| Total cost | Coefficient | 31,266 | 1,017.42 | 3.25 |
| | t-Stat | 80.09 | 1.84 | |
| | Signifi. | *** | * | |
| Pesticide cost | Coefficient | 11,022 | –3,104.41 | –28.17 |
| | t-Stat | 83.45 | –16.62 | |
| | Signifi. | *** | *** | |
| Seed cost | Coefficient | 1,227.94 | 2,364.18 | 192.53 |
| | t-Stat | 20.47 | 27.87 | |
| | Signifi. | *** | *** | |
| Price | Coefficient | 17.35 | 0 | 0 |
| | t-Stat | 2,975.13 | 0 | |
| | Signifi. | *** | NS | |
| Profit | Coefficient | 9,940.99 | 7,772.21 | 78.18 |
| | t-Stat | 7.42 | 4.1 | |
| | Signifi. | *** | *** | |

*NS* not significant

Note: *** = significant at 99 %, ** = significant at 95 %, * = significant at 90 %

The impact on the yields as well as the value of outputs is the lowest at about 21.33 %, but the rise in total cost is also lower. The fall in the pesticide cost is the highest in Andhra Pradesh at –28.17 %, but the rise in the seed cost is also the highest at 192.53 %. This is supportive of the fact why opposition to the seed prices may be the highest in Andhra Pradesh. The rise in the profits is statistically highly significant and amounts to a 78.18 %, which is in between Gujarat and Maharashtra. The absolute level of profitability of cotton in Andhra Pradesh is lowest amongst the three states. Note that similar analysis could not be carried out for Tamil Nadu since observation-wise data was not available from the AERC there.

# Chapter 13
# Farmers' Perceptions on Various Features of Bt Cotton in Andhra Pradesh

An attempt is made in this section to gauge the experience of the farmers in relation to availability and quality of Bt seeds, demand on inputs like fertilizers, pesticides, labour, irrigation etc., their costs when compared to non Bt varieties, quality and marketability of Bt cotton.

## 13.1 Some Basic Features of Bt Cotton: Views of Bt Cotton Sample Farmer Households

Licensed seed companies spend substantial amounts for marketing. Their field agents visit all the villages well before the season and contact village elders and farmers who can in turn influence others. Fifty per cent in Bt-G and 22 % in Bt-NC categories said they were approached by agents of the seed companies and by farmers on their behalf to promote the seed. Propaganda was stepped by the second year by all the agencies. Details are presented in Table 13.1.

Seed companies are vigorously promoting sales in Warangal district also. The field team came across many billboards in the villages. Mahyco was active in the first year. But, the farmers were disappointed with the outturn of their variety. They found that Rasi was performing better in the region. So, the next year Rasi stepped up their campaign to woo many more farmers (Table 13.2).

As already discussed earlier, the cost of licensed Bt seed is nearly four times to non Bt cotton. There are no complaints about the germination of the seed. Seed rate per hectare is also lowest. However 8 % of the farmers said they were using slightly higher quantity (see Table 13.2).

Non-confirmed Bt seed is sold roughly at half the price of the genuine seed. Half of the farmers cultivating this variety are using 10–15 % higher quantity.


V.P. Gandhi, D. Jain, *Introduction of Biotechnology in India's Agriculture*, DOI 10.1007/978-981-10-1091-0_13

**Table 13.1** Sample households approached by private agencies for promoting Bt cotton percentages

| Particulars | Guntur | | | Warangal | | |
|---|---|---|---|---|---|---|
| | Bt cotton-G | Bt cotton-NC | Non-Bt cotton | Bt cotton-G | Bt cotton-NC | Non-Bt cotton |
| Percentage of farmers whom private agencies approached for promoting Bt cotton | 51 | 31 | 5 | 91 | – | 10 |
| **Name of agency** | | | | | | |
| 1. Mahyco | 40 | – | – | 54 | – | – |
| 2. Rasi | 11 | – | – | 45 | – | – |
| 3. Farmers | – | 31 | – | 10 | – | – |
| 4. Others | – | – | 5 | 6 | – | 10 |
| **Agency 1: Mahyco** | | | | | | |
| Year when the agency approached | | | | | | |
| 1. This year | 44 | – | – | 43 | – | – |
| 2. Last year | 24 | – | – | 56 | – | – |
| 3. Before 2 years | – | – | – | – | – | – |
| **Agency 2: Rasi** | | | | | | |
| Year when the agency approached | | | | | | |
| 1. This year | 40 | – | – | 61 | – | – |
| 2. Last year | 17 | – | – | 22 | – | – |
| 3. Before 2 years | – | – | – | – | – | – |
| **Agency 3: Farmers** | | | | | | |
| Year when the agency approached | | | | | | |
| 1. This year | – | 31 | 5 | – | – | – |
| 2. Last year | – | 8 | – | – | – | – |
| 3. Before 2 years | – | – | – | – | – | – |
| **Agency 4: Others** | | | | | | |
| Year when the agency approached | | | | | | |
| 1. This year | – | – | – | – | – | 10 |
| 2. Last year | – | – | – | – | – | 15 |
| 3. Before 2 years | – | – | – | – | – | – |

In Warangal seed rate per hectare is not that high as in Guntur, but still 20 % of farmers use slightly higher quantities. In Bt category 41 % of farmers have said that they have reduced the seed rate by 10–15 % (Table 13.3).

All the farmers of Bt-G and 89 % of Bt-NC are convinced that Bt varieties have greater resistance to pests like boll worms. In fact that is the reason for them to switch over from non Bt even though the seed is costlier many times.

All the Bt cultivators, no matter whether they belong to genuine or not confirmed category, are buying seed every year from company dealers or unofficial seed

**Table 13.2** Seed rate of Bt cotton vis-à-vis non-Bt cotton (Guntur)

| Particulars | Bt cotton-G | | |
|---|---|---|---|
| | Irrigated | Unirrigated | Overall |
| | Percentage of households | | |
| Percentage of sample households reporting | | | |
| **Higher** seed rate for Bt cotton-**G** | | | |
| | Percentage of reporting households | | |
| 1. Below 25 % | 8 | 8 | 8 |
| 2. 25–50 % | | | |
| 3. 50–75 % | | | |
| 4. Above 75 % | | | |
| **Percentage of sample households reporting: Lower seed rate for Bt cotton-G** | | | |
| | Percentage of reporting households | | |
| 1. Below 25 % | 84 | 90 | 87 |
| 2. 25–50 % | | | |
| 3. 50–75 % | | | |
| 4. Above 75 % | | | |

**Table 13.3** Seed rate of Bt cotton vis-à-vis non-Bt cotton (Warangal)

| Particulars | Bt cotton-G | | |
|---|---|---|---|
| | Irrigated | Unirrigated | Overall |
| | Percentage of households | | |
| **Higher** seed rate for Bt cotton-**G** | 16 | 24 | 20 |
| **Lower** seed rate for Bt cotton-**G** | 47 | 35 | 41 |

breeders or their agents. Monsanto, Mahyco and Rasi are the licensed Bt seed dealers in the district. No farmer raising these seeds complains about any difficulty in purchasing them. On the other hand licensed seed is procured from unofficial seed breeders from Kurnool district. Some farmers from the village go to them in crop season and order for seed for the next season. Sometimes they also sell the seed to fellow farmers. Unless planned in advance procuring this NC seed is difficult because the whole procedure is illegal and prone to raids by Agricultural officials who are entrusted with the job of curbing its propagation.

There seems to be no dispute over the efficacy of genuine Bt seed as every farmer has agreed that it is more resistant to boll worm attack than non Bt varieties. But this view is shared by only 89 % of non-confirmed Bt cultivators. Seventy four per cent of Bt-G and 63 % of Bt-NC farmers have said that the resistance to boll worm is as high as 25–50 % when compared to non Bt (Table 13.4).

Even in Warangal district, farmers have good faith in Genuine Bt seed. Ninety eight per cent have said that the variety is superior in resisting the bollworm. Among those, 8 % felt that the superiority is as high as 50–75 %. The others felt that it is below 25 %. For other pests it has no superiority, they added (Table 13.5).

**Table 13.4** Resistance of Bt cotton to pest/insects as compared to non-Bt cotton (Guntur)

| Particulars | Bt cotton-G | Bt cotton-NC |
|---|---|---|
| | Percentage of total sample households | |
| Percentage of sample households reporting that Bt cotton is more resistant to **boll worm** than non-Bt cotton | 100 | 89 |
| **(A) Boll worm** | | |
| Extent of resistance | Percentage of reporting households | |
| 1. Below 25 % | 26 | 37 |
| 2. 25–50 % | 74 | 63 |
| 3. 50–75 % | – | – |
| 4. 75–99 % | – | – |
| 5. 100 % | – | – |
| Percentage of sample households reporting that Bt cotton is more resistant to other pests than non-Bt cotton | Nil | |

**Table 13.5** Resistance of Bt cotton to pest/insects as compared to non-Bt cotton (Warangal)

| Particulars | Bt cotton-G | Bt cotton-NC |
|---|---|---|
| | Percentage of total sample households | |
| Percentage of sample households reporting that Bt cotton is more resistant to **boll worm** than non-Bt cotton | 98 | – |
| **(A) Boll worm** | | |
| Extent of resistance | Percentage of reporting households | |
| 1. Below 25 % | 92 | – |
| 2. 25–50 % | 8 | – |
| 3. 50–75 % | – | – |
| 4. 75–99 % | – | – |
| 5. 100 % | – | – |
| Percentage of sample households reporting that Bt cotton is more resistant to other pests than non-Bt cotton | Nil | |

Marketing of Bt cotton was no problem to the cultivators. On the other hand they have said that Bt cotton looks better as the bolls are less damaged. Usually they mix the Bt cotton with other non-Bt cotton while selling. The traders also do not make any distinction between these two varieties in fixing the price.

Cultivation of licensed Bt like Mahyco and Rasi was started in the year 2003–2004. Incidentally the same year has witnessed the beginning of cultivation of unconfirmed Bt varieties. In the first year 33 % of the cotton cultivators have started raising Bt-G while 18 % opted for the cultivation of Bt-NC simultaneously. In the following year, i.e., the reference year additional 27 % have joined Bt-G cultivation whereas 22 % chose to try Bt-NC cultivation afresh (Table 13.6). Bt cultivation is

**Table 13.6** Experience of sample farmers in the cultivation of Bt cotton: percentages

| Year when the sample farms started growing Bt-cotton | Percentage of total sample households growing Bt cotton | | | | | | | |
|---|---|---|---|---|---|---|---|---|
| | Guntur | | | | Warangal | | | |
| | Bt cotton-G | | Bt cotton-NC | | Bt cotton-G | | Bt cotton-NC | |
| (a) From this year | 12 | 26.66 | 10 | 22.22 | 44 | 97.78 | – | – |
| (b) Since last year | 15 | 33.33 | 8 | 17.78 | 1 | 2.22 | – | – |

**Table 13.7** Date of flowering of Bt cotton vis-à-vis Non-Bt cotton (Guntur)

| Particulars | Bt cotton-G | | | Bt cotton-NC | | |
|---|---|---|---|---|---|---|
| | Percentage of total sample households | | | | | |
| | U | UI | Overall | I | UI | Overall |
| Early flowering of Bt cotton | 89 | 70 | 79 | 85 | 71 | 78 |
| | Percentage of reporting households | | | | | |
| 1. 1–10 days | 100 | 100 | 100 | 100 | 100 | 100 |
| 2. 10–20 days | – | – | – | – | – | – |
| 3. 20–30 days | – | – | – | – | – | – |
| 4. More than 30 days | – | – | – | – | – | |
| No change in flowering date | 11 | 30 | 21 | 15 | 29 | 22 |

said to have taken off in Warangal district only in the reference year as 98 % said that their experience extends only to the current season (Table 13.6).

## 13.2 Agronomic Features of Bt Cotton: Response of Bt Growers

### *13.2.1 Flowering*

In Guntur district, all the farmers, both Bt-G and Bt-NC, have agreed that the plants are flowering 1–10 days early. This they attribute to the pest resistance of the plants in the earlier phase of growth (Table 13.7). Similarly farmers in Warangal district have also reported early flowering of Bt cotton. Fifty nine per cent said that plant flowered early by 10 days and another 41 have reported early flowering by nearly 15–20 days (Table 13.8).

### *13.2.2 Plant Size, Boll Size and Relative Number of Pickings*

Farmers' response with respect to plant size, boll size and relative number of pickings are given in Tables 13.9 and 13.10.

**Table 13.8** Date of flowering of Bt cotton vis-à-vis non-Bt cotton (Warangal)

| Particulars | Bt cotton-G | | | Bt cotton-NC | | |
|---|---|---|---|---|---|---|
| | Percentage of total sample households | | | | | |
| | U | UI | Total | I | UI | Total |
| Early flowering of Bt cotton | 81 | 100 | 91 | – | – | – |
| | Percentage of reporting households | | | | | |
| 1. 1–10 days | 50 | 67 | 59 | – | – | – |
| 2. 10–20 days | 50 | 33 | 41 | – | – | – |
| 3. 20 –30 days | – | – | – | – | – | – |
| 4. More than 30 days | – | – | – | – | – | – |
| Percentage of sample households reporting no change in flowering date | 19 | 0 | 19 | – | – | – |

**Table 13.9** Plant size, boll size and number of picking of Bt cotton as compared to non-Bt cotton reported by Bt cotton growers: percentage (Guntur)

| Particulars | Bt cotton-G | Bt cotton-NC |
|---|---|---|
| **Percentage of sample households reporting on Bt cotton plant size:** | | |
| (a) Shorter | 75 | 22 |
| (b) Taller | – | – |
| (c) Same | 25 | 78 |
| **Percentage of sample households reporting on Bt cotton plant boll size:** | | |
| (a) Bigger | 100 | 44 |
| (b) Smaller | – | – |
| (c) Same | – | 56 |
| **Percentage of sample households reporting on number of Bt cotton picking** | | |
| (a) More | – | – |
| (b) Less | 100 | 56 |
| (c) Same | – | 44 |

Bt plants are shorter in size according to 75 % of Bt-G farmers and 22 % of Bt-NC farmers. All the farmers of Bt-G reported slightly bigger size of bolls whereas only 44 % of Bt-NC has concurred with them. All the Bt-G cultivators have observed lower No. of pickings by one or two. But 44 % of Bt-NC have felt that there is no difference in No. of pickings (Table 13.9).

Ninety six per cent of Bt -G farmers in Warangal district have said that plants are slightly shorter in size and 72 % felt that cotton bolls are bigger in size when compared to non-Bt. Almost all of them (95 %) reported that cotton pickings are less in number and are limited to 3 or 4. None of the Bt farmers have reported any inspection of their crop by agriculture department or other government personnel (see Table 13.10).

**Table 13.10** Plant size, boll size and number of picking of Bt cotton as compared to non-Bt cotton reported by Bt cotton growers: percentage (Warangal)

| Particulars | Bt cotton-G | Bt cotton-NC |
|---|---|---|
| **Percentage of sample households reporting on Bt cotton plant size:** | | |
| (a) Shorter | 96 | – |
| (b) Taller | – | – |
| (c) Same | 4 | – |
| **Percentage of sample households reporting on Bt cotton plant boll size:** | | |
| (a) Bigger | 72 | – |
| (b) Smaller | 22 | – |
| (c) Same | 6 | – |
| **Percentage of sample households reporting on number of Bt cotton picking** | | |
| (a) More | – | – |
| (b) Less | 95 | – |
| (c) Same | 5 | – |

## 13.3 Farmers' Perception on Advantages or Disadvantages of Bt Cotton vis-à-vis Non-Bt Cotton

Licensed Bt seed is sold by very few dealers in Guntur. So a large number of farmers, 93 %, have expressed a disadvantage with availability of seed. The cost of Bt-G seed is nearly four times higher than traditional varieties. There is no surprise that 41 % felt it a disadvantage and 30 % a strong disadvantage. But 74 % felt advantage with the quality of seed. When asked about pest incidence on the crop 4 % felt strong advantage while 85 % felt advantage over other varieties. None felt a disadvantage. Similarly 89 % have reported lesser use of pesticide and thereby incurred lesser cost. About 4 % of farmers felt strong advantage when it comes to cotton quality and staple length. With lessened boll damage the colour of the fibre is also very good and it has resulted in 15 % of farmers reporting strong advantage and 78 % saying advantage.

Farmers expressed overall satisfaction with Bt cotton yields. 19 % felt strong advantage and 74 % said it is advantageous. When queried about higher profits 52 % conveyed strong advantage and 41 % felt advantage over traditional varieties (Table 13.11).

A vast majority, 89 % of Bt-NC farmers did not feel much difference as far as availability of seeds is concerned. Eighty three per cent felt a certain disadvantage with cost of the seed. Regarding the quality of seeds, 72 % felt an advantage.

Lesser pest incidence is reported by 89 % and an equal number of farmers expressed advantage with expenditure on pesticides. Advantage over fertilizer use and lower labour need is reported by 44 % and 39 % respectively. The quality of

**Table 13.11** Advantages or disadvantages of Bt cotton vis-à-vis non-Bt cotton reported by Bt cotton-G growers: percentages

| | | Average rating (%) | | | | | | | | | |
|---|---|---|---|---|---|---|---|---|---|---|---|
| | | Bt cotton – G | | | | | Bt cotton – NC | | | | |
| Sl. no. | Particulars | 2 | 1 | 0 | −1 | −2 | 2 | 1 | 0 | −1 | −2 |
| 1 | Availability of seeds | – | – | 7 | 93 | – | – | – | 89 | 11 | – |
| 2 | Seed cost/price | – | – | 30 | 41 | 30 | – | 17 | 83 | – | – |
| 3 | Quality of avail. seeds | – | 74 | 26 | – | – | – | 72 | 28 | – | – |
| 4 | Pest incidence/problem | 4 | 85 | 11 | – | – | – | 89 | 11 | – | – |
| 5 | Pesticide need/cost | 4 | 89 | 7 | – | – | – | 89 | 6 | 5 | – |
| 6 | Fertilizer need/cost | 4 | 30 | 67 | – | – | – | 44 | 56 | – | – |
| 7 | Labour need/cost | – | 48 | 52 | – | – | – | 39 | 61 | – | – |
| 8 | Machine need/cost | – | – | 100 | – | – | – | 11 | 89 | – | – |
| 9 | Irrigation need/cost | – | 15 | 85 | – | – | – | – | 100 | – | – |
| 10 | Harvesting cost | – | 26 | 74 | – | – | – | 11 | 89 | – | – |
| 11 | Cotton quality | 4 | 85 | 11 | – | – | – | 94 | 6 | – | – |
| 12 | Market preference | – | 26 | 74 | – | – | – | 50 | 50 | – | – |
| 13 | Staple length | 4 | 96 | – | – | – | – | 83 | 17 | – | – |
| 14 | Fibre colour | 15 | 78 | 7 | – | – | – | 100 | – | – | – |
| 15 | Cotton price | – | 19 | 81 | – | – | – | 22 | 78 | – | – |
| 16 | Easy marketing | – | 26 | 74 | – | – | – | 6 | 94 | – | – |
| 17 | By-product output | – | – | 100 | – | – | – | 6 | 94 | – | – |
| 18 | Yield | 19 | 74 | 7 | – | – | 33 | 67 | – | – | – |
| 19 | Profit | 52 | 41 | 7 | – | – | 28 | 67 | 5 | – | – |
| 20 | Livestock feeding | – | 7 | 93 | – | – | – | – | 100 | – | – |
| 21 | Water saving | – | 15 | 85 | – | – | – | 11 | 89 | – | – |
| 22 | Suitable for early sowing | – | 4 | 96 | – | – | – | 11 | 78 | 11 | – |
| 23 | Suitable for late sowing | – | 7 | 93 | – | – | – | 11 | 78 | 11 | – |
| 24 | Others (specify) | – | – | – | – | – | – | – | – | – | – |

Rating: +2 – Strong Advantage, +1 – Advantage, 0 – No Difference, −1 – Disadvantage, −2 – Strong Disadvantage

cotton seems to be surely an advantage with 94 % farmers and market preference for Bt cotton is reported by 50 %. It has good staple length and everyone has agreed that the colour of the fibre is superior. When it comes to price and marketing aspects majority felt not much difference. Thirty three per cent had better yields whereas 28 % had better profit (Table 13.11).

Opinions of Bt Farmers of Warangal district are presented in Table 13.12. Eighty seven per cent have felt slightly disadvantage with availability of Bt G seed as not every dealer sells those seeds (see Table 13.12). The cost of Bt Seed is nearly four times higher. Fifty six per cent felt the cost as disadvantage and another 38 % felt it as strong disadvantage. Fifty eight per cent farmers have expressed faith in its pest resistance capability and 53 % felt the advantage in the cost on pesticides. The quality of cotton was appreciated by 87 % of the farmers and roughly an equal number said the colour was also good. Ninety three per cent felt advantage with the yield

**Table 13.12** Advantages or disadvantages of Bt cotton-G vis-à-vis non-Bt cotton reported by Bt cotton-G growers: percentage (Warangal)

| Sl. no. | Particulars | Average rating (%) | | | | |
|---|---|---|---|---|---|---|
| | | 2 | 1 | 0 | −1 | −2 |
| 1 | Availability of seeds | – | – | 9 | 87 | 4 |
| 2 | Seed cost/price | – | – | 7 | 56 | 38 |
| 3 | Quality of avail. Seeds | 2 | 96 | 2 | – | – |
| 4 | Pest incidence/problem | – | 58 | 36 | 7 | – |
| 5 | Pesticide need/cost | 36 | 53 | 9 | 2 | – |
| 6 | Fertilizer need/cost | – | 24 | 76 | – | – |
| 7 | Labour need/cost | – | 44 | 56 | – | – |
| 8 | Machine need/cost | – | 4 | 96 | – | – |
| 9 | Irrigation need/cost | 2 | 42 | 56 | – | – |
| 10 | Harvesting cost | 4 | 47 | 49 | – | – |
| 11 | Cotton quality | 36 | 51 | 13 | – | – |
| 12 | Market preference | 4 | 11 | 84 | – | – |
| 13 | Staple length | 4 | 84 | 9 | 2 | – |
| 14 | Fibre colour | 2 | 89 | 7 | 2 | – |
| 15 | Cotton price | 2 | 4 | 93 | – | – |
| 16 | Easy marketing | 2 | 2 | 96 | – | – |
| 17 | By-product output | – | – | 100 | – | – |
| 18 | Yield | – | 93 | 7 | – | – |
| 19 | Profit | – | 89 | 4 | 7 | – |
| 20 | Livestock feeding | – | – | 100 | – | – |
| 21 | Water saving | – | 11 | 89 | – | – |
| 22 | Suitable for early sowing | – | 13 | 87 | – | – |
| 23 | Suitable for late sowing | – | 7 | 93 | – | – |
| 24 | Others (specify) | – | – | – | – | – |

Rating: +2 – Strong Advantage, +1 – Advantage, 0 – No Difference, −1 – Disadvantage, −2 – Strong Disadvantage

and 89 % said they could have got better profits if not for the bad rainfall (see Table 13.12).

Cultivation of Bt cotton itself is a nascent technology that the cultivators have experienced for the last 2 years. Here, an attempt is made to gauge their opinions about the impact the technology has made across the cross sections of castes, size groups of farmers, economic class and on the village itself. Table 13.13 gives the responses of sample households on the impact of Bt Technology on economic development of various sections of society. A vast majority, 93 % feel that higher income families have benefited by Bt and 7 % feel that Bt has positive impact on the poor as well.

While everyone has accepted that Bt has positive impact on large and medium farmers, 13 % said landless are also helped by way of better employment as a result of better harvest. Same analogy as given in income levels has emerged on the caste

**Table 13.13** Direct/indirect impact of Bt cotton technology on the economic development of the following reported by Bt cotton growers

| Sl. no. | Particulars | Average rating (%) | | | | | | | | | |
|---|---|---|---|---|---|---|---|---|---|---|---|
| | | Guntur | | | | | Warangal | | | | |
| | | 2 | 1 | 0 | –1 | –2 | 2 | 1 | 0 | –1 | –2 |
| 1. | Village as a whole | 2.22 | 75.56 | 22.22 | – | – | – | – | 100.00 | | |
| 2. | Any particular religious group (specify) | – | 4.44 | 95.56 | – | – | – | – | 100.00 | | |
| 3. | Any particular caste (specify) | – | – | 100.00 | – | – | – | – | 100.00 | | |
| 4. | Any other group (specify) | – | – | 100.00 | – | – | – | – | 100.00 | | |
| 5. | Women | – | – | 100.00 | – | – | – | – | 100.00 | | |
| 6. | Poor | – | 6.67 | 93.33 | – | – | | 19.60 | 80.40 | | |
| 7. | Upper income | – | 93.33 | 6.67 | – | – | | 35.00 | 65.00 | | |
| 8. | Middle income | – | 82.22 | 17.78 | – | – | | 22.00 | 78.00 | | |
| 9. | Large/medium farmers | – | 100.00 | – | – | – | | 43.00 | 57.00 | | |
| 10. | Small/marginal farmers | – | 93.33 | 6.67 | – | – | | 18.65 | 81.35 | | |
| 11. | Landless | – | 13.33 | 86.67 | – | – | | – | 100.00 | | |
| 12. | Labour/wage earners | – | 6.67 | 93.33 | – | – | | 16.00 | 84.00 | | |
| 13. | Livestock owners | – | 6.67 | 93.33 | – | – | | – | 100.00 | | |
| 14. | Tribes | – | – | 100.00 | – | – | | – | – | | |
| 15. | Upper caste | – | 60.00 | 40.00 | – | – | | 35.00 | 65.00 | | |
| 16. | Lower caste | – | 46.67 | 53.33 | – | – | | 26.00 | 74.00 | | |
| 17. | Scheduled castes | – | 20.00 | 80.00 | – | – | | 11.00 | 89.00 | | |
| 18. | Youth | – | – | 100.00 | – | – | | – | 100.00 | | |
| 19. | Any other occupational group (specify) | | – | 100.00 | | | | | | | |
| 20. | Any other (specify) | | – | – | | | | | | | |

Rating: +2 – Strong Advantage, +1 – Advantage, 0 – No Difference, –1 – Disadvantage, –2 – Strong Disadvantage

front as 60 % expressed that upper castes have had a positive impact, only 20 % felt that scheduled castes were also the beneficiaries.

Bt farmers in Warangal district have reported that the advantage with Bt cultivation accrued more towards upper caste, upper income and large farmers and it has tapered down along the caste and income groups (Table 13.13).

A number of advantages were put forward by seed suppliers in support of Bt cultivation. Among these the three main reasons were lesser need for pesticides, higher yield and better quality cotton. The other reasons are higher marketability

**Table 13.14** Advantages of Bt cotton conveyed by the seed suppliers: percentages

| | Percentage of total sample households | | | | | | | |
|---|---|---|---|---|---|---|---|---|
| | Guntur | | | | Warangal | | | |
| | Bt cotton-G | | Bt cotton-NC | | Bt cotton-G | | Bt cotton-NC | |
| Advantages | No. | % | No. | % | No. | % | No. | % |
| 1. Lesser use of pesticide | 20 | 44 | 12 | 27 | 30 | 67 | – | – |
| 2. Increase in yield | 21 | 47 | 13 | 29 | 22 | 49 | – | – |
| 3. Good quality cotton | 18 | 40 | 12 | 27 | 25 | 56 | – | – |
| 4. Higher marketability | 10 | 22 | 8 | 18 | 25 | 56 | – | – |
| 5. Better price | 8 | 18 | 9 | 20 | 23 | 51 | – | – |

**Table 13.15** Measures to facilitate Bt cotton cultivation as suggested Bt sample respondents (Guntur)

| | Bt cotton-G | Bt cotton-NC |
|---|---|---|
| Particulars | Percentage of reporting farm households | |
| **Government** | | |
| 1. Supply of genuine seed at low cost | 83 | 95 |
| 2. Paying of higher price for Bt | 79 | 82 |
| **Extension agencies** | | |
| More supervision in the crop season | 26 | |
| **Private traders/companies** | | |
| 1. Supply of good quality seed | 25 | **88** |

and better price. Forty four per cent in Bt-G and 27 % in Bt-NC have reported lesser use of pesticide as the reason given by the suppliers. Similarly 47 % in Bt-G and 29 % in Bt-NC quoted increase in yield as advantage. Other advantages propagated by suppliers are given in Table 13.14.

Main benefit of the Bt cotton as conveyed by the seed suppliers in Warangal district to a large number of farmers (67 %) is the lesser use of pesticide. Other benefits that are projected to farmers are good quality cotton (56 %), higher marketability (56 %), better price (51 %) and increased yield (49 %). Details are given in Table 13.14.

## 13.4 Measures Suggested by Sample Farmers to Improve the Acceptance of Bt Cotton

An overwhelming majority of Bt farmers expressed that the government must consider supplying genuine Bt seed at reasonable cost and paying good price for Bt cotton in the market (see Table 13.15. A quarter of the farmers have also indicated that the extension agencies must step up their role in giving timely advice in crop practices as the technology itself is new to the farmer . Eighty four per cent of

**Table 13.16** Measures to facilitate Bt cotton cultivation as suggested Bt sample respondents (Warangal)

| Particulars | Bt cotton-G | Bt cotton-NC |
|---|---|---|
| | Percentage of reporting farm households | |
| **Government:** | % | |
| 1. Supply of genuine seed at low cost | 76 | – |
| 2. Paying of higher price for Bt cotton | 84 | – |
| **Extension agencies** | | |
| 1. More supervision in the crop season | 47 | – |
| **Private traders/companies** | | |
| 1. Seed cost must be brought down | 91 | – |
| 2. Suitable varieties to the local conditions should be sold | 39 | – |

**Table 13.17** Opinion of the sample respondent on pest attack on Bt cotton as compared to non-Bt cotton (Guntur)

| Particulars | Bt cotton-G | Bt cotton-NC |
|---|---|---|
| | Percentage of total sample households | |
| **(A) Boll worm** | | |
| Sample respondents reporting HIGER Pest Attack | – | – |
| Sample respondents reporting LOWER pest attack | 75 | 56 |
| Sample respondents reporting NO CHANGE in pest attack | 25 | 44 |
| Reason for HIGER pest attack | Percentage of respondents reporting higher | |
| Reason for LOWER pest attack | | |
| 1. Could be pest resistance of the plant | 100 | 100 |
| **(B) Other pests** | | |
| Sample respondents reporting HIGER pest attack | | |
| Sample respondents reporting LOWER pest attack | | |
| Sample respondents reporting NO CHANGE in pest attack | 100 | 100 |

farmers in Warangal district demanded that the government should pay higher price for Bt cotton (see Table 13.16).

Majority of farmers, 75 % in Bt G and 56 % in non-Bt category have felt that Bt varieties have superior pest resistance qualities (Table 13.17). In the other district Warangal, 7 % farmers have expressed that Bt variety has lower pest resistance and attributed it due to weather (Table 13.18).

**Table 13.18** Opinion of the sample respondent on pest attack on Bt cotton as compared to non-Bt cotton (Warangal)

| Particulars | Bt cotton-G | Bt cotton-NC |
|---|---|---|
| | Percentage of total sample households | |
| **(A) Boll worm** | | |
| Sample respondents reporting HIGER pest attack | 7 | NA |
| Sample respondents reporting LOWER pest attack | 73 | NA |
| Sample respondents reporting NO CHANGE in pest attack | 20 | NA |
| Reason for HIGER pest attack | Percentage of respondents reporting Higher | |
| 1. weather conditions | 100 | NA |
| Reason for LOWER pest attack | Percentage | |
| 1. Due to high pest resistance quality | 100 | NA |
| **(B) Other pests** | | |
| Sample respondents reporting HIGER pest attack | | |
| Sample respondents reporting LOWER pest attack | | |
| Sample respondents reporting NO CHANGE in pest attack | 100 | NA |
| Reason for HIGER pest attack | | |
| Reason for LOWER pest attack | | |

**Table 13.19** The impact of Bt cotton cultivation on environment

| Particulars | Bt cotton-G | Bt cotton-NC | Non-Bt cotton |
|---|---|---|---|
| | Percentage of total sample households | | |
| Sample households reporting adverse impact on environment due to Bt cotton cultivation | 4 | NA | NA |
| Observations made by the reported households | Percentage of households reported YES | | |
| Cotton, when stored in the house skin itching is observed | 4 | NA | NA |

Though the farmers are apprehensive about the ill effects of Bt cotton on environment they could not pin point any for the present. They themselves agree that their experience is too little in coming to any conclusion. About 4 % of sample farmers have complained skin irritation when cotton was stored in the house (see Table 13.19).

**Table 13.20** Willingness/non willingness to continue with Bt cotton cultivation (Guntur)

| | Bt cotton-G | Bt cotton-NC |
|---|---|---|
| Particulars | Percentage of total sample households | |
| Percentage of sample households reporting their willingness to continue with of Bt cotton cultivation | 80 | 100 |
| Reason for continuing Bt cultivation | Percentage of households shown willingness to continue Bt cotton | |
| More resistant to Boll worm and better yield | 80 | 100 |
| Reason for discontinuing Bt cultivation | Percentage of households shown non willingness to continue Bt cotton | |
| 1. Seed cost is prohibitive, wish to try Bt NC | 20 | – |
| 2. Did not perform well | 20 | |

**Table 13.21** Willingness/non willingness to continue with Bt cotton cultivation (Warangal)

| | Bt cotton-G | Bt cotton-NC |
|---|---|---|
| Particulars | Percentage of total sample households | |
| Percentage of sample households reporting their willingness to continue with of Bt cotton cultivation | 92 | – |
| Reason for continuing Bt cultivation | Percentage of households shown willingness to continue Bt cotton | |
| 1. More resistant to Boll worm and better yield | 92 | – |
| Reason for discontinuing Bt cultivation | Percentage of households shown non willingness to continue Bt cotton | |
| 1. Seed cost is prohibitive | 8 | – |
| 2. Did not perform well | 8 | – |
| 3. The hybrid is not suitable to the region | 8 | |

An overall favourable perception is noticed among the sample farmers towards Bt as 80 % of Bt-G and 100 % of Bt-NC categories expressed willingness to raise the crop again. Twenty per cent of Bt-G farmers have incurred losses and they attribute that to non-performance of particular varieties of hybrids (Table 13.20).

Even more percentage of farmers in Warangal are willing to continue, (96 %)/ with Bt cultivation citing its superior pest resistance. The others complained about high cost of seed and wrong selection of hybrid for Bt by the seed companies as the reasons for discontinuing (Table 13.21).

# Part IV
# State Study 2: Gujarat

# Chapter 14
# Cotton Cultivation in Gujarat

After Independence, Gujarat achieved remarkable success in boosting agricultural sector mainly in terms of crop production and productivity of food and commercial cash crops including cotton. It is obvious that growth of agricultural sector is directly linked with profitability of cultivation of crops. And, realization of profit from cultivation in a particular region is a function of many agricultural characteristics and practices of that region. Therefore, to assess more precisely the overall impact and profitability of Bt cotton in selected regions, broad picture of cotton crop related agricultural characteristics of the selected regions will be helpful. With this in view, brief information about climatic conditions, rainfall, crop pattern, irrigation, trend in area, production and productivity in cotton etc. for State and selected districts has been provided in this chapter.

## 14.1 Climate and Soil Type

Gujarat has tropical climate. The climate of the State as well as of selected districts is extreme and subject to significant variations in temperature. The range of minimum and maximum temperature of the State was 2.3–47° C. The temperature in Vadodara district normally varies between 8° C in winter and 43° C in summer. In Rajkot district, it normally varies between 7 and 44.2° C. During the reference year 2004–2005, the climatic conditions remained more or less most favourable to cotton crop in the selected areas of the study.

The soils of southern part of Vadodara district including selected Karjan taluka is moderately deep black cotton soils and it is suitable for cotton crop. The soils of western part of Vadodara district is alluvial sandy loam locally known as 'goradu'. These soils are loamy and well suited for irrigated farming including hybrid cotton. The soils of Rajkot district are shallow medium black and calcareous. Nearly 74 % area of the district falls into semi-arid zone.


V.P. Gandhi, D. Jain, *Introduction of Biotechnology in India's Agriculture*, DOI 10.1007/978-981-10-1091-0_14

**Table 14.1** District-wise average rainfall (in millimetres)

| | | | Years | | | |
|---|---|---|---|---|---|---|
| Sr. no. | District | Average rainfall (1994–2003) | 2001 | 2002 | 2003 | 2004 |
| 1 | Ahmedabad | 771 | 658 | 397 | 883 | 814 |
| 2 | Amreli | 509 | 509 | 573 | 667 | 413 |
| 3 | Anand | 743 | 525 | 547 | 1063 | 821 |
| 4 | Banaskantha | 638 | 694 | 251 | 807 | 426 |
| 5 | Bharuch | 845 | 885 | 1023 | 889 | 1085 |
| 6 | Bhavnagar | 593 | 622 | 831 | 537 | 549 |
| 7 | Dahod | 705 | 530 | 750 | 1015 | 1127 |
| 8 | Dangs | 2703 | 2200 | 2442 | 2129 | 2642 |
| 9 | Gandhinagar | 610 | 421 | 373 | 1031 | 806 |
| 10 | Jamnagar | 612 | 583 | 433 | 1429 | 869 |
| 11 | Junagadh | 839 | 896 | 589 | 1178 | 890 |
| 12 | Kheda | 718 | 715 | 479 | 1091 | 759 |
| 13 | Kuchchh | 316 | 243 | 78 | 712 | 223 |
| 14 | Mehsana | 733 | 930 | 437 | 720 | 636 |
| 15 | Narmada | 1147 | 822 | 802 | 1275 | 1255 |
| 16 | Navsari | 1631 | 2014 | 1396 | 2933 | 2492 |
| 17 | Panchmahal | 946 | 677 | 735 | 1353 | 929 |
| 18 | Patan | 626 | 540 | 282 | 720 | 603 |
| 19 | Porbandar | 586 | 623 | 271 | 666 | 477 |
| 20 | Rajkot | 562 | 429 | 373 | 989 | 719 |
| 21 | Sabarkantha | 723 | 593 | 431 | 957 | 499 |
| 22 | Surat | 1185 | 1126 | 1130 | 1730 | 1962 |
| 23 | Surendranagar | 484 | 671 | 636 | 525 | 550 |
| 24 | Vadodara | 873 | 826 | 835 | 1014 | 1154 |
| 25 | Valsad | 1851 | 2104 | 1624 | 2100 | 2246 |
| 26 | Gujarat State | 853 | 817 | 636 | 1078 | 959 |

Source: Directorate of Agriculture, Gujarat State

In Gujarat State, nearly 64 % cotton cultivation is rainfed. Therefore, cotton output of the State and farmers' net return from cotton cultivation in a particular year is directly associated with suitability of the rainfall pattern to the crop. The district-wise rainfall recorded for the years 2001–2004 and average rainfall of last 10 years ending 2003 has been displayed in Table 14.1.

Gujarat State normally receives rainfall through south-west monsoon which mostly commences between mid of June to end of June and withdraw by the end of September. The rainfall in the State is erratic and scanty. Therefore, several parts of State experience frequent drought or drought like situations which have led to poor harvests and incomes. For State as a whole, the average rainfall of last 10 years ending 2003 was 853 mm and across districts it varied from 2703 mm for Dangs to 316 mm for Kutchh. This wide variation in annual rainfall across different years and across districts causes significant fluctuations in crops productivity.

Table 14.2 Irrigation and sources of irrigation in selected districts

| Sr. no. | Particulars | | Districts/State Vadodara | Rajkot | Gujarat State |
|---|---|---|---|---|---|
| A | Irrigation: (area in 00' ha) | | | | |
| | 1 | Net Irrigated Area (NIA) | 1924 | 1512 | 29,573 |
| | 2 | Gross Irrigated Area (GIA) | 2284 | 1860 | 36,031 |
| | 3 | % of NIA to Net Sown Area | 36.62 | 20.79 | 30.65 |
| | 4 | % of GIA to Gross Crop. Area (GCA) | 40.66 | 23.89 | 32.85 |
| | 5 | Irrigation intensity | 118.71 | 123.01 | 121.84 |
| B | Source-wise net irrigation (%) | | | | |
| | 1 | Govt. Canals | 2.01 | 20.63 | 16.26 |
| | 2 | Well-Tube wells | 95.83 | 77.46 | 82.51 |
| | 3 | Other sources | 2.16 | 1.91 | 1.23 |

Figures are triennial average ending year 2001
Source: Directorate of Agriculture, Gujarat State

During the reference year 2003–2004, average annual rainfall of the State was 959 mm, whereas it was 1154 mm for Vadodara district and 719 mm for Rajkot district. During the period from last week of July, 2004 to second week of August, heavy rain affected the cotton crop in some parts of Vadodara district. It also caused problem of pest and disease namely spodoptera in selected Rajkot district. However, during the reference year other pest diseases problems remained below threshold level.

## 14.2 Irrigation and Sources of Irrigation

Irrigation is a basic need and one of the most crucial factors affecting the productivity of cotton and other crops. The groundwater, surface water and irrigation scenario of Gujarat State is not so encouraging as compared to national level. Of the total, 80 % of the surface water is available only in south and central Gujarat. The data on irrigation and sources of irrigation in selected districts is exhibited in Table 14.2. These data show that average gross irrigated area of the State was about 33 % of State GCA and year-wise, it ranged between 31 % and 35 %. It was around 41 % for Vadodara district and only 24 % for Rajkot district.

In Gujarat including selected districts, wells and tube wells are the principal sources of irrigation (see Table 14.2). In total net irrigation, the share of well and tube well irrigation was 82.5 % for State, 82.51 % for Rajkot district and 95.83 % for Vadodara district (see Table 14.2). The public sector canal was the second most important source of irrigation claiming 16.26 % of State total irrigated area.

## 14.3 Cropping Pattern

Agricultural output of the State as well as of the selected districts is highly dependent on rainfall. Therefore, cropping pattern of selected districts is directly affected by rainfall behaviour and availability of irrigation. Hence, general practice is to grow single crop during the year on areas where scope of irrigation is negligible and rabi crops are grown on limited areas where irrigation sources are available.

The cropping pattern data shown in Table 14.3 exhibit that groundnut, cotton, bajra, rice, wheat and maize were the main crops of the State. The area under cotton alone in the State was to the extent of 16.45 % of GCA. In the State cropping pattern, groundnut, cotton and bajra crops together accounted for more than 44 % of GCA.

In Vadodara district, cotton, tur, rice and maize were the main kharif crops. Generally, hybrid and Bt cotton are grown as irrigated crops. Vadodara district is a leading medium and long staple cotton producer. In the district, cotton and tur crops are most important and cotton accounted for nearly 30 % and tur 15.45 % of the district GCA.

In Rajkot district, groundnut, cotton, wheat and bajra were important crops. Majority area under groundnut is rainfed. The hybrid cotton and Bt cotton (confirmed and non-confirmed) is mostly grown in areas where assured irrigation is available and non-HB cotton is grown as rainfed as well as irrigated crop. The district is a leading producer of groundnut and cotton. Groundnut alone occupied 46.79 % of district GCA. Cotton is the second most important crop claiming 21.94 % of district GCA. Nearly 70 % area of district GCA was thus occupied by groundnut and cotton.

In both selected districts, area under Bt cotton was very high during the reference year 2004–2005. In total Bt cotton area, share of non-approved Bt cotton was also reported as significant.

## 14.4 Hybrid/HYV and Non-hybrid Varieties of Cotton in Gujarat

Along with irrigation, coverage of areas under hybrid/HYVs varieties is a crucial factor influencing the yield of cotton crop. With this in view, year-wise data of area under prominent hybrid and non-hybrid varieties of cotton in the State have been presented in Table 14.4.

Data presented in Table 14.4 show that Gujarat cotton-13 and V-797 are the two main non-hybrid cotton varieties of the State, whereas H-6, S-8, S-10, Bt approved and other hybrid varieties (mainly non-approved Bt including Navbharat-151 and others) are the main hybrid cotton varieties in the State. The data clearly show that after the approval of Bt cotton cultivation in the State in 2002, the area under cotton H-4 declined significantly and touched negligible level in 2004–2005. The area under other hybrid (mainly non-confirmed Bt cotton varieties) varieties shows continuous upward trend since 1999–2000. It moved up from 90,426 ha in 1998–1999 to 413,763 ha in 2004–2005 showing net increase of 357 %.

**Table 14.3** Crop-pattern and crop productivity in selected districts and State (Biennium ending 2004–2005). Area in "00" Hect, Production in "00" MT, yield in kg/ha

| | | Vadodara district | | | | Rajkot district | | | | Gujarat state | | | |
|---|---|---|---|---|---|---|---|---|---|---|---|---|---|
| Sr. no | Crop | Area | % Share of area | Production | Yield | Area | % Share of area | Production | Yield | Area | % Share of area | Production | Yield |
| 1 | Rice | 497 | 8.96 | 569 | 1145 | – | – | – | – | 6774 | 6.28 | 1237 | 1821 |
| 2 | Wheat | 161 | 2.9 | 374 | 2327 | 491 | 5.81 | 1889 | 3851 | 7435 | 6.9 | 19,210 | 2583 |
| 3 | Jowar | 139 | 2.5 | 191 | 1374 | 78 | 0.92 | 60 | 788 | 1472 | 1.37 | 1540 | 1044 |
| 4 | Bajra | 124 | 2.24 | 164 | 1323 | 365 | 4.32 | 557 | 1525 | 9932 | 9.21 | 13,445 | 1345 |
| 5 | Maize | 512 | 9.23 | 703 | 1372 | 10 | 0.1 | 16 | 1684 | 4720 | 4.38 | 6222 | 1322 |
| 6 | Total Cereals | 1492 | 26.9 | 2023 | 1356 | 943 | 11.16 | 2522 | 2676 | 30,760 | 28.53 | 53,103 | 1727 |
| 7 | Tur | 8,57 | 15.45 | 691 | 806 | 8 | 0.1 | 7 | 889 | 2756 | 2.56 | 2470 | 897 |
| 8 | Gram | 14 | 0.2 | 9 | 630 | 64 | 0.7 | 62 | 969 | 1362 | 1.26 | 1154 | 846 |
| 9 | Total Pulses | 1020 | 18.39 | 804 | 788 | 197 | 2.33 | 162 | 822 | 7712 | 7.15 | 5515 | 715 |
| 10 | Total Food grains | 2512 | 45.29 | 2827 | 1125 | 1140 | 13.49 | 2685 | 2356 | 38,472 | 35.69 | 58,617 | 1524 |
| 11 | Groundnut | 147 | 2.65 | 240 | 1639 | 3955 | 46.79 | 5945 | 1503 | 19,942 | 18.5 | 31,448 | 1577 |
| 12 | Total Oilseeds | 244 | 4.4 | 346 | 1416 | 4408 | 52.15 | 6358 | 1443 | 29,720 | 27.57 | 42,769 | 1439 |
| 13 | Cotton | 1655 | 29.84 | 4832 | 445 | 1855 | 21.94 | 10,400 | 953 | 17,737 | 16.45 | 47,353 | 454 |
| 14 | Gross Cropped Area(GCA) of State | 5547 | 100 | – | – | 8453 | 100 | – | – | 107,801 | 100 | – | – |

(1) Cotton Production in "00" bales of lint, each 170 kg
(2) Yield of cotton in lint kg/ha
(3) % of share of area indicates percentage to State GCA
Source: Directorate of Agriculture, Gujarat State

**Table 14.4** Prominent varieties of cotton grown in Gujarat State. (Area in ha)

| Cotton variety | Years | | | | | | |
|---|---|---|---|---|---|---|---|
| | 1998–1999 | 1999–2000 | 2000–2001 | 2001–2002 | 2002–2003 | 2003–2004 | 2004–2005 |
| Non-hybrids | | | | | | | |
| Digvijay | 84,818 | 59,984 | 20,232 | 17,926 | 51,195 | 10,292 | 37,918 |
| G-Cot.–10 | 65,535 | 12,760 | 9517 | 5821 | 4300 | 3150 | 3200 |
| G-Cot.–12 | 36,822 | 32,200 | 21,250 | 7928 | 0 | 16,000 | 16,100 |
| G-Cot.–13 | 313,114 | 259,044 | 314,248 | 232,437 | 294,282 | 228,452 | 187,850 |
| V-797 | 311,934 | 221,076 | 262,365 | 202,879 | 217,810 | 229,855 | 276,090 |
| Deviraj | 8409 | 18,936 | 22,210 | 53,557 | 26,944 | 47,410 | 39,530 |
| Other varieties | 134,496 | 51,941 | 155,808 | 98,379 | 182,183 | 184,933 | 200,560 |
| Total non-hybrid (ha) | 955,128 | 655,941 | 805,630 | 618,927 | 776,714 | 719,822 | 761,248 |
| % to total cotton | 57.60 | 42.36 | 50.86 | 35.00 | 46.39 | 43.70 | 38.16 |
| Hybrids | | | | | | | |
| H-4 (S-4) | 16,951 | 14,090 | 32,850 | 32,369 | 1440 | 1990 | 900 |
| H-6, H-8, H-10 (S-6, 8,10) | 595,669 | 695,386 | 579,055 | 902,410 | 655,227 | 507,199 | 610,590 |
| Bt (approved) | – | – | – | – | 16,900(p) | 80,000 (p) | 208,502 (p) |
| Others hybrid (inclusive non-app. Bt) | 90,426 | 182,949 | 166,337 | 214,749 | 224,006 | 338,068 | 413,763 |
| Total hybrid area | 703,046 | 892,425 | 778,236 | 1,149,528 | 897,573 | 927,257 | 1,233,755 |
| % to HB. to grand total | 42.40 | 57.64 | 49.14 | 65.00 | 53.61 | 56.30 | 61.84 |
| Grand total area (non-HB + HB) % | 1,658,174 | 1,548,366 | 1,583,866 | 1,768,455 | 1,674,287 | 1,647,079 | 1,995,000 |
| | 100.00 | 100.00 | 100.00 | 100.00 | 100.00 | 100.00 | 100.00 |

Source: Directorate of Agriculture, Gujarat State, Gandhinagar

*P* Provisional

At present, of the total cotton area of the State, nearly 62 % area is under hybrid varieties and only 38 % under non-hybrid varieties. In the total cotton area of the State, the share of non-hybrid varieties declined from 57.60 % in 1998–1999 to 38.16 % in 2004–2005. This clearly suggests that cotton farmers are shifting from non-hybrid to hybrid cotton and particularly to Bt cotton at a very fast pace.

## 14.5 Profile of Cotton Varieties Grown in Gujarat

As stated earlier, many varieties of hybrids, HYVs and non-hybrid cotton are grown in different tracts of the State. Based on attributes, these varieties may be classified into two main groups. Cotton varieties S-6, H-8 and H-10 may be put into one group and varieties V-797, Guj.Cot-13, Guj-Cot-21 (Wagad cotton), GS-23 (Bharuch) may be categorized into another group. The salient features and profile of each group of varieties are given in Table 14.5. Also map of Gujarat State showing variety-wise districts/tracts where it is grown is given in Fig. 14.1 below.

From the Table 14.5, it is evident that non-HB varieties like V-797, G-Cot-13 etc. are not grown at all in selected Vadodara district. However, it is grown in some parts of selected Rajkot district.

Generally, Bt cotton and others hybrid varieties of cotton are sown in the State during the period 15th June to 15th July. But the farmers who have adequate irrigation facilities and hoping to take rabi crop are opting for sowing of Bt cotton in the month of May.

**Table 14.5** Cotton profile of Gujarat

| Variety/hybrid | S-6,H-8,H-10,MECH 12, MECH 184,MECH-162,RCH-2 | V-797,G Cot-13,G Cot-21(Wagad Cotton) GS23(Bharuch) |
|---|---|---|
| Sowing period | June-July MECH-12,184,RCH-2 in May-June Under assured Irrigated cond. | July–August |
| Harvesting period | Oct.–March | Feb.–March |
| District/tracts | All districts of Gujarat except Valsad and Dang | Ahmedabad, Mehsana, Rajkot, Jamnagar, Kutchh, Junagadh, Banaskantha, Surendranagar, Panchmahal |
| Potential lint yield (kg per hectare) | 500–600 | 600–700 |
| Staple length (2.5 % S.L.mm) | 28–30 | 22–23 |
| Micronaire value | 3.5–4.5 | 5–5.5 |
| Tenacity (g/tex.) At 3.2 mm gl | 22–24 | 20–21 |

Source: (i) Directorate of Agriculture, Gujarat State, Gandhinagar
(ii) Indian Cotton – A Profile 2003–2004, CCI, Navi Mumbai

VARIETY /HYBRID

- S6(H6),H8,H10, MECH 12, MECH 184, RCH-2
- V797, G.COT 13, G.COT 21
- GS 23 (BHARUCH)

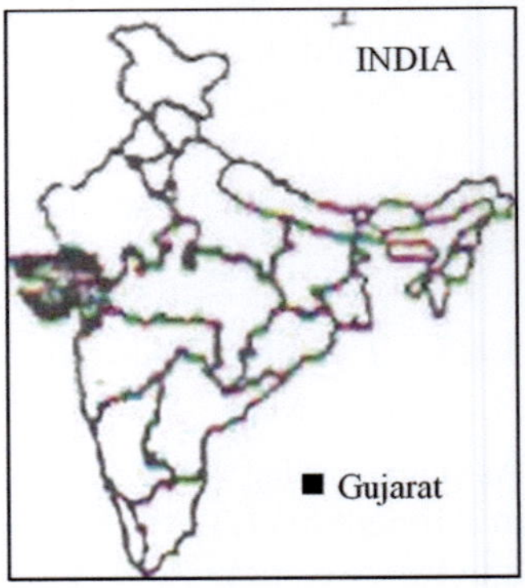


**Fig. 14.1** Gujarat's map showing tract-wise variety grown

**Gujarat**

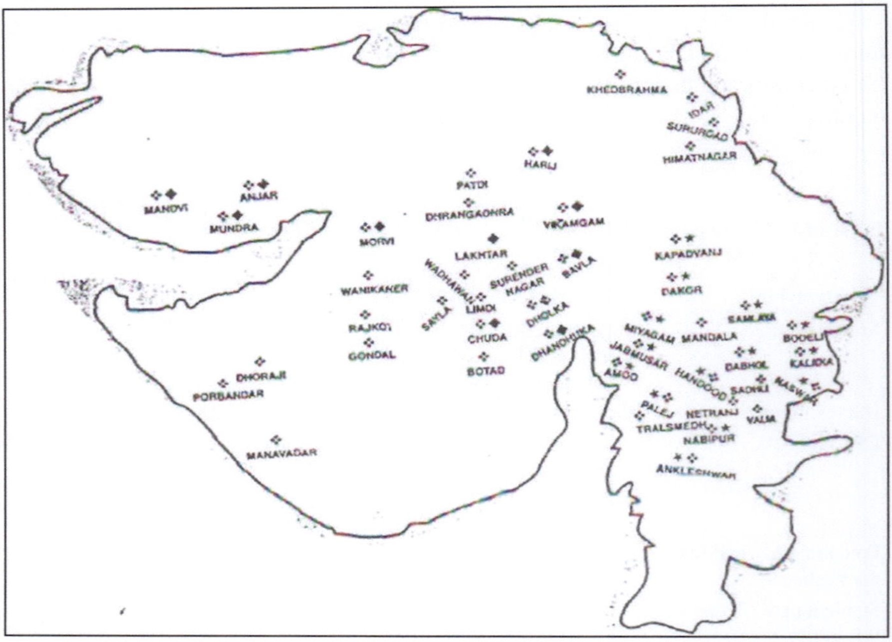

**Table 14.6** Area and yield of cotton in Gujarat – 1990–1991 to 2005–2006

| Year | Area (lakh ha) | Yield (kg/ha) | Production (lakh bales) |
|---|---|---|---|
| 1990–1991 | 9.2 | 274.8 | 13.2 |
| 1991–1992 | 11.4 | 223.6 | 14.9 |
| 1992–1993 | 11.5 | 329.3 | 22.3 |
| 1993–1994 | 11.3 | 208.3 | 19.8 |
| 1994–1995 | 12.1 | 375.1 | 26.6 |
| 1995–1996 | 14.1 | 388.0 | 32.2 |
| 1996–1997 | 14.9 | 392.0 | 34.3 |
| 1997–1998 | 15.2 | 470.0 | 42.0 |
| 1998–1999 | 16.1 | 497.0 | 47.0 |
| 1999–2000 | 15.4 | 230.0 | 20.8 |
| 2000–2001 | 16.2 | 112.0 | 11.6 |
| 2001–2002 | 17.5 | 165.0 | 16.9 |
| 2002–2003 | 16.4 | 175.0 | 16.9 |
| 2003–2004 | 16.5 | 417.0 | 40.3 |
| 2004–2005[a] | 19.1 | 651.0 | 73.0 |
| 2005–2006[a] | 20.8 | 728.0 | 89.0 |
| Annual compound growth rate (percent) | | | |
| 1990–1991 to 2001–2002 | 5.31 | −3.20 | 1.55 |
| 1990–1991 to 2005–2006 | 4.47 | 2.39 | 6.51 |
| 2000–2001 to 2005–2006 | 4.33 | 40.0 | 44.1 |

Source: Gandhi and N.V. Namboodri (2007)
[a]Revised estimates

## 14.6 Cotton Scenario in Gujarat

The total area under cotton in Gujarat in 2005–2006 was about 20.8 lakh hectares which was nearly 23 % of the cotton area of the country. After the introduction of Bt cotton, area under cotton in Gujarat is increasing every year at a fast pace. It moved up from 16.4 lakh hectares in 2002–2003 to around 20.8 lakh hectares in 2005–2006 (see Table 14.6 and graphs).

Today, Gujarat State contributes about 30 % to the national cotton production. The main cotton cultivating districts are Surendranagar (23 %), Bhavnagar (12 %), Rajkot, Ahmedabad and Vadodara (about 10 % each). As regards annual output of cotton in the State, it ranged from 13.23 lakh bales (each of 170 kg) in 1990–1991 to 40.27 lakh bales in 2003–2004 and as per revised estimate, it is likely to hit target of 89 lakh bales in 2005–2006 (see Table 14.6). The annual compound growth rate worked for the cotton yield during 1990–1991 to 2001–2002 showed negative growth of −3.2 %. However, it showed positive growth of 2.4 % for extended period 1990–1991 to 2005–2006 (see Table 14.6). The annual compound growth rates worked out for area, yield and production for the period 2000–2001 to 2005–2006 showed a very rapid growth rate of 40 %, 4 % and 44 % respectively (Figs. 14.2 and 14.3).

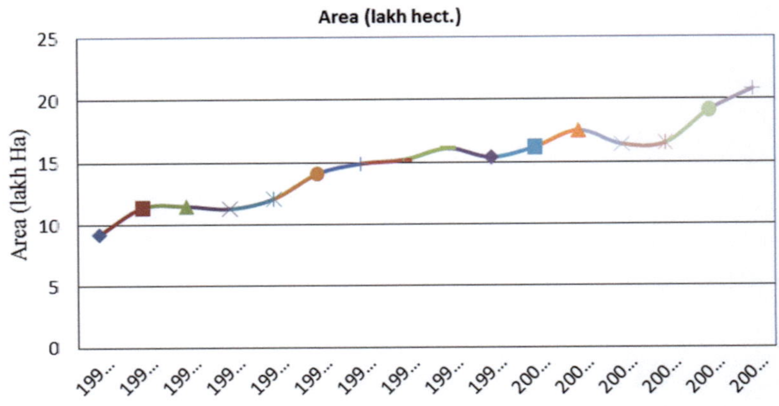


**Fig. 14.2** Area under cotton in Gujarat: 1990–1991 to 2005–2006

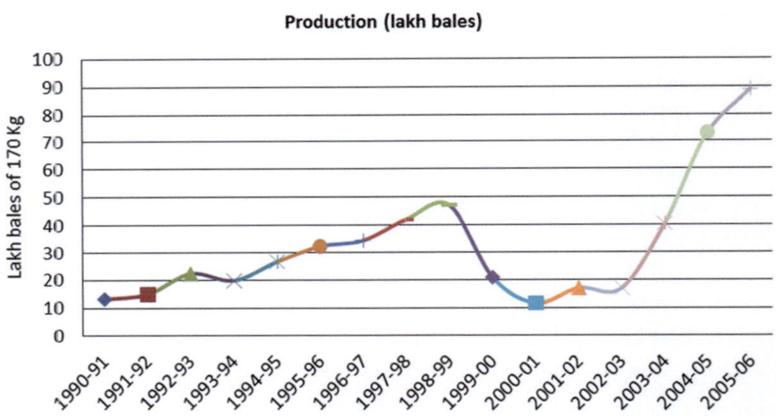


**Fig. 14.3** Production of cotton in Gujarat: 1990–1991 to 2005–2006

## *14.6.1 District-Wise Cotton Area in Gujarat*

The district-wise cotton area is given in Table 14.7 for the years 1990–1991 to 2004–2005. Though, cotton crop is being grown in all the districts of the State except Valsad and Dangs, the prominent cotton growing districts are Surendranagar, Rajkot, Bhavnagar, Ahmedabad, Vadodara, Bharuch, Mehsana (including Patan) and Amreli. The Table reveals that in the early 2000, the cotton area in the State as well as in almost all districts has shown an upward trend as compared to period of early 1990s. After the government approval of commercial cultivation of Bt cotton, in each district and State as a whole, the cotton area increased sharply as Bt cotton made inroad into area under groundnut and other crops cultivation in the State. The

**Table 14.7** District-wise area under cotton in Gujarat State during 1990–1991 to 2004–2005. Area in "00" Hect

| No. | District | 1990–1991 | | 1995–1996 | | 2000–2001 | | 2001–2002 | | 2002–2003 | | 2003–2004 | | 2004–2005 (p) | |
|---|---|---|---|---|---|---|---|---|---|---|---|---|---|---|---|
| | | Area | % to state | Area | % to state | Area | % to state | Area | % to state | Area | % to state | Area | % to state | Area | % to state |
| 1 | Ahmedabad | 1214 | 13.18 | 1452 | 10.29 | 1680 | 10.40 | 1826 | 10.44 | 1643 | 10.05 | 1680 | 10.24 | 1973 | 9.89 |
| 2 | Vadodara | 1196 | 12.99 | 1406 | 9.97 | 1577 | 9.76 | 1781 | 10.18 | 1583 | 9.68 | 1581 | 9.63 | 1780 | 8.92 |
| 3 | Bharuch[a] | 497 | 5.40 | 1014 | 7.19 | 1574 | 9.74 | 1546 | 8.84 | 1611 | 9.85 | 1681 | 10.24 | 1762 | 8.83 |
| 4 | Gandhinagar | 22 | 0.24 | 42 | 0.30 | 24 | 0.15 | 38 | 0.22 | 32 | 0.20 | 180 | 1.10 | 325 | 1.63 |
| 5 | Mehsana[a] | 735 | 7.98 | 1091 | 7.74 | 1089 | 6.74 | 1127 | 6.44 | 1020 | 6.24 | 1147 | 6.99 | 1549 | 7.77 |
| 6 | Sabarkantha | 165 | 1.79 | 138 | 0.98 | 347 | 2.15 | 351 | 2.01 | 313 | 1.91 | 377 | 2.30 | 905 | 4.54 |
| 7 | Amreli | 203 | 2.20 | 648 | 4.59 | 731 | 4.53 | 843 | 4.82 | 888 | 5.43 | 1040 | 6.34 | 1199 | 6.01 |
| 8 | Bhavnagar | 673 | 7.31 | 1437 | 10.19 | 1870 | 11.58 | 1987 | 11.36 | 1884 | 11.52 | 1901 | 11.58 | 2101 | 10.53 |
| 9 | Jamnagar | 232 | 2.52 | 385 | 2.73 | 318 | 1.97 | 372 | 2.13 | 298 | 1.82 | 320 | 1.95 | 552 | 2.77 |
| 10 | Kutch | 596 | 6.47 | 624 | 4.42 | 443 | 2.74 | 583 | 3.33 | 409 | 2.50 | 402 | 2.45 | 514 | 2.58 |
| 11 | Rajkot | 956 | 10.38 | 1222 | 8.66 | 1805 | 11.17 | 2023 | 11.56 | 1781 | 10.89 | 1680 | 10.24 | 2422 | 12.14 |
| 12 | Surendranagar | 2026 | 22.00 | 3799 | 26.94 | 3775 | 23.37 | 4232 | 24.19 | 4201 | 25.70 | 3845 | 23.43 | 3875 | 19.43 |
| 13 | Gujarat State | 9210 | 100 | 14,104 | 100 | 16,154 | 100 | 17,498 | 100 | 16,349 | 100 | 16,410 | 100 | 19,945 | 100 |

% shows district's share in total Cotton area of state
Source: Directorate of Agriculture, Govt. of Gujarat Gandhinagar
(*p*) Provisional Figure
[a]Bharuch Includes Newly formed Narmada District and Mehsana Includes Patan district

increase in cotton area of the State in 2004–2005 was around 21.5 % as compared to previous year 2003–2004. The fast momentum of adoption of Bt cotton was mainly responsible for such a sharp increase in cotton area.

Data presented in Table 14.7 shows that the distribution of cotton area is highly varying across districts. Surendranagar is an important cotton producing district with cotton area around 400,000 ha. The district grew mainly Vagad deshi (short staple) cotton under dry farming conditions due to very poor irrigation facilities. Ahmedabad is also famous for deshi cotton (V-797 and Vagad) producing district and cotton area of the district during 1990–2005 ranged between 121,000 and 197,000 ha. In Vadodara district, mainly medium staple Digvijay, long staple American cotton and Bt cotton are grown and area under cotton increased from 120,000 ha in 1990–1991 to 178,000 ha in 2004–2005. In terms of cotton area, Rajkot occupied 2nd rank in the State. The cotton area in the Rajkot district increased at a very fast pace and it touched 244,000 ha in 2004–2005 as against only 96,000 ha in 1990–1991. The most favourable economics of Bt cotton cultivation was the main reason for such significant increase in cotton area. Bhavnagar, Bharuch, Amreli and Sabarkantha are other traditionally cotton belt districts. Out of 25 districts of the State, 7 districts (Ahmedabad, Vadodara, Bharuch, Rajkot, Surendranagar, Mehsana and Bhavnagar) together accounted for as much as 78 % of State total cotton area in 2004–2005.

## *14.6.2 District-Wise Cotton Production in Gujarat*

The district-wise data on cotton production are presented in Table 14.8. In post-introduction period of Bt cotton, State witnessed sharp rise in cotton production. This is mainly due to higher area under Bt cotton, higher irrigation coverage and favourable climatic and rainfall conditions.

District-wise data on cotton production show that districts like Surendranagar and Ahmedabad accounted for relatively lower share of production as compared to their share in area. This is mainly due to lower productivity as cotton grown here has been rainfed and low yielding non-hybrid. In Surendranagar, which has the highest area under cotton among all districts, production has been fluctuating from year to year and no definite trend is noticeable. Similar trends prevailed for Rajkot and Ahmedabad districts too. However, districts like Rajkot, Vadodara, Bhavnagar, Bharuch and Sabarkantha generally recorded higher production as compared to their share in area. This is mainly because of higher yield achieved by these districts. The data clearly reveals that cotton production in Vadodara, Bharuch, Gandhinagar, Rajkot, Bhavnagar, Sabarkantha, Jamnagar and Amreli districts, jumped significantly after 2001–2002 which are coincides with post-introduction period of Bt cotton.

It is seen from the data presented in Table 14.8 that cotton production in almost all the districts of the State has been showing fluctuations across the years. However, in case of Surendranagar, Rajkot and Ahmedabad districts, fluctuations in cotton

**Table 14.8** District-wise production of cotton in Gujarat State. (Production in "00" bales, each of 170 kg lint.)

| No. | District | 1990–1991 | | 1995–1996 | | 2000–2001 | | 2001–2002 | | 2002–2003 | | 2003–2004 | | 2004–2005 (p) | |
|---|---|---|---|---|---|---|---|---|---|---|---|---|---|---|---|
| | | Area | % to state | Area | % to state | Area | % to state | Area | % to state | Area | % to state | Area | % to state | Area | % to state |
| 1 | Ahmedabad | 2080 | 15.72 | 1452 | 10.29 | 928 | 7.99 | 1045 | 6.14 | 1135 | 6.74 | 2463 | 6.12 | 4294 | 9.42 |
| 2 | Vadodara | 1541 | 11.65 | 1406 | 9.97 | 1735 | 14.94 | 1584 | 9.30 | 2439 | 14.47 | 3164 | 7.86 | 4146 | 9.10 |
| 3 | Bharuch[a] | 521 | 3.94 | 1014 | 7.19 | 1480 | 12.74 | 1660 | 9.75 | 2131 | 12.65 | 2540 | 6.31 | 4709 | 10.33 |
| 4 | Gandhinagar | 60 | 0.45 | 42 | 0.30 | 32 | 0.28 | 60 | 0.35 | 107 | 0.64 | 506 | 1.26 | 820 | 1.80 |
| 5 | Mehsana[a] | 799 | 6.04 | 1091 | 7.74 | 1044 | 8.99 | 1446 | 8.49 | 1225 | 7.27 | 1578 | 3.92 | 3304 | 7.25 |
| 6 | Sabarkantha | 244 | 1.84 | 138 | 0.98 | 393 | 3.38 | 517 | 3.04 | 291 | 1.73 | 1401 | 3.48 | 2763 | 6.06 |
| 7 | Amreli | 292 | 2.21 | 648 | 4.59 | 587 | 5.05 | 830 | 4.87 | 1404 | 8.33 | 3392 | 8.42 | 2024 | 4.44 |
| 8 | Bhavnagar | 1033 | 7.81 | 1437 | 10.19 | 861 | 7.41 | 1301 | 7.64 | 2891 | 17.16 | 4116 | 10.22 | 4696 | 10.31 |
| 9 | Jamnagar | 435 | 3.29 | 345 | 2.45 | 222 | 1.91 | 508 | 2.98 | 247 | 1.47 | 1879 | 4.67 | 1344 | 2.95 |
| 10 | Kutch | 728 | 5.50 | 624 | 4.42 | 702 | 6.04 | 957 | 5.62 | 617 | 3.66 | 994 | 2.47 | 626 | 1.37 |
| 11 | Rajkot | 1631 | 12.33 | 1222 | 8.66 | 700 | 6.03 | 2117 | 12.43 | 735 | 4.36 | 10,118 | 25.13 | 7763 | 17.04 |
| 12 | Surendra-nagar | 2600 | 19.65 | 3799 | 26.94 | 1813 | 15.61 | 3961 | 23.26 | 2540 | 15.07 | 6125 | 15.21 | 6610 | 14.51 |
| 13 | Gujarat State | 13,229 | 100 | 14,104 | 100 | 11,614 | 100 | 17,032 | 100 | 16,850 | 100 | 40,269 | 100 | 45,568 | 100 |

% shows district's share in total Cotton area of state

Source: Directorate of Agriculture, Govt. of Gujarat Gandhinagar

(*p*) Provisional Figure

[a]Bharuch Includes Newly formed Narmada District and Mehsana Includes Patan district

production were very wide and significant. It varied in Surendranagar district between 181,300 bales and 661,000 bales, in Ahmedabad it varied between 92,800 bales and 42,940 bales and for Rajkot between 70,000 bales and 1,011,800 bales.

In districts like Vadodara, Gandhinagar etc. which are well endowed with irrigation, the fluctuations in cotton production were found to be in relatively narrow range.

### *14.6.3 District-Wise Lint Cotton Yield in Gujarat*

The district-wise lint cotton yield figures per hectare are presented in Table 14.9. The year-wise cotton yield in lint form for the State during considered time periods varied from as low as only 122 kg in 2000–2001 to 454 kg in 2003–2004. Likewise in Rajkot district, it varied from 66 kg in 2000–2001 to 1023 kg in 2003–2004. In Vadodara district, year-wise fluctuations in cotton yield were there, but varied in a relatively narrow range of 151 kg in 2001–2002 to 396 kg in 2004–2005.

For the same time period, wide fluctuations in lint cotton yield were observed across the districts. In 2003–2004, it ranged from a low of 234 kg per hectare in Mehsana district to 1023 kg per hectare in Rajkot district.

It is seen from the Table that Vadodara, Bharuch, Gandhinagar, Sabarkantha, Jamnagar and Rajkot districts recorded lint cotton yield higher than the State average yield of 388 kg during 2004–2005. In spite of highest area under cotton in the State,

**Table 14.9** District-wise yield of cotton in Gujarat State

| | | Yield of Cotton (Kg.lint/Hectare) | | | | | | | |
|---|---|---|---|---|---|---|---|---|---|
| No. | District | 1990–1991 | 1995–1996 | 1999–2000 | 2000–2001 | 2001–2002 | 2002–2003 | 2003–2004 | 2004–2005 (p) |
| 1 | Ahmedabad | 291 | 175 | 206 | 94 | 97 | 117 | 249 | 370 |
| 2 | Vadodara | 219 | 257 | 276 | 187 | 151 | 262 | 340 | 396 |
| 3 | Bharuch[a] | 178 | 196 | 318 | 160 | 182 | 225 | 257 | 454 |
| 4 | Gandhinagar | 457 | 503 | 582 | 233 | 268 | 567 | 478 | 429 |
| 5 | Mehsana | 185 | 279 | 213 | 163 | 221 | 204 | 234 | 363 |
| 6 | Sabarkantha | 251 | 505 | 252 | 193 | 250 | 159 | 632 | 519 |
| 7 | Amreli | 245 | 391 | 226 | 137 | 167 | 269 | 554 | 287 |
| 8 | Bhavnagar | 261 | 371 | 187 | 78 | 111 | 261 | 368 | 380 |
| 9 | Jamnagar | 319 | 401 | 227 | 119 | 232 | 141 | 998 | 414 |
| 10 | Kuchchh | 208 | 269 | 438 | 269 | 279 | 256 | 423 | 207 |
| 11 | Rajkot | 289 | 442 | 262 | 66 | 177 | 70 | 1023 | 540 |
| 12 | Surendranagar | 218 | 166 | 184 | 82 | 159 | 103 | 271 | 290 |
| 13 | Gujarat State | 244 | 265 | 230 | 122 | 165 | 175 | 454 | 388 |

% shows district's share in total Cotton area of state

Source: Directorate of Agriculture, Govt. of Gujarat Gandhinagar

[a]Bharuch Includes Newly formed Narmada District and Mehsana Includes Patan district

Surendranagar district recorded much lower yield in almost all the years mainly owing to negligible coverage of irrigation to cotton crop. Similar situation was observed in Ahmedabad district too.

It was observed that districts which had higher coverage of irrigation to cotton crop recorded relatively higher cotton yield. This suggests that coverage of irrigation is a key variable for increasing the productivity of Bt cotton and non-Bt cotton crop.

### 14.6.4 District-Wise Cotton Yield in Irrigated and Unirrigated Area and Coverage of Irrigation

It is an established fact that irrigation and variety of cotton grown are two most crucial factors impacting the cotton yield in a big way. With this in view, district-wise data on yield of irrigated cotton, unirrigated cotton and irrigation coverage have been furnished in Table 14.10.

In the State, irrigation coverage to cotton was to the extent of 40%. Across districts, irrigation coverage to cotton crop showed significantly wide variations and it ranged from as low as 8.22% for Ahmedabad district to cent per cent for Gandhinagar district. It was more than 50% for Sabarkantha, Rajkot, Junagadh, Amreli, Bhavnagar, Kheda and Vadodara districts. This high variation in irrigation coverage for cotton across districts is causing high variations in cotton productivity.

In all the districts of the State, the yield of irrigated cotton was significantly higher than that of non-irrigated cotton. The ratio of yield of irrigated cotton and unirrigated cotton for the State worked out to 2.95, whereas among districts it varied from 1.54 for Kheda district to 5.55 for Kutchh district. As compared to non-irrigated cotton, yield of irrigated cotton was observed more than three times in Banaskantha, Mehsana, Surat, Bhavnagar, Junagadh, Rajkot, Jamnagar and Kutch districts. The very high yield of irrigated cotton clearly suggests very strong positive association between irrigation coverage and yield of cotton crop. This further suggests that cotton production can be increased further by bringing more areas of cotton under irrigation.

### 14.6.5 District-Wise Sales of Approved Bt Cotton Seeds in Gujarat During 2004–2005

The cultivation of illegal Bt cotton in Gujarat in 2001–2002 raised controversy and forced Government of India to think over whether or not allow cultivation of Bt cotton in country. Hence, after long consideration on all issues related with Bt cotton, GEAC approved in 2002 three Bt hybrids viz., MECH-12, MECH-162 and MECH-184 for cultivation in Gujarat and other States. Later on, two more hybrids from Rasi seeds were approved. The authorized Bt cotton seeds are sold in a bag and each seed bag contains 450 g Bt cotton seeds and 120 g non-Bt cotton seeds. The

**Table 14.10** District-wise Yield of irrigated and unirrigated cotton in Gujarat. Yield in Lint kg/ha, Area in 00' Hect, Prod. in 00, Bales

| Sr. no. | District | Irrigated cotton | | | Unirrigated cotton | | | Total cotton | | | % of Irri. coverage to cotton | Ratio of IY and UY |
|---|---|---|---|---|---|---|---|---|---|---|---|---|
| | | Area | Prod. | Yield | Area | Prod. | Yield | Area | Prod. | Yield | | |
| 1 | Ahmedabad | 141 | 297 | 358 | 1576 | 1250 | 135 | 1717 | 1547 | 153 | 8.22 | 2.65 |
| 2 | Banaskantha | 44 | 126 | 487 | 102 | 55 | 92 | 146 | 181 | 211 | 29.93 | 5.29 |
| 3 | Vadodara | 884 | 1544 | 297 | 765 | 852 | 189 | 1649 | 2396 | 247 | 53.61 | 1.57 |
| 4 | Bharuch | 454 | 1083 | 406 | 1159 | 1027 | 151 | 1613 | 2110 | 222 | 28.15 | 2.69 |
| 5 | Gandhinagar | 83 | 224 | 459 | 0 | 0 | 0 | 83 | 224 | 459 | 100.00 | N.A. |
| 6 | Kheda | 93 | 225 | 411 | 63 | 99 | 267 | 156 | 324 | 353 | 59.62 | 1.54 |
| 7 | Mehsana | 298 | 820 | 468 | 796 | 603 | 129 | 1044 | 1423 | 232 | 27.41 | 3.63 |
| 8 | Panchmahal | 33 | 84 | 433 | 41 | 35 | 145 | 74 | 119 | 273 | 44.59 | 2.99 |
| 9 | Sabarkantha | 323 | 715 | 376 | 24 | 21 | 149 | 347 | 736 | 361 | 93.08 | 2.52 |
| 10 | Surat | 15 | 37 | 419 | 32 | 26 | 138 | 47 | 63 | 228 | 32.61 | 3.04 |
| 11 | Amreli | 471 | 1240 | 448 | 453 | 636 | 239 | 924 | 1876 | 345 | 50.97 | 1.87 |
| 12 | Bhavnagar | 979 | 2198 | 382 | 945 | 571 | 103 | 1924 | 2769 | 245 | 50.88 | 3.70 |
| 13 | Jamnagar | 220 | 794 | 614 | 110 | 84 | 130 | 330 | 878 | 452 | 66.67 | 4.72 |
| 14 | Junagadh | 183 | 618 | 574 | 81 | 64 | 134 | 264 | 682 | 439 | 50.29 | 4.28 |
| 15 | Kutch | 224 | 717 | 544 | 241 | 139 | 98 | 465 | 856 | 313 | 48.17 | 5.55 |
| 16 | Rajkot | 1435 | 4036 | 478 | 393 | 285 | 123 | 1828 | 4321 | 402 | 78.50 | 3.82 |
| 17 | Surendranagar | 834 | 1770 | 361 | 3259 | 2438 | 127 | 4093 | 4208 | 175 | 20.38 | 2.84 |
| 18 | Gujarat State | 6712 | 16,527 | 419 | 10,040 | 8186 | 139 | 16,752 | 24,713 | 251 | 40.07 | 2.95 |

Yield figures shown are an average of 2001–2002 to 2003–2004
Source: Directorate of Agriculture, Govt. of Gujarat, Gandhinagar
*IY* Irrigated Cotton Yield, *UY* Unirrigated Cotton Yield

suggested seed rate is one bag for 1 acre. The approved Bt cotton seed price was around Rs.1600 per bag in 2004–2005 which was two to three times as compared to price of unauthorized non-approved varieties of Bt cotton seed. Among Bt cotton growers of the State, many farmers preferred these low priced unauthorized local variants of Bt seeds (Navbharat-151 and others) as farmers found it almost same in respect of quality, yield and income.

District-wise sales of approved Bt cotton seeds in 2004–2005 is given in Table 14.11. The data reveals that sales of Bt cotton seeds across districts showed wide variations and it was higher in Bhavnagar, Rajkot, Vadodara and Jamnagar districts. The year-wise examination of approved sales of Bt cotton seeds shows significant increase in 2004–2005 over previous year.

It was observed that farmers in Gujarat are not adopting recommended system of growing refuge crops on boundary of Bt cotton plots. Moreover, sale of non-approved Bt cotton seeds was much higher than approved Bt cotton seeds.

**Table 14.11** District-wise sales of approved Bt cotton seeds in 2004–2005

| | | Nos. of bags (each of 450 g Bt) | | |
|---|---|---|---|---|
| Sr. no. | District | MECH-12 MECH-162 MECH-184 | Rasi's RCH-2 | Total |
| 1 | Ahmedabad | 2330 | – | 2330 |
| 2 | Amreli | 22,420 | 10,550 | 32,970 |
| 3 | Bharuch | 14,000 | 6010 | 20,010 |
| 4 | Bhavnagar | 92,000 | 13,500 | 105,500 |
| 5 | Vadodara | 43,800 | 12,810 | 56,610 |
| 6 | Gandhinagar | 3600 | – | 3600 |
| 7 | Jamnagar | 34,000 | 12,850 | 46,850 |
| 8 | Junagadh | 27,000 | 11,130 | 38,130 |
| 9 | Kheda and Anand | 5850 | 4642 | 10,492 |
| 10 | Kutch | 16,000 | 3750 | 19,750 |
| 11 | Mehsana | 14,500 | – | 14,500 |
| 12 | Panchmahal | 6000 | 2110 | 8110 |
| 13 | Rajkot | 77,600 | 23,670 | 101,270 |
| 14 | Surendranagar | 6800 | 4910 | 11,710 |
| 15 | Sabarkantha | 24,600 | – | 24,600 |
| 16 | Surat | 8000 | 4230 | 12,230 |
| 17 | North Gujarat | 1500 | 11,095 | 12,595 |
| 18 | Total | 400,000 | 121,257 | 521,257 |

Source: (1) Directorate of Agriculture, Gujarat State Gandhinagar

(2) Data Provided by Rasi Seeds, Salem

## 14.7 Minimum Support Prices and Annual Average Market Prices of Cotton

Before the start of cotton season, every year the Government of India, based on the recommendation of the CACP, announces Minimum Support Prices (MSP) for the two basic varieties of Cotton viz., F.414/H-777/J-34 (medium staples) and H-4 (long staple). The prices for other varieties of cotton are subsequently fixed by the Textile Commissioner based on market differentials and discussions with various organizations related with cotton such as CCI, East India Cotton Association, farmers' co-operative organizations etc.

In Table 14.12, support prices of Cotton for those varieties of cotton which are mainly grown in Gujarat State have been shown.

Vagad and V-797 varieties are very popular in the rainfed farming areas of Gujarat and particularly in Surendranagar, Ahmedabad, Rajkot, Kutch, Patan and Mehsana districts even though, government has not been announcing MSP for Vagad Cotton since 1999–2000.

In recent years, the global market prices of cotton have been impacting highly on domestic market prices of cotton. Hence, during the cotton season 2004–2005, owing to depressed global prices of raw cotton (Cotton) domestic market also witnessed continuous down trend in Cotton prices. During the first picking of H-4 cotton, the market price of Cotton was around Rs.2250 per quintal which subsequently declined to around Rs. 1900–1950/qtl. during 4th–7th picking period and stabilized near to MSP level. In some interior rural areas of the State, the prices of Cotton ruled even below support prices. The plausible reasons given for such decline in Cotton prices were large volume of carried over stock, good harvest of Cotton and lower global prices of cotton.

**Table 14.12** Support prices for varieties of cotton of FAQ – 2000–2001 to 2004–2005. (As announced by Govt. of India). (Rs. Qtl.)

| Sr. no | Variety | Basic staple length (2.5 % span length) mm | Micronnaire value | 2000–2001 | 2001–2002 | 2002–2003 | 2003–2004 | 2004–2005 |
|---|---|---|---|---|---|---|---|---|
| 1 | V-797 | 22.00 | 4.2–4.8 | 1515 | 1560 | 1560 | 1605 | 1640 |
| 2 | G.Cot.12,13,21 | 23.00 | 4.2–5.0 | 1440 | 1485 | 1485 | 1530 | 1560 |
| 3 | H-4 /H-6/H-10 | 30.00 | 3.6–4.2 | 1825 | 1875 | 1875 | 1925 | 1960 |
| 4 | Sankar -6/10 (Saurashtra) | 29.00 | 3.7–4.3 | 1840 | 1840 | 1860 | 1910 | 1960 |
| 5 | RCH-2 (Bt) | 30.00 | 3.5–4.5 | – | – | – | – | 1960 |
| 6 | MECH-12/162/184 (Bt) | 29.00 | 3.5–4.5 | – | – | 1875 | 1925 | 1960 |
| 7 | MCU-5 | 33.00 | 3.0–3.5 | 1875 | 1950 | 1950 | 2000 | 2035 |

Source: Indian cotton, a profile 2003–2004, published by CCI, Mumbai

*FAQ* fair average quality

**Table 14.13** Annual average prices of cotton and lint candy for years of 2000–2001 to 2004–2005

| Sr. no. | Year | MSP of cotton (Rs./qtl.) | | Average market rate of cotton (Rs. /qtl.) | | Market price as % over MSP | | Annual average market price of lint candy (Rs./candy) | |
|---|---|---|---|---|---|---|---|---|---|
| | | H-4 | S-6 | H-4 | S-6 | H-4 | S-6 | H-4 | S-6 |
| 1 | 2000–2001 | 1825 | 1840 | 2207 | 2310 | 120.93 | 125.54 | 19,676 | 20,863 |
| 2 | 2001–2002 | 1875 | 1840 | 1891 | 1900 | 100.85 | 103.26 | 15,559 | 16,659 |
| 3 | 2002–2003 | 1875 | 1860 | 2215 | 2323 | 118.13 | 124.89 | 19,910 | 20,720 |
| 4 | 2003–2004 | 1925 | 1910 | 2533 | 2632 | 131.58 | 137.8 | 22,139 | 21,309 |
| 5 | 2004–2005[a] | 1960 | 1960 | 2025 | 2076 | 105.92 | 105.92 | 16,353 | 17,190 |

Cotton Prices of Bt Cotton normally are very near or slightly higher than Price of H-4/S-6
*MSP* minimum support price. 1 Candy = 355.62 kg
[a]Up to January 2005

The profit level of cultivation of cotton is highly influenced by the prevailing market prices of Cotton. Hence, average annual rate of H-4 and S-6 Cotton and lint candy for the years 2000–2001 to 2004–2005 is shown in Table 14.13. The average prices of H-4 and S-6 Cotton in each year had ruled above the level of MSP. However, in 2001–2002 and 2004–2005 Cotton market prices were quite fluctuating and uneven and reached near to MSP. The average prices of Cotton S-6 for the year 2004–2005 was lower by 21 % compared to previous year 2003–2004. The market prices of Cotton showed wide fluctuation in different years and hence it had impacted the economy of cotton cultivation.

Overall review of the chapter reveals that Gujarat is a leading cotton producing State of India and in recent years, particularly after introduction of Bt cotton, productivity of cotton in the State witnessed spectacular rise. However, Bt cotton in Gujarat State was found ideal only for assured irrigation situations and non-suitable for rainfed areas. In majority areas of the State, Bt cotton proved cost efficient and generated higher profit and productivity. No striking difference noticed between market price of Bt cotton and non-Bt hybrid cotton. The coverage under Bt cotton in the State is moving up at a fast pace. In total Bt cotton area of the State, share of non-approved varieties of Bt cotton was higher as compared to share of approved varieties of Bt cotton. In Gujarat, farmers are not following recommended practice of planting refuge plants on boundary of Bt cotton plots. Bt cotton was found suitable to State soils and climate and hence majority of cotton growers believed that Bt cotton cultivation would be helpful in raising productivity as well as the net income.

# Chapter 15
# Sampling and Methodology

The study is based on both primary as well as secondary data. For primary data collection, the selection of sample districts, tehsils, villages and sample farmers was made in the following manner:

## 15.1 Sample Design

### *15.1.1 Selection of Districts*

For the selection of sample districts, major cotton growing districts of the State were classified into two categories considering criteria given below:

- **Category I**: Districts having sufficiently large area under Bt cotton and also characterized with higher irrigation facilities.
- **Category II**: Districts having sufficiently large cultivation of Bt cotton but having relatively lower irrigation facilities.

From the above two categories, **Vadodara** district from category I and **Rajkot** district from category II were selected purposively. While selecting two districts care was exercised to select districts from two different agro climatic zones.

### *15.1.2 Selection of Sample Tehsils*

From each of selected district, one tehsil having cotton as a major crop and having significantly large area under Bt cotton was selected purposively. Accordingly, **Karajan** tehsil from **Vadodara** district and **Gondal** tehsil from **Rajkot** district were selected (see Table 15.1).


V.P. Gandhi, D. Jain, *Introduction of Biotechnology in India's Agriculture*, DOI 10.1007/978-981-10-1091-0_15

**Table 15.1** List of selected sample districts, tehsils, villages and category-wise selected households

| Selected tehsils (district) | Selected villages | Nos. of sample households | | | | | | | | | | | | | | | |
|---|---|---|---|---|---|---|---|---|---|---|---|---|---|---|---|---|---|
| | | Genuine Bt cotton (G) | | | | Non-confirm Bt cotton (NC) | | | | (G + NC) total Bt cotton | | | | Non-Bt hybrid (HB) cotton | | | |
| | | SF | MF | LF | Total | SF | MF | LF | Total | SF | MF | LF | Total | SF | MF | LF | Total |
| Gondal (Rajkot) | Charkhadi | 0 | 3 | 1 | 4 | 7 | 2 | 2 | 11 | 7 | 5 | 3 | 15 | 7 | 5 | 3 | 15 |
| | Devla | 4 | 3 | 1 | 8 | 3 | 2 | 2 | 7 | 7 | 5 | 3 | 15 | 7 | 5 | 3 | 15 |
| | Patidar | 6 | 1 | 3 | 10 | 1 | 4 | 0 | 5 | 7 | 5 | 3 | 15 | 7 | 5 | 3 | 15 |
| | Total | 10 | 7 | 5 | 22 | 11 | 8 | 4 | 23 | 21 | 15 | 9 | 45 | 21 | 15 | 9 | 45 |
| Karjan (Vadodara) | Dhavat | 4 | 4 | 1 | 9 | 3 | 1 | 2 | 6 | 7 | 5 | 3 | 15 | 7 | 5 | 3 | 15 |
| | Mangrol | 0 | 3 | 1 | 4 | 7 | 2 | 2 | 11 | 7 | 5 | 3 | 15 | 7 | 5 | 3 | 15 |
| | Simli | 0 | 2 | 0 | 2 | 7 | 3 | 3 | 13 | 7 | 5 | 3 | 15 | 7 | 5 | 3 | 15 |
| | Total | 4 | 9 | 2 | 15 | 17 | 6 | 7 | 30 | 21 | 15 | 9 | 45 | 21 | 15 | 9 | 45 |
| All | – | 14 | 16 | 7 | 37 | 28 | 14 | 11 | 53 | 42 | 30 | 18 | 90 | 42 | 30 | 18 | 90 |

*SF* small farmers (Below 2.00 ha), *MF* medium farmers (2.00–4.00 ha), *LF* large farmers (Above 4.00 ha)

### 15.1.3 Selection of Sample Villages

Three sample villages from each of selected tehsil were selected purposively considering criteria shown below:

- Selected village must have adequate number of Bt cotton growers as well as non-Bt HB cotton growers.[1]
- One village near to town place as well as cotton market yard.
- Second village at least 15 km away from town place and has road linkage
- Third village in which agriculture is relatively progressive.

The list of selected villages and selected tehsils is given in Table 15.1.

### 15.1.4 Selection of Sample Farmers

It was decided to select 15 Bt cotton growers and 15 non-Bt HB cotton growers sample households from each selected village in the following manner:

Firstly, all the cotton cultivating households of the selected village were classified into two categories as Bt cotton growers and non-Bt HB cotton growers. The farmers growing non-confirmed Bt cotton (Navbharat-151 or local Bt cotton) were considered on par with Bt cotton growers. The farmers growing both, Bt cotton as well as non-Bt cotton, were selected as Bt cotton grower when his area under Bt

[1] Different types of cotton are grown by farmers. Non Bt HB Cotton stands for Hybrid Cotton while Cotton-G stands for Genuine or registered Bt Cotton. Cotton- NC stands for Non-Confirmed or Unregistered BT Cotton.

cotton was higher than area under non-Bt cotton and vice-versa. Again on the basis of their landholding, Bt cotton growers and non-Bt HB cotton growers were classified into three categories, namely, small farmers (below 2.00 ha), medium farmers (2.00–4.00 ha) and large farmers (above 4.00 ha). From each sample village, 15 Bt cotton growers comprising seven small farmers, five medium farmers and three large farmers were selected at random. Following a similar selection procedure, 15 non-Bt HB cotton growers were also selected from each sample villages. Thus, overall, from each selected villages, total 30 sample farmers (15 Bt cotton growers and 15 non-Bt cotton growers) were selected. For the study, altogether, a sample of 180 (30×3=90 from each district) sample households was selected for collecting primary data for the study (see Table 15.1). The circumstances listed below compelled us to deviate slightly from the prescribed selection procedure:

- Bt cotton is a hybrid variety and hence in Gujarat including selected study areas it is mostly grown as irrigated crop. Therefore, it was not possible for us to find unirrigated Bt cotton growers in selected villages. Hence, in view of non-availability of unirrigated Bt cotton growers, we selected all sample farmers having irrigated Bt cotton (HB). In order to make meaningful comparison of data, as a counterpart, only irrigated non-Bt cotton (hybrid) growers were selected. In short, all the selected sample farmers were of irrigated hybrid cotton.
- When adequate number of non-Bt cotton/Bt cotton growers belonging to a particular size class in one selected village was not available, we selected the shortfall number of farmers of same size class from another selected village and kept the total sample size unchanged.

### *15.1.5 Method of Primary Data Collection*

The primary survey instruments was prepared and finalized by CMA, IIM, Ahmedabad after consultation with associated AERCs. The season-wise primary data were collected by recall method from the selected sample households by interviewing the decision makers of the households. Quantitative/qualitative information was collected in the schedule on various study related aspects such as educational background, landholding, irrigation, season-wise crop pattern, variety-wise area under Bt cotton and non-Bt cotton in last 5 years, seed rate, sources of cottonseeds, variety-wise cost of cultivation of Bt cotton and non-Bt cotton, intensity of pest attack in cotton, insecticides/pesticides used, qualitative impact of Bt cotton as compared to non-Bt cotton etc. Also opinions of farmers on various aspects such as germplasm, quality of seeds, other inputs, support system, environmental effect etc. were collected.

In addition to the field survey, important information was also obtained through personal discussion and contacts with concerned officers of State government, CCI and progressive farmers growing Bt cotton.

### 15.1.6 Secondary Data Collection

The secondary data required for the study were collected from the State government offices including Directorate of Agriculture, Gujarat, regional office of CCI, websites of CCI and cotton Advisory Board and Central/State government publications. The time series data on area, yield, production of cotton (variety-wise) for selected districts and State, annual average prices of Cotton lint for different years, staple length of important cotton varieties etc. were collected from above mentioned sources.

### 15.1.7 Analytical Framework

The main objective of the study is to examine the advantages and disadvantages of Bt cotton as compared to non-Bt cotton. Therefore, average cost of cultivation, yield per unit, cost of insecticides/pesticides, input use pattern, net return etc. were worked out size group-wise for genuine Bt cotton, NC Bt cotton and non-Bt cotton separately and by comparing these results conclusions are drawn about difference between Bt cotton and non-Bt cotton. Also using this analysis, an attempt is made to identify reasons for difference in cost of cultivation and net return. The general views collected from farmers were analysed to study the impacts perceived by the farmers such as pest incidence, quality of seeds, impact on other crops and environment, availability of seeds, suitability of Bt cotton, usefulness of technology, continuation of sowing of Bt cotton in years to come etc. There are a few cases where selected genuine Bt cotton growers had grown non-confirmed Bt or non-Bt cotton and selected non-Bt cotton growers had grown Bt cotton. In such situation for the analysis of genuine Bt cotton growers, only genuine Bt cotton plots are considered and other cotton plots are ignored. Similarly for analysis of non-Bt cotton growers, only non-Bt cotton plots are considered and other cotton plots ignored.

### 15.1.8 Reference Year

It was decided to select agricultural year 2004–2005 as reference year. As final harvesting of cotton crop ended around February, it was decided to conduct field survey during March to June 2005.

# Chapter 16
# Nature and Performance of Bt Cotton vs. Non-Bt Cotton in Gujarat

## 16.1 Introduction

In this chapter, the socio-agro economic profile of selected 180 sample households is given based on the analysis of farm level survey data collected for cotton season 2004–2005. Two districts namely, Rajkot and Vadodara were selected and from each selected district, 90 sample households comprising 45 Bt cotton growers and 45 non-Bt HB cotton growers covering small, medium and large farmers were selected.

## 16.2 Cropping Pattern

Since our main objective is to study the effect of Bt cotton on crop income, profit level, yield level and climatic and soil suitability, it is pertinent to examine cropping pattern adopted by sample households and particularly proportion of area devoted to different varieties of cotton crop. It is also worthwhile to examine and identify difference in the cropping pattern of Bt and non-Bt cotton growers. The cropping pattern of the sample households is presented in Table 16.1. The Table exhibits the per farm area under the important crops as a percentage of GCA.

Overall (both districts together) as well as in both districts, the GCA was higher for Bt cotton households as compared to non-Bt cotton households. Further, examination of overall cropping pattern reveals that in both the districts, cotton was the most important crop and sample farmers had allocated major share of total cropped area to it. The Bt cotton growers have relatively higher share to cotton crop (50.00 % of GCA) as compared to the non-Bt cotton growers (44.67 % of GCA). Besides cotton, groundnut, tur, wheat and sugarcane were other important crops of the selected districts. Cotton and these other crops together accounted for more than 80 % of GCA for both group (Bt and non-Bt) of households.


V.P. Gandhi, D. Jain, *Introduction of Biotechnology in India's Agriculture*, DOI 10.1007/978-981-10-1091-0_16

**Table 16.1** Cropping pattern of sample households (Figures are per cent to GCA)

| Sr. no | Crop | Rajkot (R) | | Vadodara (V) | | Overall (R+V) | |
|---|---|---|---|---|---|---|---|
| | | Bt cotton hhs | Non-Bt cotton hhs | Bt cotton hhs | Non-Bt cotton hhs | Bt cotton hhs | Non-Bt cotton hhs |
| A | Kharif season | | | | | | |
| | Bt cotton (G) | 21.55 | – | 20.27 | 1.24 | 20.92 | 0.57 |
| | Bt cotton (NC) | 17.43 | – | 36.72 | 2.14 | 26.46 | 1.89 |
| | Non-Bt cotton | 4.36 | 35.09 | 0.71 | 50.62 | 2.62 | 42.21 |
| | Total cotton | 43.34 | 35.09 | 57.70 | 54.00 | 50.00 | 44.67 |
| | Groundnut | 28.21 | 36.86 | – | – | 15.20 | 19.96 |
| | Tur | – | – | 11.02 | 12.26 | 5.15 | 5.62 |
| | Vegetables | 0.80 | 0.22 | 2.47 | 3.15 | 1.22 | 1.57 |
| B | Rabi season | | | | | | |
| | Wheat | 8.72 | 5.13 | 2.10 | 4.89 | 5.66 | 5.02 |
| | Garlic | 2.98 | 8.50 | | | 1.54 | 4.60 |
| | Fodder | 5.05 | 1.10 | 0.52 | 0.23 | 2.95 | 0.70 |
| | Vegetables | 1.15 | 2.89 | | | 0.74 | 1.57 |
| C | Summer season | | | | | | |
| | Sugarcane | | | 11.29 | 12.21 | 5.39 | 5.60 |
| | Vegetables | 0.08 | 0.04 | 0.24 | | 0.15 | 0.02 |
| | Fodder | 2.52 | 1.20 | 4.99 | 2.02 | 3.68 | 1.58 |
| D | Gross cropped area (GCA) % | 100.00 | 100.00 | 100.00 | 100.00 | 100.00 | 100.00 |
| | Hect.[a] | 4.36 | 4.11 | 3.81 | 3.48 | 4.08 | 3.79 |

[a]Average per farm

In Rajkot district, cotton, groundnut and garlic crops dominated the cropping pattern and together covered at least 75 % of GCA in both categories (Bt and non-Bt) of households. Wheat, garlic and fodder crops were important rabi crops and for Bt cotton households they claimed about 8.72 %, 2.98 % and 5.05 % area of GCA respectively. Garlic accounted for 8.50 % area of GCA for non-Bt cotton households. As compared to non-Bt cotton households, 8.25 % more area was devoted to cotton crop by the Bt cotton households. The data further reveals that farming economy in Rajkot district is highly relying on output of the crops viz., cotton, groundnut and garlic.

On account of non-availability of adequate water for irrigation in rabi season, majority of cotton farmers of Rajkot district had harvested cotton only up to five to six pickings and thereafter they utilized it either for growing late rabi/summer crops or keeping it idle for next year sowing.

In Vadodara district, cotton, tur and sugarcane were most important crops and data clearly reveal that economy of farmers of this region is heavily relying on the prospect of these three crops. Bt cotton households devoted as much as 57.70 % of GCA to only cotton crop. The corresponding figure for non-Bt households was

54 %. Tur and sugarcane together covered around 22 % for Bt cotton households, whereas it was around 24 % for non-Bt households.

In Vadodara region, on account of adequate availability of water for irrigation throughout the year, the general practice of cotton farmers is to give irrigation to cotton after each picking operation and to take at least seven to nine pickings of cotton and hence, the picking operation of cotton last up to end of February. This practice coupled with good area under long duration sugarcane crop reduces the availability of land for growing rabi/summer crops and thereby causing low cropping intensity.

## 16.3 Area Under Cotton

Size of operational holding influences the cost of cultivation of crops, adoption of modern and new agriculture technology and capital investment in agriculture. In this context, size of operational holding is examined in Table 16.2. The Table presents the category-wise average operational area of the sample households. The overall average operated area per household worked out to 3.45 ha for Bt cotton growers and 3.13 ha for non-Bt cotton growers. In Rajkot district, it was 3.26 ha for Bt growers and 3.05 ha for non-Bt cotton growers. In Vadodara district, it was 3.65 ha for Bt growers and 3.20 ha for non-Bt cotton growers. For different size groups, the overall (Rajkot and Vadodara together) average operational area per household for Bt cotton growers worked out to 1.58 ha for SF, 3.15 ha for MF (medium) and 8.33 ha for LF. For non-Bt cotton growers, it worked out to 1.28 ha for SF, 3.09 ha for MF and 7.49 ha for LF (see Table 16.2).

As majority of the sample households have small to moderate size of operational holding, the adoption of modern agriculture technology and capital investment in agriculture is bound to be at low to medium level.

## 16.4 Cotton Varieties Used by the Sample Households

The overall distribution of sample farmers according to type of cotton grown during agricultural year 2001–2002 to 2004–2005 has been presented in Table 16.3. The following important points are emerging from the Table.

The commercial cultivation of Bt cotton was allowed officially in the year 2002–2003. In this year, only 30 sample farmers had grown Bt cotton. Surprisingly, out of these 30 Bt cotton growers, only one sample farmer grew approved variety of Bt cotton, whereas rest 29 farmers opted for non-approved varieties of Bt cotton. In the year 2001–2002 (prior year of approval of Bt cotton) not a single farmer had grown Bt cotton.

In year 2001–2002, out of 180 sample farmers only 87 farmers (48.33 %) had grown conventional non-Bt cotton varieties.

**Table 16.2** Per household average operated area

| District | Farm size | Bt cotton (G) | | | Bt cotton (NC) | | | Total Bt cotton(G+NC) | | | Non-Bt cotton | | |
|---|---|---|---|---|---|---|---|---|---|---|---|---|---|
| | | Owned land | L.I. land | Total op. area | Owned land | L.I. land | Total op. area | Owned land | L.I. land | Total op. area | Owned land | L.I. land | Total op. area |
| Rajkot | Small (SF) | 1.66 | 0 | 1.66 | 1.67 | 0 | 1.67 | 1.67 | 0 | 1.67 | 1.38 | 0 | 1.38 |
| | Medium (MF) | 3.25 | 0 | 3.25 | 3.06 | 0 | 3.06 | 3.15 | 0 | 3.15 | 2.89 | 0 | 2.89 |
| | Large (LF) | 8.34 | 0 | 8.34 | 3.65 | 2.02 | 5.67 | 6.25 | 0.9 | 7.15 | 7.19 | 0.04 | 7.23 |
| | All | 3.69 | 0 | 3.69 | 2.5 | 0.35 | 2.85 | 3.08 | 0.18 | 3.26 | 3.04 | 0.01 | 3.05 |
| Vadodara | Small (SF) | 1.5 | 0 | 1.5 | 1.39 | 0.1 | 1.49 | 1.42 | 0.08 | 1.5 | 1.18 | 0 | 1.18 |
| | Medium (MF) | 3.24 | 0 | 3.24 | 3.02 | 0 | 3.02 | 3.15 | 0 | 3.15 | 3.31 | 0 | 3.31 |
| | Large (LF) | 10.8 | 6.35 | 17.15 | 6.93 | 0.4 | 7.33 | 7.79 | 1.74 | 9.53 | 6.69 | 1.07 | 7.76 |
| | All | 3.79 | 0.84 | 4.63 | 3.01 | 0.15 | 3.16 | 3.27 | 0.38 | 3.65 | 2.99 | 0.21 | 3.2 |
| Overall | Small (SF) | 1.62 | 0 | 1.62 | 1.5 | 0.06 | 1.56 | 1.54 | 0.04 | 1.58 | 1.28 | 0 | 1.28 |
| | Medium (MF) | 3.25 | 0 | 3.25 | 3.04 | 0 | 3.04 | 3.15 | 0 | 3.15 | 3.09 | 0 | 3.09 |
| | Large (LF) | 9.04 | 1.81 | 10.85 | 5.73 | 0.99 | 6.72 | 7.02 | 1.31 | 8.33 | 6.94 | 0.55 | 7.49 |
| | All | 3.72 | 0.34 | 4.06 | 2.79 | 0.24 | 3.03 | 3.17 | 0.28 | 3.45 | 3.02 | 0.11 | 3.13 |

L.I. Land = Leased-in land, Op. = Operated, SF = Below 2.00 ha, MF = between 2.00 and 4.00 ha, LF = Above 4.00 ha

**Table 16.3** Overall distribution of sample farmers according to type of cotton grown during 2001–2002 to 2004–2005

| Year | Nos. of selected farmers | Nos. of sample farmers growing cotton | | | | | | Total cotton growers (Bt + non-Bt) | Nos. of farmers growing cotton with staple length | | |
|---|---|---|---|---|---|---|---|---|---|---|---|
| | | Bt cotton (Bt) | | | Non-Bt cotton (NBt) | | | | | | |
| | | Genuine | Non-confirmed | Total Bt growers | Hybrid | Non-hybrid | Total non-Bt growers | | M | SM | L |
| 1 | 2 | 3 | 4 | 5 | 6 | 7 | 8 | 9 | 10 | 11 | 12 |
| 2001–2002 | 180 | 0 | 0 | 0 | 73 | 14 | 87 | 87 | 4 | 14 | 69 |
| | 100 | 0 | 0 | 0 | 40.56 | 7.77 | 48.33 | 48.33 | 4.6 | 16.09 | 79.31 |
| 2002–2003 | 180 | 1 | 29 | 30 | 96 | 28 | 124 | 154 | 15 | 53 | 86 |
| | 100 | 0.56 | 16.11 | 16.67 | 53.33 | 15.55 | 68.88 | 85.55 | 9.74 | 34.42 | 55.84 |
| 2003–2004 | 180 | 24 | 53 | 77 | 73 | 30 | 103 | 180 | 28 | 35 | 117 |
| | 100 | 13.33 | 29.44 | 42.77 | 40.56 | 16.67 | 57.23 | 100 | 15.56 | 19.44 | 65 |
| 2004–2005 | 180 | 137 | 53 | 90 | 56 | 34 | 90 | 180 | 21 | 49 | 110 |
| | 100 | 29.44 | 29.44 | 50 | 31.11 | 18.89 | 50 | 100 | 11.67 | 27.22 | 61.11 |

(i) In column 1–9, figures in brackets denote percentage to total selected farmers for the respective variety of cotton
(ii) In column 10–12, figures in brackets denote percentage to total cotton growers in respective year i.e., column 9
*M* medium staple (20.50–25.50 mm), *SM* super medium (25.51–27.50 mm), *L* long (28.00–33.00 mm)

In each year during 2002–2003 to 2004–2005, number of farmers growing non-approved varieties of Bt cotton outclassed the number of farmers growing approved varieties of Bt cotton. This suggests popularity of non-approved Bt cotton among farmers and easy availability of non-approved Bt seeds in the market.

In each year during 2001–2002 to 2004–2005, the number of sample farmers growing cotton has shown significant upward trend which suggest that introduction of Bt cotton caused shift in crop pattern in favour of cotton and brought farmers back again to cotton cultivation. Majority (around 88 %) sample farmers had grown cotton varieties which had staple length either super medium (25.50–27.50 mm) or long (>28 mm).

## 16.5 Average Cost of Seed Used

Bt cotton cultivation is cost-intensive and Bt seed is very costly as compared to seed of non-Bt conventional hybrid cotton varieties. For achieving optimum crop productivity, timely and adequate use of all related inputs as per recommendation is most essential. Keeping in view high cost of Bt seeds, an attempt is made here to examine whether or not Bt cotton growers followed the recommended level of seed rate.

From Table 16.4, it is evident that actual seed rate applied by Bt cotton growers (G), Bt cotton growers (NC) and non-Bt hybrid cotton growers was slightly higher

**Table 16.4** Average seed price and average seed rate for Bt and non-Bt cotton

| Cotton type | Particulars | Year 2001–2002 | 2002–2003 | 2003–2004 | 2004–2005 |
|---|---|---|---|---|---|
| Bt cotton (G) | Average seed rate (kg/ha) | – | – | 1.077 | 1.210 |
| | Average price (Rs/kg) | – | – | 2944 | 3371 |
| Bt cotton (NC) | Average seed rate (kg/ha) | – | 1.557 | 1.613 | 1.660 |
| | Average price (Rs/kg) | – | 827 | 1195 | 1396 |
| Overall Bt cotton (G+NC) | Average seed rate (kg/ha) | – | 1.557 | 1.432 | 1.473 |
| | Average price (Rs/kg) | – | 827 | 1698 | 2092 |
| Non-Bt cotton (HB) | Average seed rate (kg/ha) | 1.781 | 2.232 | 1.722 | 1.968 |
| | Average price (Rs/kg) | 526 | 564 | 642 | 673 |

than the recommended seed rate of 1.125 kg/ha (one bag of 450 g/acre), 1.5 kg/ha and 1.75 kg/ha respectively. Despite high cost of seeds, marginally higher seed rate applied by sample farmers was mainly attributed to following reasons:

- Few resowing cases;
- Bt cotton seed is available mostly in a bag of 450 g. Hence, farmers who require seed quantity less than 450 g. or multiple of 450 g. have no alternative but to purchase compulsory higher quantity of seed and use it.

The average seed price per kg. paid by farmers of Bt cotton (G), Bt cotton (NC) and non-Bt hybrid cotton was Rs.3371, Rs.1396 and Rs.673 respectively. The slightly higher application of seed rate in Bt cotton by sample farmers clearly reveals their willingness to invest on new technology even though it is cost intensive, and also to use all recommended inputs adequately for enhancing productivity and generating higher income.

## 16.6 Application of Pesticides

The present non-Bt hybrid cotton varietie s are highly suffering from bollworm, sucking pests, leaf curl virus and other pests/insects attack. Among these pests, bollworm is a major danger and devastating. The loss due to bollworm is estimated at around 40–50 % of cotton yield. To control the bollworm and prevent yield loss, farmers are spraying more doses of pesticides. Therefore, it is not only escalating the pesticides cost and in turn cost of cultivation, but also causing yield and quality loss too. Subsequently, this affects the rate of return from cotton cultivation. Moreover, constant and continuous exposure to pesticides could result in severe health impacts for farmers and farm workers.

A new cotton variety using BT technology was introduced to provide solution of making cotton plants free from bollworm infestation and particularly American bollworm. The introducer of Bt cotton claimed that it is resistant to pests, particularly bollworm and hence prevents the attack of bollworm. With a view to ascertain this claim, information on type of pests/insects attack and their intensity level was collected from sample farmers and presented in Table 16.5.

Out of total 37 genuine Bt cotton growers, only two farmers reported mild American/pink bollworm attack, whereas 61 % of non-Bt hybrid cotton growers reported moderate to light American/pink bollworm attack. From the total non-confirmed Bt cotton growers, only 20.75 % reported bollworm infestation. This result clearly supports the claim that BT technology in cotton is highly effective in providing strong resistance to American bollworm for virtually Bt cotton growers. Now, it remains to be seen whether or not this intensity of effectiveness will sustain in the years to come. Some farmers expressed the apprehension that due to non-growing of suggested refugia plants surrounding Bt cotton, the effectiveness of Bt cotton in preventing bollworm attack is bound to be weaker in the years to come as bollworm will gradually develop resistance power to Bt gene.

Both, Bt cotton and non-Bt cotton suffered attack of soil pests, sucking pests and foliage feeding pests. However, intensity of attack of these pests was found slightly lower in Bt cotton as compared to non-Bt cotton. This clearly suggests that Bt cotton is substantially effective in protecting cotton plants from only bollworm infestation. It is found not so effective in preventing sucking pests and other pests/insects infestation. However, for other than bollworm pests, Bt variety appears to have slightly less infestation as compared to non-Bt varieties. Hence, farmers must be told that Bt cotton does not control all pests. Consequently, there may be need of spraying insecticides in Bt cotton too.

Pesticides are one of the important and crucial inputs of cotton cultivation. In total cost of production, share of pesticides is significant. The primary aim of introducing Bt cotton was to make cotton plants free from bollworm infestation and thereby to effect reduction in pesticides consumption and cost. With a view to ascertain effectiveness of Bt technology in reducing use and cost of pesticides in cotton, information on average quantity of pesticides used per hectare and cost of pesticides per hectare is presented for Bt and non-Bt cotton in Table 16.6.

In field survey, farmers reported a number of pesticides having different quality brands, form and power. The pesticides used by farmers were either in liquid or powder form. The names of major pesticides used by farmers were BHC, cholorophyrifos, cholorofenac, carboxin, carbosulfan, endosulfan, sevin, fenitrothion, monocrotophos, quinalphos, thiodan, sulphur etc. Owing to difference in quality, power and brand, very wide difference was noticed in prices of pesticides. Hence, comparison of number of sprays or quantity of pesticides used in Bt and non-Bt cotton will not explain fully the real impact of Bt technology on pesticides consumption.

Overall (both districts together) non-Bt hybrid cotton required as many as 5.44 number of pesticides sprays per hectare, while approved and non-approved Bt required an average of 4.83 and 5.00 pesticides sprays respectively. This shows that approved Bt cotton growers have used 11.21 % less number of pesticides sprays as compared to hybrid non-Bt cotton. Similar trend was also noticed in Rajkot and Vadodara districts too (see Table 16.6).

**Table 16.5** Percentage distributions of sample farmers according to type of pests/insects attack in cotton

| | | Percentage of sample farmers reporting pest attack | | | | | | | | | | | | | | | |
|---|---|---|---|---|---|---|---|---|---|---|---|---|---|---|---|---|---|
| | | Bt cotton (G) | | | | Bt cotton (NC) | | | | Total Bt cotton (G+NC) | | | | Non-Bt cotton (HB) | | | |
| | | H | M | L | Total | H | M | L | Total | H | M | L | Total | H | M | L | Total |
| Boll worm | American | – | – | – | – | 1.89 | 1.89 | 15.09 | 18.87 | 1.11 | 1.11 | 8.89 | 11.11 | 7.78 | 28.89 | 18.89 | 55.56 |
| | Pink | – | – | – | – | – | – | 1.89 | 1.89 | – | – | 1.11 | 1.11 | – | 3.33 | 2.22 | 5.55 |
| | Spotted | – | – | 5.41 | 5.41 | 5.66 | – | 5.66 | 11.32 | 3.33 | – | 5.55 | 8.88 | – | 3.33 | 4.44 | 7.77 |
| | Mites | – | – | – | – | – | 1.89 | – | 1.89 | – | 1.11 | – | 1.11 | – | 1.11 | – | 1.11 |
| Sucking pests | Trips | – | 2.70 | 5.41 | 8.11 | 3.77 | 1.89 | 1.89 | 7.55 | 2.22 | 2.22 | 3.33 | 7.77 | 3.33 | 5.56 | 2.22 | 11.11 |
| | Leaf hopper | – | 8.11 | 32.43 | 40.54 | 5.66 | 24.53 | 20.75 | 50.94 | 3.33 | 17.78 | 25.56 | 46.67 | 7.78 | 25.56 | 10.00 | 43.34 |
| | Whitefly | 13.51 | 2.70 | 16.22 | 32.43 | 11.32 | 18.87 | 3.77 | 33.96 | 12.22 | 12.22 | 8.89 | 33.33 | 15.56 | 15.56 | 10.00 | 41.12 |
| | Aphids and others | 16.22 | 2.70 | 43.24 | 62.16 | 24.53 | 16.98 | 45.28 | 86.79 | 21.11 | 11.11 | 44.44 | 76.66 | 15.56 | 32.22 | 23.33 | 71.11 |
| Foliage feeding pests | Leaf roller | – | 2.70 | 5.41 | 8.11 | 1.89 | – | 9.43 | 11.32 | 1.11 | 1.11 | 7.78 | 10.00 | – | 3.33 | 10.00 | 13.33 |
| | Cater pillar | – | 10.81 | 5.41 | 16.22 | 22.64 | 24.53 | 11.32 | 58.49 | 13.33 | 18.89 | 8.89 | 41.11 | 13.33 | 22.22 | 10.00 | 45.55 |
| Soil-pests | Root weevil and others | – | 2.70 | 0.00 | 2.70 | – | – | 1.89 | 1.89 | – | 1.11 | 1.11 | 2.22 | 1.11 | – | – | 1.11 |

*H* heavy pest attack, *M* moderate pest attack, *L* light pest attack, *G* genuine (approved) Bt, *NC* non-confirmed Bt

**Table 16.6** Average quantity of pesticides used and number of spray in Bt and non-Bt cotton

| District | Particulars | Bt (G) | Bt (NC) | Bt (G+NC) | Non-Bt |
|---|---|---|---|---|---|
| Rajkot (R) | Average number of spray | 4.22 (−16.27) | 4.19 (−16.87) | 4.20 (−16.67) | 5.04 (0.00) |
| | Qty. per sprays (ml/ha) | 364 (−52.60) | 725 (−5.60) | 549 (−28.52) | 768 (0.00) |
| | Total qty. of pesticides used (L/ha) | 1.53 (−60.47) | 3.04 (−21.45) | 2.30 (−40.57) | 3.87 (0.00) |
| | Total cost of pesticides (Rs/ha) | 2337 (−28.75) | 3699 (12.77) | 2965 (−9.60) | 3280 (0.00) |
| Vadodara (V) | Average number of spray | 5.72 (−3.05) | 5.62 (−4.75) | 5.65 (−4.24) | 5.90 (0.00) |
| | Qty. per sprays (ml / ha) | 527 (−21.58) | 469 (−30.21) | 489 (−27.23) | 672 (0.00) |
| | Total qty. of pesticides used (L / ha) | 3.02 (−23.93) | 2.63 (−33.75) | 2.76 (−30.48) | 3.97 (0.00) |
| | Total cost of pesticides (Rs/ha) | 3152 (2.30) | 2202 (−28.53) | 2530 (−17.88) | 3081 (0.00) |
| Overall (R + V) | Average number of spray | 4.83 (−11.21) | 5.00 (−8.09) | 4.92 (−9.56) | 5.44 (0.00) |
| | Qty. per sprays (ml/ha) | 430 (−40.11) | 580 (−19.22) | 516 (−28.13) | 718 (0.00) |
| | Total qty. of pesticides used (L/ha) | 2.13 (−45.38) | 2.79 (−28.46) | 2.53 (−35.13) | 3.90 (0.00) |
| | Total cost of pesticides (Rs/ha) | 2682 (−15.34) | 2771 (−12.53) | 2732 (−13.76) | 3168 (0.00) |

Figures in bracket are percentage increase/decrease over non-Bt cotton
*G* genuine (approved) Bt, *NC* non-confirmed Bt

In respect of quantity of pesticides used per spray and total quantity used per hectare, it was reported that Bt cotton growers (approved and non-approved together) utilized less quantity of pesticides as compared to non-Bt cotton growers in both the districts. Owing to lower pesticides usage, expenses incurred on pesticides by Bt cotton growers was also found lower as compared to that of non-Bt cotton growers in both the districts. In Rajkot, it was Rs.2337 per hectare for approved Bt cotton, whereas it was Rs.3280 for non-Bt cotton. The contrary to expectation, the unapproved Bt (NC) growers in Rajkot district sprayed slightly less quantity of pesticides as compared to non-Bt cotton, but total cost of pesticides sprayed for Bt (NC) was found somewhat higher. This has happened mainly due to wide difference in quality, brand and power (concentration) of pesticides used. As Bt cotton seed is more costly, they used high power costly pesticides, whereas they used less relatively low price, low power pesticides for non-Bt cotton. In Vadodara, it was Rs.2530 for total Bt (G+NC), Rs.2202 for non-approved Bt and Rs.3081 for non-Bt cotton. Overall, on an average, Bt cotton growers (G+NC) spent nearly 13.76 % less amount on pesticides as compared to non-Bt cotton farmers.

According to sample farmers, Bt cotton protects plants from only bollworm attack. It is equally susceptible to other pests/diseases, which are appearing at various stages of cotton crop. Therefore, Bt cotton also requires the application of good quantity of pesticides/insecticides. The data presented in Table 16.6 clearly suggest that quantum of reduction in pesticides consumption in Bt cotton was very low and not in line with claim made by Bt promoters. The major reasons for not achieving expected reduction in pesticides consumption in Bt cotton are:

- On account of higher investment on seed and other inputs farmers devoted more care to crop and hence with the notice of any sign of pests/diseases in cotton, as a precautionary measures, farmers sprayed pesticides/insecticides, even though it is not required. When pests' appearance is below threshold level, a pesticide spraying is not needed. Even though, due to fear, farmers sprayed pesticides on cotton plants. This faulty practice reduced the savings on pesticides.
- In reference year, due to more favourable climatic conditions, bollworm pressure was relatively low in all the cotton varieties including non-Bt cotton. Therefore, cost of pesticides sprayed in non-Bt cotton was relatively low for the reference year. This brought down the savings on pesticides for Bt cotton.
- As Bt cotton suffered attack from sucking pests, leaf curling virus, foliage feeding pests, soil pests and other diseases compelled farmers to spray pesticides/insecticides in Bt cotton. This raised the cost of pesticides for Bt cotton.
- In a few cases, substandard quality of pesticides pushed up the consumption and subsequently cost of pesticides.
- Few farmers were not able to distinguish between Bt and non-Bt cotton. And, therefore, continued to spray same level of pesticides/insecticides in Bt and non-Bt cotton.

Overall, one can conclude that Bt technology is pesticides savings but the quantum of savings is very low and far from level of expectation.

# Chapter 17
# Economics of Bt Cotton vs. Non-Bt Cotton in Gujarat

## 17.1 Introduction

In this chapter, an attempt has been made to examine operation-wise/item-wise cost of cultivation of Bt cotton and non-Bt hybrid cotton (only for irrigated) using field level data collected from sample farmers. Productivity, output-input ratio, cost efficiency and net and gross return per unit of land for Bt vs. non-Bt cotton has also been worked out. The intensity of pests attack and pesticides cost differentials between Bt and non-Bt cotton have also been examined. In order to know farmers' views and their stand on various aspects associated with the cultivation of Bt vis-à-vis non-Bt, data were collected from 90 Bt cotton growing sample farmers. Analysis based on these perception data has also been attempted here.

## 17.2 Average Bt and Non-Bt Cotton Yields

According to pro Bt cotton group, yield superiority over non-Bt cotton is one of the most important positive impacts of Bt cotton. For verification of this claim, yield differential between Bt and non-Bt cotton is examined here. Figure 17.1 and Table 17.1 provide comparative yield performance of Bt and non-Bt cotton across districts and farm sizes.

From Table 17.1 and Fig. 17.1, it is obvious that the overall yield performance of Bt cotton was far better than that for non-Bt cotton. For both districts together, an average yield of total Bt cotton (G+NC) was 32.20 qtls/ha, which was 28.44% higher than that of 25.07 qtls/ha for non-Bt cotton. The corresponding figures for approved Bt cotton (G) were 36.34 qtls/ha and for non-approved Bt cotton (NC) was 28.99 qtls/ha. This clearly suggests yield superiority of genuine/approved Bt cotton varieties over the non-confirmed Bt cotton and non-Bt cotton varieties. Almost similar pattern of yield trend was observed in both the selected districts.


V.P. Gandhi, D. Jain, *Introduction of Biotechnology in India's Agriculture*, DOI 10.1007/978-981-10-1091-0_17

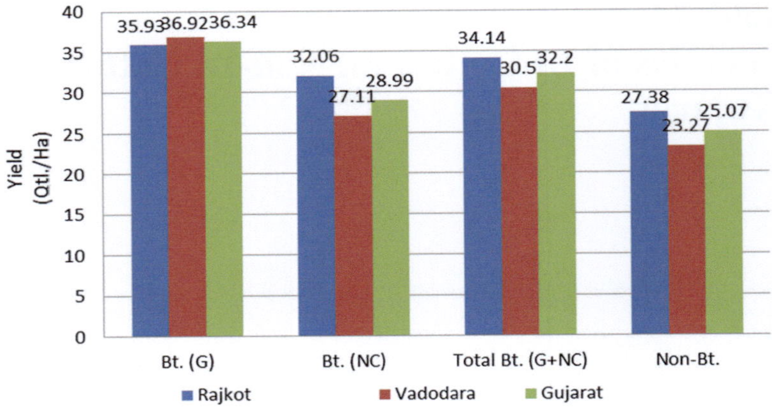


**Fig. 17.1** Yield of Bt and non-Bt cotton

**Table 17.1** Average yields, revenue and price realization for Bt and non-Bt cotton

| District | Particulars | Cotton | | | |
|---|---|---|---|---|---|
| | | Bt (G) | Bt (NC) | Total Bt (G+NC) | Non-Bt |
| Rajkot (R) | Average area per farm (ha) | 1.93 (134.03) | 1.58 (109.72) | 1.75 (121.53) | 1.44 (100.00) |
| | Yield of cotton (kg/ha) | 3593 (131.23) | 3206 (117.09) | 3414 (124.69) | 2738 (100.00) |
| | Value of cotton (Rs/ha) | 72,335 (133.79) | 63,493 (117.44) | 68,260 (126.25) | 54,066 (100.00) |
| | Average price realized per qtl (Rs) | 2013 (101.95) | 1980 (100.29) | 1999 (101.25) | 1975 (100.00) |
| Vadodara (V) | Average area per farm (ha) | 2.09 (112.97) | 1.97 (106.49) | 2.01 (108.65) | 1.85 (100.00) |
| | Yield of cotton (kg/ha) | 3692 (158.66) | 2711 (116.50) | 3050 (131.07) | 2327 (100.00) |
| | Value of cotton (Rs/ha) | 72,984 (163.48) | 53,166 (119.09) | 60,102 (134.63) | 44,643 (100.00) |
| | Average price realized per qtl (Rs) | 1977 (103.04) | 1961 (102.22) | 1971 (102.71) | 1918 (100.00) |
| Overall (R + V) | Average area per farm (ha) | 1.99 (120.61) | 1.8 (109.09) | 1.88 (113.94) | 1.65 (100.00) |
| | Yield of cotton (kg/ha) | 3634 (144.95) | 2899 (115.64) | 3220 (128.44) | 2507 (100.00) |
| | Value of cotton (Rs/ha) | 72,490 (148.32) | 57,056 (116.74) | 63,942 (130.83) | 48,873 (100.00) |
| | Average price realized per qtl (Rs) | 1995 (102.32) | 1968 (100.96) | 1986 (101.86) | 1949 (100.00) |

Figures in bracket denote percentage with respect to non-Bt
*G* genuine, *NC* non-confirmed Bt

Examination of yield across districts clearly reveals inter-district variations. In both the districts, the yield obtained for Bt cotton was substantially higher than that for non-Bt cotton. The overall average yield obtained by total Bt(G+NC) farmers in Rajkot district was 34.14 qtls/ha, which was nearly 12 % higher than 30.50 qtls/ha for Vadodara district. The non-Bt cotton farmers of Rajkot district obtained about 18 % higher yield than that for Vadodara district. This shows that sample households of Rajkot achieved notable increase in the yield of Bt and non-Bt cotton as compared to their counterpart in Vadodara district (see Table 17.1). Favourable rainfall and climatic condition helped cotton farmers of Rajkot district to achieve higher yield. The damage caused to cotton crop (Bt and non-Bt) due to excessive rainfall led to relatively lower cotton yield in Vadodara district.

Examination of yield data across different farm categories shows that yield level of Bt cotton for all the three categories of farmers were found to be higher than for corresponding categories of non-Bt cotton (see Table 17.2). The yield level of Bt cotton (G+NC) was found highest (34.34 qtls/ha) for medium landholding farmers, wherein it was observed lowest (29.24 qtls/ha) for small landholding farmers (see Table 17.2).

**Table 17.2** Farm size-wise average yields, revenue and price realization for Bt and non-Bt cotton

| Farm size | Particulars | Bt cotton (G) | Bt cotton (NC) | Total Bt cotton (G+NC) | Non-Bt cotton |
|---|---|---|---|---|---|
| Small | Area in ha (per farm) | 0.98 | 1.05 | 1.03 | 0.87 |
| | Yield of cotton (kg/ha) | 3270 | 2762 | 2924 | 2107 |
| | Value of cotton (Rs/ha) | 64,993 | 52,778 | 57,952 | 40,699 |
| | Average price realized (Rs/qtl) | 1988 | 1911 | 1982 | 1932 |
| Medium | Area in ha (per farm) | 2.17 | 1.93 | 2.05 | 1.59 |
| | Yield of cotton (kg/ha) | 3955 | 2764 | 3434 | 2561 |
| | Value of cotton (Rs/ha) | 76,545 | 54,748 | 68,138 | 49,949 |
| | Average price realized (Rs/qtl) | 1935 | 1981 | 1984 | 1950 |
| Large | Area in ha (per farm) | 3.61 | 3.55 | 3.57 | 3.56 |
| | Yield of cotton (kg/ha) | 3394 | 3096 | 3213 | 2695 |
| | Value of cotton (Rs/ha) | 68,380 | 60,625 | 63,969 | 53,345 |
| | Average price realized (Rs/qtl) | 2015 | 1958 | 1991 | 1979 |
| Overall | Area in ha (per farm) | 1.99 | 1.80 | 1.88 | 1.65 |
| | Yield of cotton (kg/ha) | 3634 | 2899 | 3220 | 2507 |
| | Value of cotton (Rs/ha) | 72,490 | 57,056 | 63,942 | 48,873 |
| | Average price realized (Rs/qtl) | 1995 | 1968 | 1986 | 1949 |

*G* genuine, *NC* non-confirmed Bt

**Table 17.3** Yield variations in Bt and non-Bt cotton

| Cotton type | Range of yield (qtl/ha) | | Mean | Standard deviation | CV (%) |
|---|---|---|---|---|---|
| | Minimum | Maximum | | | |
| Total Bt (G+NC) | 12.80 | 53.54 | 32.20 | 9.11 | 28.29 |
| Non-Bt | 6.18 | 41.18 | 25.07 | 6.31 | 25.17 |

*CV* coefficient of variation

The coefficient of variation (CV) for Bt cotton was found 28.29 %, slightly higher compared to 25.17 % for non-Bt cotton (see Table 17.3). Moreover, the yield range of Bt cotton was 12.80–53.54 qtls/ha, whereas for non-Bt cotton, it was 6.18–41.18 qtls/ha. The higher CV and higher range of variability clearly suggests slightly higher inter-farm yield variability for Bt cotton. This result does not support the claim that Bt cotton has less inter-farm variations and more stability in yield as compared to non-Bt cotton.

### 17.2.1 Picking-Wise Break-Up of Productivity and Maturity Period of Bt and Non-Bt Cotton

It is said that Bt cotton has relatively short maturity period and higher number of pickings. For verification of these claims, the data of production obtained in each picking has been collected and presented in Table 17.4.

The data in Table 17.4 show that the average gap between sowing and first harvesting of Cotton (maturity period) was 123 days for total Bt cotton (G+NC), which was shorter by about 10 days as compared to 133 days for non-Bt cotton varieties. The average gap between two successive pickings was found in the range of 18–20 days and was observed to be almost uniform for both Bt and non-Bt cotton. The number of pickings shows absolutely no variation between Bt and non-Bt convention hybrid varieties. However, Rajkot farmers completed harvesting of Cotton in six pickings, whereas Vadodara farmers continued it up to eight and more pickings. Owing to limited water resources, the number of cotton pickings in Rajkot was less. In Rajkot district, about 95 % cotton was harvested at the end of fourth picking for both Bt and non-Bt cotton. In Vadodara district at the end of fifth pickings, about 84 % of Bt production and about 90 % of non-Bt cotton production was obtained. Further, it can be concluded that number of pickings in Bt and non-Bt cotton varied from one region to another depending upon the availability of water and other agro climatic factors.

## 17.3 Cost of Production, Value of Output and Profit

The conflicting assertions made by pro and anti Bt cotton groups about the impact of Bt technology with special reference to various operational costs and gross cost of cultivation indicates that a clear picture is yet to emerge. In this context, it is

**Table 17.4** Picking-wise break-up of productivity of Bt and non-Bt cotton (kg/ha)

| Picking | Rajkot (R) | | | Vadodara (V) | | | Overall (R + V) | | |
|---|---|---|---|---|---|---|---|---|---|
| number | Bt (G) | Bt (NC) | Non-Bt | Bt (G) | Bt (NC) | Non-Bt | Bt (G) | Bt (NC) | Non-Bt |
| First | 621 (17.27) | 679 (21.18) | 528 (19.30) | 237 (6.42) | 221 (8.15) | 198 (8.51) | 463 (12.72) | 448 (15.45) | 468 (14.53) |
| Second | 1196 (33.29) | 1159 (36.15) | 995 (36.35) | 961 (26.02) | 446 (16.45) | 467 (20.07) | 1105 (30.34) | 783 (27.01) | 927 (28.79) |
| Third | 1024 (28.51) | 811 (25.30) | 771 (28.16) | 661 (17.90) | 692 (25.53) | 687 (29.52) | 877 (24.09) | 719 (24.81) | 973 (30.22) |
| Fourth | 547 (15.21) | 382 (11.92) | 385 (14.05) | 753 (20.40) | 601 (22.17) | 471 (20.24) | 638 (17.52) | 484 (16.69) | 504 (15.65) |
| Fifth | 178 (4.97) | 139 (4.34) | 58 (2.11) | 496 (13.45) | 352 (12.99) | 267 (11.47) | 308 (8.46) | 246 (8.49) | 206 (6.41) |
| Sixth | 27 (0.75) | 36 (1.12) | 1 (0.04) | 334 (9.04) | 226 (8.33) | 141 (6.05) | 145 (3.98) | 135 (4.66) | 89 (2.76) |
| Seventh | – | – | – | 156 (4.23) | 113 (4.17) | 62 (2.66) | 65 (1.79) | 61 (2.10) | 39 (1.21) |
| Eight and more | – | – | – | 97 (2.55) | 60 (2.21) | 34 (1.48) | 40 (1.10) | 23 (0.79) | 14 (0.43) |
| Total production (kg/ha) | 3593 (100.00) | 3206 (100.00) | 2738 (100.00) | 3692 (100.00) | 2711 (100.00) | 2327 (100.00) | 3641 (100.00) | 2899 (100.00) | 3220 (100.00) |
| Gap between sowing and first picking (days) | 122 | 122 | 131 | 123 | 125 | 134 | 123 | 124 | 133 |

Figures in brackets denote percentages to total production
*G* genuine Bt, *NC* non-confirmed Bt

pertinent to examine impact of Bt technology in cotton on each operational cost and gross cost of cultivation. Table 17.5 examines item-wise cost data of Bt cotton vs. non-Bt cotton.

The data presented in Table 17.5 clearly shows that the average gross cost of cultivation for total Bt cotton (G+NC) was higher than that for non-Bt hybrid cotton in both the districts. Overall (both districts together), the average cost of cultivation per hectare comes to Rs.31,815 for Bt(G), Rs.28,145 for Bt(NC), Rs.29,743 for total Bt (G+NC) and Rs.26,993 for non-Bt cotton. The overall average cost of cultivation for total Bt cotton is higher by Rs.2750, an increase of 10.19 % compared to that for non-Bt cotton (see Table 17.5). For Vadodara district, the cost of cultivation for Bt cotton was up by 13.20 %; whereas it was 6.17 % up for Rajkot district (see Table 17.5). The examination of cost of cultivation data according to landholding categories do not reveals any definitive trend. However, cost of cultivation was found lowest for small size farmers for both Bt as well as non-Bt cotton.

There were several reasons which pushed up the cost of cultivation of Bt cotton. Firstly, the seed cost per hectare of Bt cotton (G) and total Bt (G+NC) cotton were Rs.4079 and Rs.3079 respectively, whereas it was only Rs.1324 for non-Bt cotton. Thus, as compared to non-Bt growers, average expenses incurred per hectare on seed for Bt(G) and total Bt(G+NC) was about 208 % and 133 % higher (see Table 17.5) respectively. Secondly, owing to substantially higher productivity, cost of picking/harvesting operation was about 22.91 % higher for total Bt(G+NC) compared to that for non-Bt cotton. Thirdly, there was a reduction in pesticides cost for Bt cotton, but quantum of reduction was much lower than expectations and hence cost saving on pesticides was lower. This saving was not enough to compensate fully for the higher seed cost.

The absolute amount spent on items like FYM, fertilisers, human labour (excluding picking) and irrigation for cultivation of Bt cotton differs only marginally from non-Bt cotton. The data given in Table 17.5 clearly suggest that Bt technology is a cost intensive and not cost saving in cotton.

### *17.3.1 Percentage Share of Inputs in Gross Cost of Cultivation*

The share of each operational cost in the gross cost of cultivation is expected to be influenced by Bt technology in cotton. Therefore, to study the composition of share of inputs in total cost of cultivation, mainly for seed, harvesting, irrigation and pesticides, the related data have been presented in Table 17.6. As expected, a considerable variation was observed in share of these costs between Bt and non-Bt cotton.

The share of seed cost in total cost was 10.35 % for total Bt cotton (G+NC) which was substantially higher than share of 4.91 % for non-Bt cotton. This is so because of the abnormally high seed prices of Bt varieties as compared to that for non-Bt cotton varieties. Though, picking operation accounts for the largest share (21.56 %) in total cost of Bt cotton, it was only marginally higher than its share (19.33 %) in total cost of non-Bt cotton. The share of pesticides cost in total cost of

**Table 17.5** Item-wise average cost of cultivation per hectare (figures in Rs)

| | | Rajkot (R) | | | | Vadodara (V) | | | | Gujarat (R + V) | | | |
|---|---|---|---|---|---|---|---|---|---|---|---|---|---|
| No | Item | Bt cotton (G) | Bt cotton (NC) | Bt cotton (G+NC) | Non-Bt cotton | Bt cotton (G) | Bt cotton (NC) | Bt cotton (G+NC) | Non-Bt cotton | Bt cotton (G) | Bt cotton (NC) | Bt cotton (G+NC) | Non-Bt cotton |
| 1 | Seed | 4031 | 2995 | 3554 | 1333 | 4145 | 1885 | 2666 | 1317 | 4079 | 2307 | 3079 | 1324 |
| 2 | Human labour | 4913 | 5376 | 5126 | 5632 | 3722 | 3837 | 3797 | 3389 | 4407 | 4421 | 4415 | 4370 |
| 3 | Bullock labour | 1873 | 2681 | 2245 | 2287 | 2361 | 2822 | 2662 | 2714 | 2080 | 2768 | 2468 | 2527 |
| 4 | Tractor | 719 | 678 | 700 | 675 | 1444 | 1282 | 1338 | 983 | 1027 | 1053 | 1042 | 848 |
| 5 | Farm yard manure | 3315 | 1739 | 2589 | 2470 | 412 | 594 | 531 | 620 | 2082 | 1029 | 1488 | 1429 |
| 6 | Fertilizer | 2626 | 3248 | 2913 | 2907 | 4265 | 3152 | 3537 | 3591 | 3321 | 3189 | 3247 | 3292 |
| 7 | Pesticides | 2337 | 3699 | 2965 | 3280 | 3152 | 2202 | 2530 | 3081 | 2682 | 2771 | 2732 | 3168 |
| 8 | Irrigation | 3908 | 4544 | 4201 | 4699 | 3974 | 3228 | 3486 | 3228 | 3935 | 3727 | 3818 | 3872 |
| 9 | Picking | 6971 | 6494 | 6751 | 5840 | 7429 | 5425 | 6117 | 4733 | 7164 | 5831 | 6412 | 5217 |
| 10 | Transport and marketing | 579 | 696 | 633 | 631 | 678 | 454 | 531 | 390 | 615 | 546 | 576 | 496 |
| 11 | Others | 345 | 416 | 378 | 437 | 524 | 556 | 545 | 460 | 421 | 503 | 467 | 450 |
| 12 | Total cost of cultivation | 31,617 | 32,565 | 32,054 | 30,190 | 32,106 | 25,438 | 27,743 | 24,507 | 31,815 | 28,145 | 29,743 | 26,993 |

*G* genuine Bt, *N* non-confirmed Bt

**Table 17.6** Percentage share of cost items in total cost of cultivation (figures in percentage)

| | | Rajkot (R) | | | | Vadodara (V) | | | | Gujarat (R + V) | | | |
|---|---|---|---|---|---|---|---|---|---|---|---|---|---|
| No | Item | Bt cotton (G) | Bt cotton (NC) | Bt cotton (G+NC) | Non-Bt cotton | Bt cotton (G) | Bt cotton (NC) | Bt cotton (G+NC) | Non-Bt cotton | Bt cotton (G) | Bt cotton (NC) | Bt cotton (G+NC) | Non-Bt cotton |
| 1 | Seed | 12.75 | 9.20 | 11.09 | 4.42 | 12.91 | 7.41 | 9.61 | 5.37 | 12.82 | 8.20 | 10.35 | 4.91 |
| 2 | Human labour | 15.54 | 16.51 | 15.99 | 18.66 | 11.59 | 15.08 | 13.69 | 13.83 | 13.85 | 15.71 | 14.84 | 16.19 |
| 3 | Bullock labour | 5.92 | 8.23 | 7.00 | 7.57 | 7.35 | 11.09 | 9.60 | 11.08 | 6.54 | 9.84 | 8.30 | 9.36 |
| 4 | Tractor | 2.27 | 2.08 | 2.19 | 2.24 | 4.50 | 5.04 | 4.82 | 4.01 | 3.23 | 3.74 | 3.50 | 3.14 |
| 5 | Farm yard manure | 10.48 | 5.34 | 8.08 | 8.18 | 1.28 | 2.34 | 1.91 | 2.53 | 6.55 | 3.66 | 5.00 | 5.30 |
| 6 | Fertilizer | 8.31 | 9.97 | 9.09 | 9.63 | 13.28 | 12.39 | 12.75 | 14.65 | 10.44 | 11.33 | 10.92 | 12.19 |
| 7 | Pesticides | 7.39 | 11.36 | 9.25 | 10.86 | 9.82 | 8.66 | 9.12 | 12.57 | 8.43 | 9.84 | 9.19 | 11.74 |
| 8 | Irrigation | 12.36 | 13.95 | 13.11 | 15.57 | 12.38 | 12.69 | 12.56 | 13.17 | 12.37 | 13.24 | 12.84 | 14.34 |
| 9 | Picking | 22.05 | 19.94 | 21.06 | 19.34 | 23.14 | 21.33 | 22.05 | 19.31 | 22.52 | 20.72 | 21.56 | 19.33 |
| 10 | Transport and marketing | 1.83 | 2.14 | 1.97 | 2.09 | 2.11 | 1.79 | 1.92 | 1.59 | 1.93 | 1.94 | 1.94 | 1.84 |
| 11 | Others | 1.09 | 1.28 | 1.18 | 1.45 | 1.63 | 2.19 | 1.97 | 1.88 | 1.32 | 1.79 | 1.57 | 1.67 |
| 12 | Total cost of cultivation | 100.00 | 100.00 | 100.00 | 100.00 | 100.00 | 100.00 | 100.00 | 100.00 | 100.00 | 100.00 | 100.00 | 100.00 |

*G* genuine Bt, *N* non-confirmed Bt

cultivation was 9.19 % for Bt cotton which was relatively lower than the share of 11.74 % of non-Bt cotton. As compared to non-confirmed (NC) Bt cotton, the share of pesticides cost was lower for approved Bt cotton (G). Surprisingly, the share of irrigation was 12.84 % for Bt cotton, which was lower than its share of 14.34 % for non-Bt cotton. Like irrigation, the share of fertiliser was also observed somewhat lower for Bt cotton (10.92 %) as compared to that for non-Bt cotton (12.19 %). This suggests that Bt cotton neither requires higher doses of fertilisers nor more irrigation than counterpart non-Bt cotton. On the whole, it is emerging from the analysis that the seed and harvesting cost are the major contributors for the increase in the cost of cultivation of Bt cotton.

### *17.3.2 Average Cost of Production of Bt and Non-Bt Cotton*

It is seen from the data presented in Table 17.7 that average cost of production per qtl. of Bt cotton (G) as well as Bt cotton (NC) in both the selected districts was found on lower side than that for non-Bt cotton. Overall, average cost of production of cotton (Cotton) per qtl. estimated for Bt (G), Bt (NC), total Bt (G+NC) and non-Bt cotton were Rs.875, Rs.971, Rs.923 and Rs.1077 respectively. Overall, the cost of production for Bt cotton (G) and total Bt cotton were found lower by about 19 % and 14 % respectively. This shows that cost efficiency of Bt cotton is higher than counterpart non-Bt cotton. Across the districts, the cost of production per qtl. for both Bt and non-Bt cotton was found slightly higher in Rajkot as compared to Vadodara district. Across size groups, the cost of cultivation per hectare for both, Bt and non-Bt cotton was found highest for medium farmers, whereas it was lowest for small farmers (see Table 17.7).

Despite higher cost of cultivation, the cost of production per qtl. for Bt cotton was found lower. The higher yield of Bt cotton out weight the higher cost of cultivation and it ultimately caused notable decline in the cost of production.

### *17.3.3 Average Price Realization and Revenue Differentials*

As per market sources, the quality difference between Bt and non-Bt cotton is very marginal. Owing to less spraying of pesticides, Bt cotton was found relatively cleaner and better in colour and hence in respect of quality, Bt cotton has slight edge over non-Bt cotton. On account of good market acceptance of the product, Bt cotton growers have not faced any problems in selling the product. On the contrary, Bt cotton growers realized somewhat better price. The data provided in Table 17.8 show that overall average price realized per quintal for total Bt cotton (G+NC) was Rs.1986 as against Rs.1949 for conventional non-Bt hybrid cotton. The market price of cotton during the reference year fluctuated widely between Rs.2300 and 1750/qtl. It was higher in October-November and then due to global and other

**Table 17.7** Average cost of production of Bt and non-Bt cotton (in Rs/qtl)

| District | Bt cotton | | | Non-Bt cotton |
|---|---|---|---|---|
| | Genuine (G) | Non-confirm (NC) | Total (G+NC) | |
| Rajkot (R) | 880 (79.78) | 1016 (92.11) | 939 (85.13) | 1103 (100.00) |
| Vadodara (V) | 870 (82.62) | 938 (89.03) | 910 (86.42) | 1053 (100.00) |
| Overall (R + V) | 875 (81.24) | 971 (90.16) | 923 (85.70) | 1077 (100.00) |

impact, it depressed gradually and at the peak of the market arrival of the product, it ranged between Rs.1800 and 1950/qtl. Due to relatively shorter maturity period and early sowing by some Bt farmers (in May), the picking of Cotton for Bt cotton started around 30 days early. This helped Bt farmers in fetching relatively higher average price of the Cotton. From Table 17.8, it is evident that in respect of average price realization for both types of cotton, the Rajkot farmers were found in a marginally better position.

Across farm sizes, no significant difference was witnessed in average price realized for Bt cotton. However, in respect of non-Bt cotton, small farmers realized lowest price of Rs.1932/qtls. where in large farmers realized highest price of Rs. 1979/qtl (see Table 17.8).

The main economic benefits of Bt cotton stems from gains on revenue side as a result of increase in the yields. The average revenue per hectare for total Bt cotton was Rs.63,942, an increase of Rs.15,069 (30.83 %) over Rs. 48,873 for non-Bt hybrid cotton (see Table 17.8). The revenue gain for Bt cotton (G) over non-Bt cotton was 48.32 %, whereas it was only 16.74 % for Bt cotton (NC). This apparently shows superiority of approved Bt cotton over non-approved Bt cotton too. For Bt cotton, the percentage of revenue gains was found higher than its yield gains mainly because of better price realization of the product.

### *17.3.4 Average Net Profit per Hectare for Bt and Non-Bt Cotton*

Since the introduction of Bt cotton, the most debated issue has been its economic viability under Indian soil and climatic condition. Therefore, economic viability of Bt cotton is studied here in terms of net profit per hectare. The average net profit per hectare is calculated by subtracting average cost of cultivation per hectare from the gross value of production per hectare. Table 17.8 presents the data on average net profit and output-input value ratio for Bt (G), Bt (NC) and non-Bt cotton.

It is seen from the data presented in Table 17.8 that cultivation of both, Bt and non-Bt cotton, was found profitable for the sample households. Further, it clearly shows that average net profit realized from cultivation of approved Bt as well as non-approved Bt cotton was significantly higher than the non-Bt hybrid cotton in both the selected districts.

**Table 17.8** Net profit and output-input value ratios for Bt and non-Bt cotton (district-wise)

| Particulars | Rajkot (R) | | | | Vadodara (V) | | | | Overall (R + V) | | | |
|---|---|---|---|---|---|---|---|---|---|---|---|---|
| | Bt (G) | Bt (NC) | Total Bt (G+NC) | Non-Bt | Bt (G) | Bt (NC) | Total Bt (G+NC) | Non-Bt | Bt (G) | Bt (NC) | Total Bt (G+NC) | Non-Bt |
| Gross value of production (Rs/ha) | 72,335 (133.79) | 63,493 (117.44) | 68,260 (126.25) | 54,066 (100.00) | 72,984 (163.48) | 53,166 (119.09) | 60,102 (134.63) | 44,643 (100.00) | 72,490 (148.32) | 57,056 (116.74) | 63,942 (130.83) | 48,873 (100.00) |
| Gross cost of cultivation (Rs/ha) | 31,617 (104.72) | 32,565 (107.86) | 32,054 (106.17) | 30,190 (100.00) | 32,106 (131.01) | 25,438 (103.80) | 27,743 (113.21) | 24,507 (100.00) | 31,815 (117.86) | 28,145 (104.27) | 29,743 (110.19) | 26,993 (100.00) |
| Net profit (Rs/ha) | 40,718 (170.55) | 30,928 (129.54) | 36,206 (151.65) | 23,876 (100.00) | 40,878 (203.00) | 27,728 (137.70) | 32,359 (160.70) | 20,136 (100.00) | 40,675 (185.90) | 28,911 (132.14) | 34,199 (156.30) | 21,880 (100.00) |
| Net profit over non-Bt (Rs/ha) | 16,843 | 7053 | 12,331 | 0 | 20,741 | 7591 | 12,222 | 0 | 18,795 | 7031 | 12,319 | 0 |
| Output-input value ratios | 2.29 | 1.95 | 2.13 | 1.79 | 2.27 | 2.09 | 2.17 | 1.82 | 2.28 | 2.03 | 2.15 | 1.81 |

Figures in brackets denotes percentage with respect to non-Bt cotton

*G* genuine Bt, *NC* non-confirmed Bt

The overall average net profit realized per hectare from cultivation of total Bt cotton (G+NC) was Rs.34,199, whereas it was only Rs.21,880 for non-Bt hybrid cotton varieties. The corresponding figures stood at Rs.40,675 for approved Bt cotton (G) and Rs.28,911 for non-confirmed Bt cotton (NC). The net increase in profit margin for total Bt cotton (G+NC) over non-Bt cotton was Rs.12,319/ha, which shows relative gain of over 56 %. The examination of data of increment benefits of Bt cotton (G) and Bt cotton (NC) over non-Bt cotton across different farm sizes (Table 17.9) suggest inverse relationship. For Bt cotton (G), it is decreasing with the increase in farm size.

**Table 17.9** Farm size-wise net profit and output-input value ratios for Bt and non-Bt cotton

| Farm size | Particulars | Bt cotton | | | Non-Bt (HB) cotton |
|---|---|---|---|---|---|
| | | G | NC | Total | |
| | Gross value of production (Rs/ha) | 64,993 | 52,778 | 57,952 | 40,699 |
| | Gross cost of cultivation (Rs/ha) | 28,267 | 27,844 | 27,978 | 25,122 |
| Small | Net profit (Rs/ha) | 36,726 | 24,934 | 29,974 | 15,577 |
| | | (235.77) | (160.06) | (192.42) | (100.00) |
| | Net profit over non-Bt (Rs/ha) | 21,149 | 9357 | 14,397 | 0 |
| | Out-input value ratio | 2.30 | 1.90 | 2.07 | 1.62 |
| | Gross value of production (Rs/ha) | 76,545 | 54,748 | 68,138 | 49,949 |
| | Gross cost of cultivation (Rs/ha) | 34,793 | 26,745 | 31,269 | 28,167 |
| Medium | Net profit (Rs/ha) | 41,752 (191.68) | 28,003 (128.56) | 36,869 (169.26) | 21,782 (100.00) |
| | Net profit over non-Bt (Rs/ha) | 19,970 | 6221 | 15,087 | 0 |
| | Out-input value ratio | 2.20 | 2.05 | 2.18 | 1.77 |
| | Gross value of production (Rs/ha) | 68,380 | 60,625 | 63,969 | 53,345 |
| | Gross cost of cultivation (Rs/ha) | 29,686 | 29,347 | 29,475 | 27,188 |
| Large | Net profit (Rs/ha) | 38,694 (147.93) | 31,278 (119.58) | 34,494 (131.87) | 26,157 (100.00) |
| | Net profit over non-Bt (Rs/ha) | 12,537 | 5121 | 8337 | 0 |
| | Out-input value ratio | 2.30 | 2.07 | 2.17 | 1.96 |
| | Gross value of production (Rs/ha) | 72,490 | 57,056 | 63,942 | 48,873 |
| | Gross cost of cultivation (Rs/ha) | 31,815 | 28,145 | 29,743 | 26,993 |
| Overall | Net profit (Rs/ha) | 40,675 (185.90) | 28,911 (132.14) | 34,199 (156.30) | 21,880 (100.00) |
| | Net profit over non-Bt (Rs/ha) | 18,795 | 7031 | 12,319 | 0 |
| | Out-input value ratio | 2.28 | 2.03 | 2.15 | 1.81 |

Figures in bracket denote percentage w.r.t. non-Bt
*G* genuine Bt, *NC* non-approved Bt

As compared to non-Bt cotton, the net profit per hectare for approved Bt cotton (G) was up by Rs.18,795, which is an increase of nearly 86 %. In both the districts, cultivation of approved Bt cotton (G) was more profitable than non-approved Bt cotton (NC). This clearly shows that Bt farmers have higher level of income as compared to non-Bt farmers.

Three factors, sharp increase in yield, better price realization of the product and lower cost of production seem responsible for sharp upsurge in the net profit from Bt cotton cultivation.

## 17.4 Econometric Analysis of the Performance of Bt Cotton vs. Non-Bt Cotton

### *17.4.1 Output-Input Value Ratio for Bt and Non-Bt Cotton*

The output-input value ratio is a ratio of gross value of production to gross cost of cultivation. It indicates cost efficiency and rate of return on investment. From Table 17.9, it is evident that overall output-input value ratio for the Bt cotton (G+NC) is 2.15 which is higher than 1.81 for non-Bt cotton. In both the selected districts, similar trend is noticed. The examination of output-input value ratio across different farm sizes also showing similar trend (see Table 17.9). The higher output-input ratio for Bt cotton clearly suggests that it is economically more viable than non-Bt cotton.

### *17.4.2 Regression Analysis*

By using dummy variable, regression analysis has been attempted to estimate impact of Bt technology on yield, profit, pesticide cost etc. and to test its statistical significance. The analysis confirms positive and significant impact of Bt cotton on yield, value of output and net profit (see Table 17.10). The impact on yield was statistically highly significant at 35.69 %, whereas it was found at 64.65 % for net profit. The analysis further indicates that pesticide cost for Bt cotton is reduced by 16.50 % (see Table 17.10). The impact on yield, profit and value of output is found highly significant at 1 and 5 % level of significance.

**Table 17.10** Regression analysis – impact of Bt cotton (total observations N = 180)

| Dependent variable | Description | Independent variable | | Impact of Bt in % |
|---|---|---|---|---|
| | | Constant | Bt | |
| Pesticide cost | Coeff. | 3042 | –502 | –16.50 |
| | 't' value | 14.89 | –2.11 | |
| | Significance | 51 | 55 | |
| Seed cost | Coeff. | 1339 | 17.20 | 128.45 |
| | 't' value | 13.30 | 11.79 | |
| | Significance | 51 | 51 | |
| Total cost of cultivation | Coeff. | 25,982 | 3493 | 13.44 |
| | 't' value | 38.16 | 3.57 | |
| | Significance | 51 | 51 | |
| Yield | Coeff. | 2332 | 832 | 35.69 |
| | 't' value | 28.19 | 7.10 | |
| | Significance | 51 | 51 | |
| Value production | Coeff. | 44,867 | 17,189 | 38.31 |
| | 't' value | 25.27 | 7.19 | |
| | Significance | 51 | 51 | |
| Profit | Coeff. | 19,605 | 12,680 | 64.65 |
| | 't' value | 16.33 | 7.72 | |
| | Significance | 51 | 51 | |

$S_1$ Significant at 1 % level, $S_5$ Significant at 5 % level, *NS* Non-significant at 1 and 5 % level

# Chapter 18
# Farmers' Perceptions on Various Features of Bt Cotton in Gujarat

In order to study type and nature of problems faced by farmers and their suggestions for improving the performance of Bt cotton, opinions and suggestions on various problems associated with Bt cotton cultivation were collected from all Bt cotton growing sample households. The following findings are based on perception data collected from these sample farmers.

## 18.1 Some Basic Features of Bt Cotton: Views of Bt Cotton Sample Farmer Households

Almost all the sample farmers reported that number of bolls per plant was observed to be higher in Bt cotton. Bt cotton plants showed vigorous growth. The size of bolls of Bt and non-Bt cotton was more or less similar. All the Bt cotton growers reported early flowering in the range of 15–25 days.

The purpose of refuge or non-Bt cotton cultivation around Bt plots is that the bollworm resistance of Bt plant is delayed. In the study, nearly 94 % Bt growers had not planted mandate refuge crop around Bt cotton field. The small farmers avoided the planting of refugia crop, mainly because of their small landholding. The large farmers deliberately avoided planting refugee crop due to fear of reduction in profit level. Moreover, the unapproved Bt growers neither received any instruction regarding growing of refuge nor did they cultivate on their own.

Many Bt cotton growers were not fully aware about the importance of refuge crop. Therefore, there is a need for sincere combine efforts from seed companies and extension agencies for convincing farmers about advantages of planting of refugee crop. The approved Bt cotton growers find seed germination rate as most satisfactory. However, few non-confirmed Bt cotton growers were not satisfied in respect of seed germination rate.


V.P. Gandhi, D. Jain, *Introduction of Biotechnology in India's Agriculture*, DOI 10.1007/978-981-10-1091-0_18

The perception of an average farmer on the future of Bt cotton farming seemed to be positive. Nearly 99 % Bt users were satisfied with Bt performance and they firmly indicated that in the years to come, they will continue the cultivation of Bt cotton. Moreover, by effecting changes in their crop pattern, they intend to increase the acreage under Bt cotton. According to sample farmers, overall Bt cotton has more advantages as compared to non-Bt hybrid cotton.

## 18.2 Farmers' Perception on Advantages or Disadvantages of Bt Cotton vis-à-vis Non-Bt Cotton

The responses received on few questions asked on advantages or disadvantages of Bt cotton vis-à-vis non-Bt cotton are presented in Table 18.1.

The data exhibit that nearly 96 % farmers reported abnormally higher seed price of Bt cotton as a strong disadvantage. Nearly 31 % sample farmers considered higher harvesting cost as a modest disadvantage. Majority of sample households found Bt cotton moderately better in terms of controlling pest incidences and saving of pesticides. As per majority sample farmers, the use level of fertilisers and irrigation in Bt cotton does not differ significantly from that for non-Bt cotton. Almost all Bt cotton growers reported handsome gains in yield and net profit. The yield gains

**Table 18.1** Farmers' perception on advantages or disadvantages of Bt cotton vis-à-vis Non-Bt cotton

| Sr. No | Particulars | % of Bt cotton growers reporting | | | | |
|---|---|---|---|---|---|---|
| | | Advantages | | No difference | Disadvantages | |
| | | Strong | Somewhat | | Strong | Somewhat |
| 1 | Seed availability | 16.67 | 24.44 | 38.89 | 20.00 | – |
| 2 | Seed price | – | – | 4.44 | 40.00 | 55.56 |
| 3 | Pest incidence | 4.44 | 86.67 | 8.89 | – | – |
| 4 | Pesticide seed/cost | 5.56 | 88.89 | 5.56 | – | – |
| 5 | Fertilizer seed/cost | 1.11 | 12.22 | 80.00 | 5.56 | 1.11 |
| 6 | Irrigation seed/cost | – | 4.44 | 91.12 | 3.33 | 1.11 |
| 7 | Labour cost/need | – | 5.56 | 84.44 | 10.00 | – |
| 8 | Harvesting cost | – | – | 68.89 | – | 31.11 |
| 9 | Cotton quality and fibre colour | 13.33 | 52.22 | 34.44 | – | – |
| 10 | Cotton price | 15.56 | 24.44 | 58.89 | 1.11 | – |
| 11 | Yield | 54.44 | 38.89 | 6.67 | – | – |
| 12 | Profit | 52.22 | 42.22 | 5.56 | – | – |
| 13 | Suitability for early sowing | 21.11 | 68.89 | 10.00 | – | – |
| 14 | Market preferences | 17.78 | 18.89 | 63.33 | – | – |
| 15 | Improvement in village economy | 8.88 | 88.89 | 2.22 | – | – |

and profit gains are two strong advantages of Bt cotton. Not a single Bt grower faced difficulty in selling the Bt produce. About 65 % farmers found quality of Bt cotton slightly better than non-Bt cotton. They found Bt cotton cleaner with better colour. Owing to better quality of Cotton, Bt growers fetched somewhat higher price for the product. All sample farmers agreed that Bt cotton is suitable for early sowing under irrigated situation. As per 98 % sample farmers, cultivation of Bt cotton impacted positively on village economy. It also generated positive impact to some extent on the income level of wage earners/village labourers.

## 18.3 Measures Suggested by Sample Farmers to Improve the Acceptance of Bt Cotton

Since Bt cotton is a relatively new crop, farmers need technical guidance and assistance prior to sowing and at different stages of crop cultivation. According to majority of sample farmers, neither government extension agencies nor representative of seed companies paid visit to Bt cotton fields. Also, they had not provided any guidance at different stages of the Bt cotton cultivation. The fellow farmers and to some extent seed dealers provided necessary help and guidance to solve the problems.

### *18.3.1 Farmers' Perception on Environmental Impact of Bt Cotton*

In response to a few questions on environmental impact of Bt cotton, all the sample farmers expressed no adverse impact of Bt cotton on other adjoining crops, insect population, farm animal health, human health and soil health. On the contrary, owing to relatively lower use of pesticides, Bt cotton is found more health friendly than non-Bt cotton.

# Part V
# State Study 3: Maharashtra

# Chapter 19
# Cotton Cultivation in Maharashtra

This section examines the status and performance of cotton in the state of Maharashtra, as well as in the districts sampled for the study. It then presents the sampling design and profile.

Cotton is traditionally one of the important crops in Maharashtra. Area under cotton is about three million hectares in the state and it accounts for about 30 % of the cotton area in the country. The area was about 2.5 million hectares until 1994–1995, and since it has expanded to about 3 million hectares. Even though the share in area is high, the share in the production is low due to the relatively low yield levels. Further, a declining trend in the 1990s, with a particularly sharp declining trend in the yields. However, a change is evident in the early 2000s with an upturn in the production and yield growth rates when these years are included.

Cotton experiences wide fluctuation in Maharashtra due to vagaries of weather as it is mainly cultivated under rain-fed conditions. The cotton production in the state shows very low levels of production during 1970–1971, 1991–1992, 1997–1998 mainly because of crop failures due to the poor rainfall or early withdrawal of monsoon. The yield levels in all these years were very low; see Table 19.1 and Figs. 19.1, 19.2 and 19.3.

## 19.1 Cotton Cultivation in the Selected Districts

The districts sampled for this study, Buldhana and Jalgaon, belong to two different agro-climatic regions of Maharashtra, and are important cotton growing districts in state. Jalgaon has greater irrigation availability for cotton whereas the cotton cultivation in Buldhana is predominantly rain-fed. Due to this, the yields are relatively low in Buldhana compared to Jalgaon. While area under cotton shows a decrease in Buldhana, it shows an increase in Jalgaon (Table 19.2 and Figs. 19.4 and 19.5). As


V.P. Gandhi, D. Jain, *Introduction of Biotechnology in India's Agriculture*, DOI 10.1007/978-981-10-1091-0_19

**Table 19.1** Area, production and yield of cotton in Maharashtra

| Year | Area (000 ha) | Production (Lint) (000 Bales) | Yield (Kg/Ha) |
|---|---|---|---|
| 1960–1961 | 2500 | 2843 | 114 |
| 1965–1966 | 2716 | 1787 | 66 |
| 1970–1971 | 2750 | 824 | 30 |
| 1975–1976 | 2307 | 1326 | 58 |
| 1980–1981 | 2550 | 2081 | 82 |
| 1985–1986 | 2709 | 3372 | 125 |
| 1990–1991 | 2721 | 3188 | 117 |
| 1991–1992 | 2759 | 1965 | 71 |
| 1992–1993 | 2574 | 3214 | 125 |
| 1993–1994 | 2481 | 4465 | 180 |
| 1994–1995 | 2760 | 4463 | 162 |
| 1995–1996 | 3078 | 4781 | 155 |
| 1996–1997 | 3085 | 3143 | 102 |
| 1997–1998 | 3139 | 1753 | 56 |
| 1998–1999 | 3199 | 2619 | 82 |
| 1999–2000 | 3254 | 3099 | 95 |
| 2000–2001 | 3077 | 3064 | 100 |
| 2001–2002 | 3105 | 4572 | 147 |
| 2002–2003 | 2800 | 4424 | 158 |
| 2003–2004 | 2766 | 3100 | 112 |
| 2004–2005 | 3049 | 5200 | 171 |
| Annual growth rate (%) | | | |
| 1990–1991 to 2000–2001 | 2.31 | −1.09 | −3.29 |
| 1990–1991 to 2004–2005 | 0.93 | 2.05 | 1.12 |

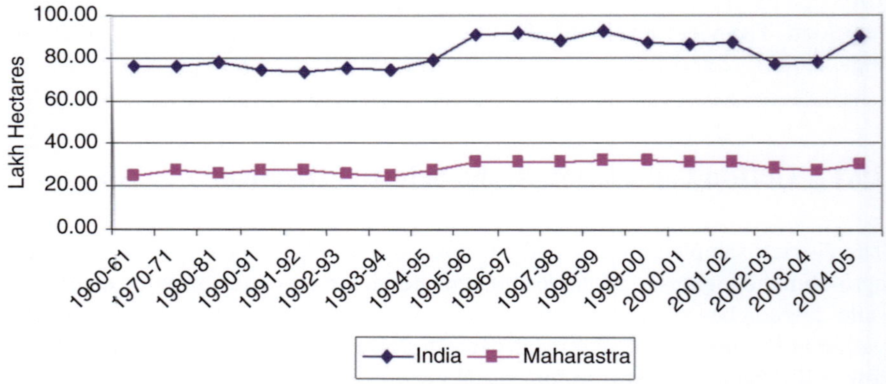


**Fig. 19.1** Area under cotton in India and Maharashtra

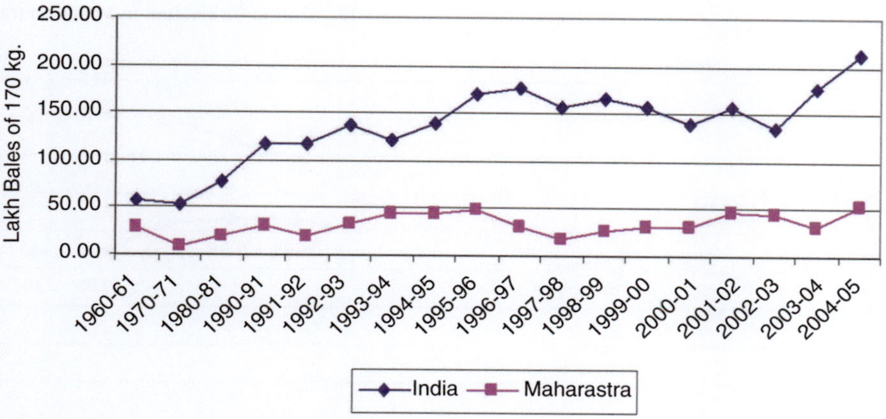


**Fig. 19.2** Production of cotton in India and Maharashtra

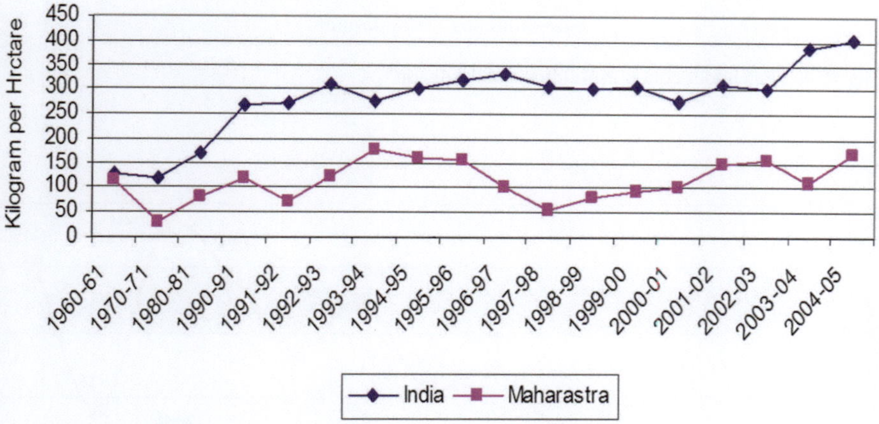


**Fig. 19.3** Yield of cotton in India and Maharashtra

**Table 19.2** Area, production and yield of cotton in sample districts

| | Area in'00 ha | | Production'00 Bales | | Yield in Kg/ha | |
|---|---|---|---|---|---|---|
| | Buldhana | Jalgaon | Buldhana | Jalgaon | Buldhana | Jalgaon |
| 1990–1991 | 2667 | 1893 | 1699 | 1909 | 108 | 171 |
| 1991–1992 | 2571 | 2040 | 717 | 1116 | 47 | 93 |
| 1992–1993 | 2186 | 2034 | 1256 | 2340 | 98 | 196 |
| 1993–1994 | 2112 | 2098 | 2075 | 2542 | 167 | 206 |
| 1994–1995 | 2436 | 2633 | 2122 | 3454 | 148 | 223 |
| 1995–1996 | 2547 | 2779 | 1861 | 3143 | 124 | 192 |
| 1996–1997 | 2524 | 2899 | 2425 | 4475 | 163 | 262 |
| 1997–1998 | 2655 | 3404 | 1199 | 3706 | 77 | 185 |
| 1998–1999 | 2658 | 3719 | 2025 | 4895 | 130 | 224 |
| 1999–1900 | 2580 | 3991 | 2635 | 5620 | 174 | 239 |
| 2000–2001 | 2198 | 4115 | 892 | 2811 | 69 | 116 |
| 2001–2002 | 2201 | 4030 | 1978 | 5049 | 153 | 213 |

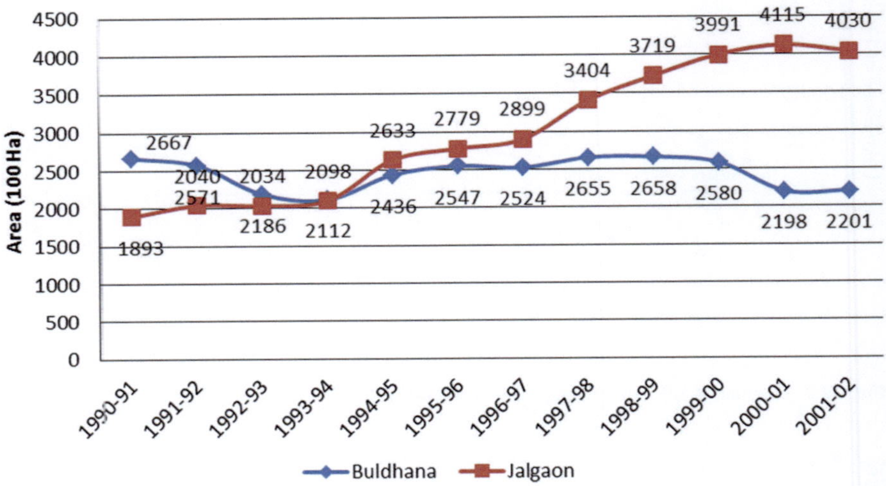


**Fig. 19.4** Area under cotton in Buldhana and Jalgaon

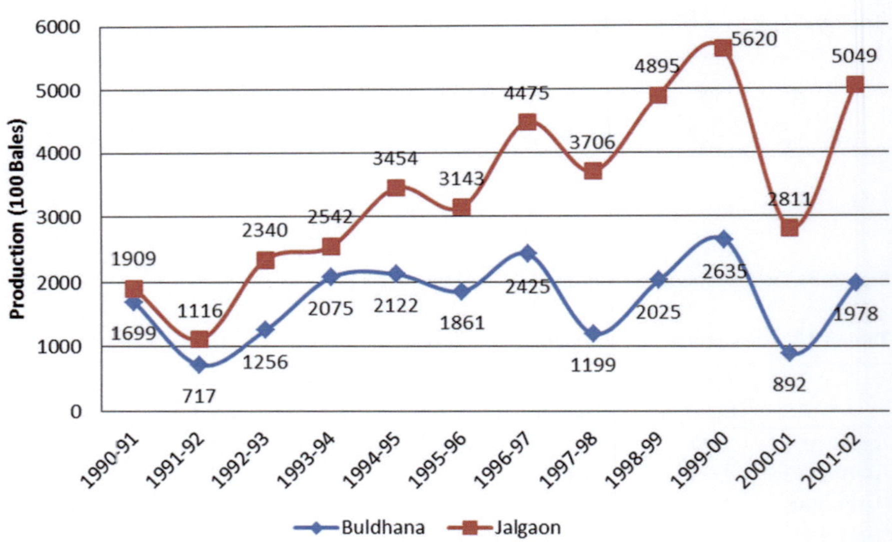


**Fig. 19.5** Production of cotton in Buldhana and Jalgaon

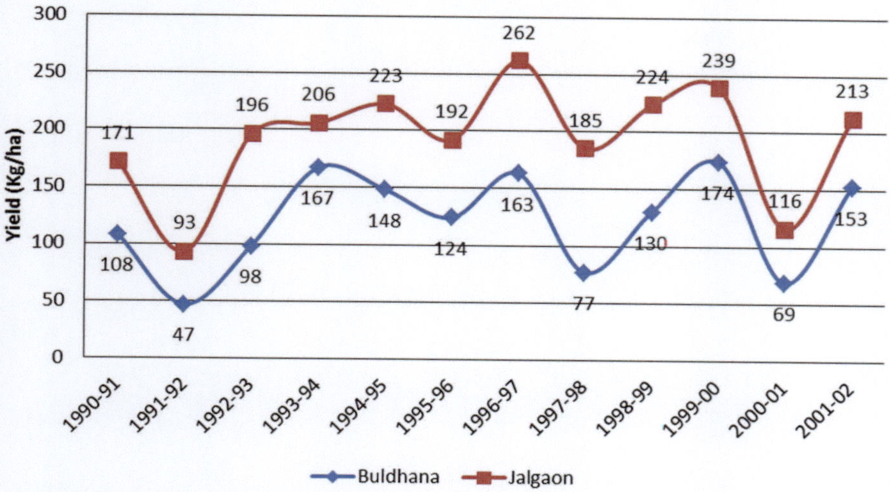


**Fig. 19.6** Yield of cotton in Buldhana and Jalgaon

elsewhere in the state of Maharashtra, both production and yield of cotton in these districts shows large fluctuations, and hardly any trend is evident until 2001–2002 (Fig. 19.6).

# Chapter 20
# Sampling and Methodology

This chapter presents the profile of the sample households on features such as age, education, experience in cotton farming, farm size, irrigation status, sources of irrigation, cropping pattern, access to cotton markets and towns, and their distribution across Bt and non-Bt farmer households. It also explores the relationship of these factors to the adoption of Bt Cotton.

## 20.1 Sampling Design

A sample survey of farmers was undertaken in two sample districts of the state of Maharashtra. Broadly, a stratified random sampling process was followed. A major consideration for the selection of the districts was sampling of different agro-climatic regions, and as indicated, the two districts belonged to two different agricultural regions of the state. The selection the districts was done on the basis of information provided by the District Agricultural Office on the cultivation of cotton varieties, particularly Bt and non-Bt cotton. Given the objectives of the study, a major consideration was the presence of Bt cotton in the districts as per the records of Commissionerate of Agriculture, Government of Maharashtra, and the District Agricultural Offices. After selecting the districts, three Talukas were selected in each district again considering the agro-ecologic diversity and the presence of Bt cotton. From each taluka one village was selected at random. Farmers growing Bt and Non-Bt cotton and across small, medium and large farm were selected through a stratified random sampling process considering representation, as far as possible, of different farm sizes, and irrigated and unirrigated conditions. In this way, a total of 12 villages and 154 farmers were covered across 2 districts and 6 talukas. The names of selected talukas, villages and the number of farmer households selected under different categories are given in the Table 20.1 below. Data were collected for the agricultural year 2004–2005. The data indicated that among the Bt farmers,


V.P. Gandhi, D. Jain, *Introduction of Biotechnology in India's Agriculture*, DOI 10.1007/978-981-10-1091-0_20

**Table 20.1** Selection of sample farmer households

| S.N. | District | Taluka | Village | Number of farmer households | | | | | |
|---|---|---|---|---|---|---|---|---|---|
| **Bt cotton** | | | | | | | | | |
| | | | | **I** | **UI** | **Total** | **Small** | **Medium** | **Large** |
| 1 | Jalgaon | Jalgaon | Nasirabad | 8 | 8 | 16 | 0 | 5 | 11 |
| 2 | Jalgaon | Bhusaval | Sakhegaon | 2 | 7 | 9 | 2 | 2 | 5 |
| 3 | Jalgaon | Jamner | Gharkheda | 8 | 2 | 10 | 0 | 2 | 8 |
| 4 | Buldhana | Jalgaon | Sungaon | 9 | 4 | 13 | 2 | 2 | 9 |
| 5 | Bildhana | Motala | Advihir | 8 | 9 | 17 | 1 | 6 | 10 |
| 6 | Buldhana | Bhuldhana | Tharatkhed | 13 | 7 | 20 | 12 | 6 | 2 |
| Total | | | | 48 | 37 | 85 | 17 | 23 | 45 |
| **Non Bt cotton** | | | | | | | | | |
| | | | | **I** | **UI** | **Total** | **Small** | **Medium** | **Large** |
| 1 | Jalgaon | Jalgaon | Nasirabad | 2 | 9 | 11 | 0 | 3 | 8 |
| 2 | Jalgaon | Bhusaval | Sakhegaon | 2 | 16 | 18 | 8 | 5 | 5 |
| 3 | Jalgaon | Jamner | Gharkheda | 3 | 6 | 9 | 4 | 4 | 1 |
| 4 | Biuldhana | Jalgaon | Sungaon | 3 | 7 | 10 | 3 | 3 | 4 |
| 5 | Buldhana | Motala | Advihir | 1 | 9 | 10 | 0 | 1 | 9 |
| 6 | Buldhana | Bhuldhana | Tharatkhed | 10 | 1 | 11 | 8 | 2 | 1 |
| Total | | | | 21 | 47 | 69 | 23 | 18 | 28 |

*I* Irrigated, *UI* Unirrigated

**Table 20.2** Irrigated and unirrigated farmer households under Bt and non-Bt cotton

| | Bt cotton | | | Non-Bt cotton | | |
|---|---|---|---|---|---|---|
| Farm size | Irrigated | Unirrigated | Total | Irrigated | Unirrigated | Total |
| Small (below 5 Acre) | 8 (47.1) | 9 (52.9) | 17 (100.0) | 9 (39.1) | 14 (60.9 | 23 (100.0) |
| Medium (5–10 Acre) | 14 (60.9) | 9 (39.1) | 23 (100.0) | 5 (27.8) | 13 (72.2) | 18 (100.0) |
| Large (Above 10 Acre) | 26 (57.8) | 19 (42.2) | 45 (100.0) | 7 (25.0) | 21 (75.0) | 28 (100.0) |
| Total | 48 (56.5) | 37 (43.5) | 85 (100.0) | 21 (30.4) | 48 (69.6) | 69 (100.0) |

about 82 % took up Bt cultivation for the first time in 2003–2004 and the rest during 2004–2005, i.e. the survey year.

The irrigation status of the sampled Bt and Non-Bt cotton farmer households given in Table 20.2 showed that while 56.5 % of Bt growers had irrigation, and this was about 30 % for the non-Bt cotton farmers. The distribution between irrigated and unirrigated and under different farm size, viz., small, medium and large is also given in Table 20.2.

**Table 20.3** Average age of head of household among Bt and non-Bt sample farmer households

| | Bt-cotton | | Non-Bt cotton | |
|---|---|---|---|---|
| Age | No. of farmers | Per cent | No. of farmers | Per cent |
| Below 30 | 8 | 9.41 | 6 | 8.70 |
| 30–40 | 17 | 20.00 | 17 | 24.64 |
| 40–50 | 25 | 29.41 | 19 | 27.54 |
| 50–60 | 26 | 30.59 | 19 | 27.54 |
| Above 60 | 9 | 10.59 | 8 | 11.59 |
| Total | 85 | 100 | 69 | 100 |

**Table 20.4** Level of education of the head of household among Bt and non-Bt farmer households

| | Bt cotton | | Non-Bt cotton | |
|---|---|---|---|---|
| Level of education | No. of farmers | Per cent | No. of farmers | Per cent |
| No formal education | 30 | 35.29 | 31 | 44.93 |
| Up to primary | 31 | 36.47 | 28 | 40.58 |
| Up to secondary | 11 | 12.94 | 6 | 8.70 |
| Graduate | 6 | 7.06 | 3 | 4.35 |
| Others | 7 | 8.24 | 1 | 1.45 |
| All households | 85 | 100 | 69 | 100 |

### *20.1.1 Age, Education and Experience in Cotton Cultivation*

The distribution of the age of the head of households across Bt and non-Bt cotton sample farmer households indicates that there is no large difference in the age of sample farmer households (Table 20.3). However, the level of education of the head of households shows that the percentage of heads of households with higher education tends to be relatively more under the Bt cotton farmers (Table 20.4). For example, about 15 % of the heads of households under Bt cotton were either graduates or had technical education as opposed to 8 % under non-Bt cotton. Similarly Bt cotton growers have more experience in the cultivation of cotton than non-Bt cotton (Table 20.5). About 60 % of heads of households under Bt cotton have more than 15 year experience in the cultivation of cotton as opposed to just over 25 % under non-Bt cotton. Thus, Bt cotton appears to be positively associated with education and experience in growing cotton.

### *20.1.2 Farm Size and Irrigation*

Next we have examined the farm size and the level of irrigation across the Bt and non-Bt cotton households. The average farm size in terms of area operated and the level of irrigation of Bt and not-Bt sample farmer households are given in Table 20.6.

**Table 20.5** Experience in cotton cultivation by the head of household among Bt and non-Bt cotton farmer households

| | Bt cotton | | Non-Bt cotton | |
|---|---|---|---|---|
| Experience in years | No. of farmers | Per cent | No. of farmers | Per cent |
| Below 5 years | 2 | 2.35 | 1 | 1.45 |
| 5–10 years | 8 | 9.41 | 12 | 17.39 |
| 10–15 years | 24 | 28.24 | 38 | 55.07 |
| Above 15 years | 51 | 60.00 | 18 | 26.09 |
| Total | 85 | 100.00 | 69 | 100.00 |

**Table 20.6** Average farm size and irrigation status of Bt and non-Bt sample farmer households (acres)

| | Bt cotton | | | | Non-Bt cotton | | | |
|---|---|---|---|---|---|---|---|---|
| | No. of farmers | Area operated | Area irrigated | % Area irrigated | No. of farmers | Area operated | Area irrigated | % Area irrigated |
| Small | 17 | 3.41 | 1.59 | 46.55 | 23 | 4.00 | 1.28 | 32.07 |
| Medium | 23 | 8.52 | 5.26 | 61.73 | 18 | 8.39 | 1.78 | 21.19 |
| Large | 45 | 22.64 | 9.84 | 43.47 | 28 | 21.39 | 6.21 | 29.05 |
| Total | 85 | 14.98 | 6.95 | 46.43 | 69 | 12.20 | 3.41 | 27.97 |

The average farm size of Bt growers is marginally higher. The farm sizes in these dry areas tend to be bigger. The percentage of area irrigated in total operated area is substantially higher on an average among the Bt growers, 46 % as compared to 28 %, and this difference exists in all farm sizes.

### 20.1.3 Sources of Irrigation

The sources of irrigation in the study area were canal, tube-wells, open wells and ponds (Table 20.7). While canal irrigation was the major source of irrigation for the Bt cotton growers, it was tube-well for the non-Bt growers. Other sources of irrigation among the sample farmer households were open wells and ponds, and more Bt cotton growers had access to these than non-Bt growers.

### 20.1.4 Access to Market for Cotton

Findings on the access to the markets for cotton for the sample farmers indicated that for about one-third of farmers the nearest cotton market was within a radius of 5 km (Table 20.8). For about 24 % of the Bt growers and 15 % of the non-Bt growers the nearest cotton market was beyond 10 km. On the whole there is no substantial difference in the access to the cotton market across the Bt and non-Bt farmers.

**Table 20.7** Sources of irrigation under Bt and non-Bt among sample farmer households

| Source of irrigation | Bt-cotton | Per cent | Non-Bt cotton | Per cent |
|---|---|---|---|---|
| Canal | 32 | 58.18 | 0 | 0.00 |
| Tube-well | 5 | 9.09 | 19 | 67.86 |
| Open well | 9 | 16.36 | 1 | 3.57 |
| Drip irrigation | 0 | 0.00 | 7 | 25.00 |
| Pond | 9 | 16.36 | 0 | 0.00 |
| Others | 0 | 0.00 | 1 | 3.57 |
| Overall | 55 | 100.00 | 28 | 100.00 |

**Table 20.8** Distance of the farmer households from the nearest cotton market

| | Bt cotton | | Non-Bt cotton | |
|---|---|---|---|---|
| Distance in Km | No. of farmers | Per cent | No. of farmers | Per cent |
| Below 2 Km | 16 | 18.82 | 11 | 15.94 |
| 2–5 Km | 17 | 20.00 | 10 | 14.49 |
| 5–8 Km | 32 | 37.65 | 47 | 68.12 |
| 8–10 | 0 | 0.00 | 0 | 0.00 |
| Above 10 | 20 | 23.53 | 11 | 15.94 |
| Overall | 85 | 100.00 | 69 | 100.00 |

# Chapter 21
# Nature and Performance of Bt Cotton vs. Non-Bt Cotton in Maharashtra

## 21.1 Introduction

This chapter provides the findings and analysis on various features such as varieties of cotton cultivated, pest resistance, pesticide use, input use, performance, and the economics, comparing Bt and Non-Bt cotton.

## 21.2 Cropping Pattern

The findings indicate that the cropping season in the survey area is predominantly the single season of kharif. Table 21.1 furnishes the season wise cropping pattern for Maharashtra.

The Table shows that about 85 % of the gross cropped area falls under the kharif season for both Bt and Non-Bt farmers. This may be due to two reasons: first, kharif is the rainy season and rain-fed agriculture accounts for a major share in the cultivated area. Secondly, cotton is a longer duration kharif crop as compared to many other crops. Thus, taking a rabi crop after cotton is difficult. The average area of Bt cotton per household was 1.59 ha with irrigation and 1.99 ha without irrigation. This was respectively 0.70 and 2.64 ha for non-Bt cotton. The other major crops followed by cotton are jowar, pulses and wheat. The cropping pattern does not show a large difference between Bt and non-Bt farmers. This is examined in the following Table through percentages.

The Table 21.2 shows that the share of area under cotton for the sample Bt growers was 43.1 %, whereas the same for the non-Bt growers was 40.1 % (Table 21.2). This indicates that cotton is the dominant crop. The share of cotton is the highest for unirrigated Non-Bt farmers indicating its great importance in this group. Other major crops grown during the kharif were jowar, maize, pulses etc. As mentioned




V.P. Gandhi, D. Jain, *Introduction of Biotechnology in India's Agriculture*, DOI 10.1007/978-981-10-1091-0_21

**Table 21.1** Cropping pattern among the Bt and non-Bt cotton sample farmer households, average acreage in hectares

| | Bt- farmers | | | Non-Bt farmers | | |
|---|---|---|---|---|---|---|
| | Irrigated | Unirrigated | Total | Irrigated | Unirrigated | Total |
| **Kharif-2004** | | | | | | |
| Cotton | 1.59 | 1.99 | 3.58 | 0.70 | 2.64 | 3.34 |
| Jowar | 0.35 | 1.07 | 1.46 | 0.09 | 1.19 | 1.27 |
| Maize | 0.13 | 0.36 | 0.49 | 0.20 | 0.30 | 0.50 |
| Soyabean | 0.21 | 0.40 | 0.61 | 0.28 | 0.32 | 0.59 |
| Urad+Mung | 0.17 | 0.69 | 0.86 | 0.23 | 0.69 | 0.92 |
| Banana | 0.49 | 0.06 | 0.55 | 0.70 | 0.00 | 0.70 |
| Sugarcane | 0.56 | 0.00 | 0.56 | 0.04 | 0.00 | 0.04 |
| Tur | 0.15 | 0.59 | 0.74 | 0.03 | 0.49 | 0.52 |
| Chilli | 0.06 | 0.37 | 0.43 | 0.20 | 0.30 | 0.49 |
| Mango | 0.00 | 0.31 | 0.31 | 0.00 | 0.05 | 0.05 |
| Total Kharif | 3.71 | 5.84 | 9.55 | 2.47 | 5.98 | 8.45 |
| **Rabi-2004–2005** | | | | | | |
| Wheat | 0.58 | 0.18 | 0.76 | 0.46 | 0.46 | 0.93 |
| Gram | 0.34 | 0.46 | 0.72 | 0.04 | 0.14 | 0.18 |
| Sunflower | 0.07 | 0.16 | 0.19 | 0.01 | 0.11 | 0.11 |
| Jowar | 0.12 | 0.32 | 0.44 | 0.07 | 0.24 | 0.32 |
| Groundnut | 0.02 | 0.04 | 0.06 | 0.00 | 0.08 | 0.08 |
| Onion | 0.06 | 0.00 | 0.06 | 0.04 | 0.04 | 0.09 |
| Total rabi | 1.19 | 1.16 | 2.23 | 0.63 | 1.08 | 1.71 |
| Grand total | 4.90 | 7.00 | 11.90 | 3.10 | 7.06 | 10.16 |

above, the share of rabi area in gross cropped area is less than 15 %, and wheat was the major crop grown during the rabi season.

## 21.3 Area Under Cotton

Table 21.3 furnishes the land holding group wise and irrigation status wise estimates of per holding area. In the study area, nearly 58 % of the total area under Bt cotton was cultivated under irrigated condition, whereas 42 % was raised under unirrigated condition. On the other hand nearly 44 % of Non Bt cotton was grown under irrigated condition and 56 % was cultivated under unirrigated condition. This indicates that between the Bt and Non Bt cotton varieties, cultivation of Bt cotton is more on irrigated lands. All these observations are more or less the same in the three land holding groups indicating that size of land holding is not a factor influencing the spread of Bt cotton cultivation.

**Table 21.2** Cropping pattern among Bt and non-Bt cotton farmer households (percentage)

| Crops | Bt- farmers | | | Non-Bt farmers | | |
|---|---|---|---|---|---|---|
| | Irrigated | Unirrigated | Total | Irrigated | Unirrigated | Total |
| **Kharif-2004** | | | | | | |
| Cotton | 45.36 | 40.70 | 43.12 | 28.97 | 45.63 | 40.85 |
| Jowar | 5.80 | 12.68 | 10.06 | 2.79 | 14.69 | 11.05 |
| Maize | 2.13 | 4.32 | 3.41 | 6.01 | 3.66 | 4.36 |
| Soyabean | 3.48 | 4.74 | 4.22 | 8.15 | 3.93 | 5.18 |
| Udad+Mung | 2.80 | 8.15 | 5.92 | 6.87 | 8.48 | 7.99 |
| Banana | 8.12 | 0.70 | 3.81 | 20.60 | 0.00 | 6.06 |
| Sugarcane | 9.28 | 0.00 | 3.89 | 1.29 | 0.00 | 0.38 |
| Tur | 2.51 | 6.97 | 5.11 | 0.86 | 6.07 | 4.55 |
| Chilli | 0.97 | 4.39 | 2.96 | 5.79 | 3.66 | 4.29 |
| Mango | 0.00 | 3.62 | 2.11 | 0.00 | 0.63 | 0.44 |
| Total Kharif | 80.46 | 86.27 | 84.62 | 81.33 | 86.74 | 85.13 |
| **Rabi 2004–2005** | | | | | | |
| Wheat | 9.48 | 2.09 | 5.27 | 13.52 | 5.71 | 8.14 |
| Gram | 5.61 | 5.44 | 4.95 | 1.29 | 1.70 | 1.58 |
| Sunflower | 1.16 | 1.95 | 1.30 | 0.43 | 1.34 | 0.95 |
| Jowar | 1.93 | 3.76 | 3.00 | 2.15 | 2.99 | 2.75 |
| Groundnut | 0.39 | 0.49 | 0.45 | 0.00 | 0.98 | 0.69 |
| Onion | 0.97 | 0.00 | 0.41 | 1.29 | 0.54 | 0.76 |
| Total rabi | 19.54 | 13.73 | 15.38 | 18.67 | 13.26 | 14.87 |
| Grand total | 100 | 100 | 100 | 100 | 100 | 100 |

**Table 21.3** Average area under Bt cotton in Maharashtra (ha)

| Farm size | Bt cotton | | | Non Bt cotton | | |
|---|---|---|---|---|---|---|
| | I | UI | Total | I | UI | Total |
| Small | 0.86 (57.72) | 0.63 (42.28) | 0.74 (100.0) | 0.85 (44.27) | 1.07 (55.73) | 0.92 (100.0) |
| Medium | 1.55 (56.16) | 1.21 (43.84) | 1.42 (100.0) | 1.38 (53.91) | 1.18 (46.09) | 1.24 (100.0) |
| Large | 1.79 (51.88) | 1.66 (48.12) | 1.74 (100.0) | 2.02 (56.11) | 1.58 (43.89) | 1.69 (100.0) |
| Overall | 1.56 (54.55) | 1.3 (45.45) | 1.45 (100.0) | 1.37 (50.93) | 1.32 (49.07) | 1.35 (100.0) |

*I* irrigated, *U* unirrigated

## 21.4 Cotton Varieties Used by the Sample Households

The sample farmers grew four varieties of Bt cotton: MECH 184, MECH 12, MECH 162 and RASHI 2 (Table 21.4). About 27 % of the sample households cultivated both RASHI2 and MECH12. RASHI 2 was more popular among large farm size

**Table 21.4** The varieties of cotton grown by the sample farmer households

| | Number of farmer households | | | | Percentage | | | |
|---|---|---|---|---|---|---|---|---|
| Varieties | Small | Medium | Large | Total | Small | Medium | Large | Total |
| **Bt-cotton** | | | | | | | | |
| MECH 184 | 6 | 9 | 10 | 25 | 35.29 | 39.13 | 22.22 | 29.41 |
| MECH 12 | 6 | 5 | 8 | 19 | 35.29 | 21.74 | 17.78 | 22.35 |
| MECH 162 | 1 | 3 | 2 | 6 | 5.88 | 13.04 | 4.44 | 7.06 |
| RASHI 2 | 1 | 1 | 10 | 12 | 5.88 | 4.35 | 22.22 | 14.12 |
| RASHI 2/MECH 12 | 3 | 5 | 15 | 23 | 17.65 | 21.74 | 33.33 | 27.06 |
| All varieties | 17 | 23 | 45 | 85 | 100 | 100 | 100 | 100 |
| Non-Bt cotton | | | | | | | | |
| Ankur | 5 | 4 | 5 | 14 | 21.74 | 22.22 | 17.86 | 20.29 |
| Bunny | 5 | 2 | 4 | 11 | 21.74 | 11.11 | 14.29 | 15.94 |
| Ajit | 9 | 4 | 6 | 19 | 39.13 | 22.22 | 21.43 | 27.54 |
| Others | 4 | 8 | 13 | 25 | 17.39 | 44.44 | 46.43 | 36.23 |
| All varieties | 23 | 18 | 28 | 69 | 100 | 100 | 100 | 100 |

**Table 21.5** Seed rate used in Bt compared to non-Bt

| | |
|---|---|
| Per cent of farmers responding | 100 |
| Percentage of farmers reporting use of lower seed rate | 100 |
| 25 % | 57.65 |
| 25–50 % | 42.45 |

and MECH varieties were more popular among medium and small farms. However there was no difference in the use of these varieties across irrigated and unirrigated conditions. In non-Bt cotton, three varieties were more popular among the growers and they were Ankur, Bunny and Ajit. There was no major difference in the use of these varieties among different sizes of farms but Ajit was more popular among small farmers.

## 21.5 Average Cost of Seeds Used

Regarding the seed rate for Bt cotton as compared to non-Bt cotton , all of them indicated that the seed rate used for Bt was lower compared to non-Bt About 57.65 % of them reported the seed rate used was lower by up to 25 %, and the rest 42.35 % reported 25–50 % lower seed rate compared to non-Bt cotton (Table 21.5). This indicates an adjustment to the high price of seeds and their more efficient use.

## 21.6 Application of Pesticides

Cotton has many pest problems and the most important reason for the adoption of Bt cotton is its resistance to pests, particularly boll worms, which can be a devastating problem for cotton. In this context, information has been collected regarding the pest problems observed by the farmers in their cotton crops. The results are reported in Table 21.6 below.

In the case of boll worms, including American, pink and spotted boll worms, no infestation is indicated by over 70 % of farmers for Bt cotton, whereas light to heavy incidence is reported in the vast majority of cases by non-Bt cotton farmers. Only about 4–6 % of the sample for Bt cotton reports moderate to heavy infestation, whereas this number was up to 60 % in non-Bt cotton. Surprisingly, there is also a difference in the sucking and foliage feeding pests, where the incidence is mainly none to light in the case of Bt cotton, whereas it is moderate to heavy in the case of non-Bt cotton. Thus, Bt cotton appears to clearly provide resistance to boll worms for a larger majority of farmers, and also to other pests for most of the farmers. However a small number of farmers indicate incidence of boll worms, particularly other kinds of boll worms.

**Table 21.6** Pest/insect attack on cotton: response of Bt cotton growers (per cent)

| | BT | | | | | Non-BT | | | | |
|---|---|---|---|---|---|---|---|---|---|---|
| | Percent | Infestation reported | | | | Percent | Infestation reported | | | |
| Pest/insect | reporting | None | Light | Moderate | Heavy | reporting | None | Light | Moderate | Heavy |
| A. Boll worm | | | | | | | | | | |
| 1. American boll worm | 96.47 | 74.39 | 21.95 | 3.66 | 0.00 | 11.76 | 30.00 | 50.00 | 20.00 | 0.00 |
| 2. Pink boll worm | 87.06 | 75.68 | 20.27 | 4.05 | 0.00 | 61.18 | 3.85 | 36.54 | 30.77 | 28.85 |
| 3. Spotted boll worm | 90.59 | 72.73 | 20.78 | 6.49 | 0.00 | 56.47 | 2.08 | 41.67 | 39.58 | 16.67 |
| 4. Others | 18.82 | 50.00 | 6.25 | 18.75 | 25.00 | 56.47 | 4.17 | 41.67 | 35.42 | 18.75 |
| B. Sucking pests | | | | | | | | | | |
| 1. Thrips | 96.47 | 4.88 | 56.10 | 35.37 | 3.66 | 11.76 | 30.00 | 0.00 | 20.00 | 50.00 |
| 2. Leafhopper | 95.29 | 3.70 | 58.02 | 30.86 | 7.41 | 58.82 | 0.00 | 18.00 | 56.00 | 26.00 |
| 3. Whitefly | 95.29 | 6.17 | 58.02 | 33.33 | 2.47 | 57.65 | 0.00 | 20.41 | 55.10 | 24.49 |
| 4. Others | 3.53 | 33.33 | 0.00 | 66.67 | 0.00 | 60.00 | 1.96 | 19.61 | 47.06 | 31.37 |
| C. Foliage feeding pests | | | | | | | | | | |
| 1. Leaf roller | 94.12 | 27.50 | 45.00 | 26.25 | 1.25 | 42.35 | 2.78 | 38.89 | 50.00 | 8.33 |
| 2. Caterpillar | 89.41 | 27.63 | 40.79 | 28.95 | 2.63 | 40.00 | 2.94 | 29.41 | 61.76 | 5.88 |
| 3. Others | 7.06 | 50.00 | 33.33 | 16.67 | 0.00 | 5.88 | 20.00 | 40.00 | 40.00 | 0.00 |
| (D) Soil pests | | | | | | | | | | |
| 1. Termite | 88.24 | 34.67 | 17.33 | 36.00 | 12.00 | 40.00 | 2.94 | 14.71 | 55.88 | 26.47 |

**Table 21.7** Application of pesticides in Bt and non-Bt cotton

| | Bt cotton | Non-Bt cotton |
|---|---|---|
| Average number of sprays | 3.37 | 5.28 |
| Cost per ha (Rs.) | 3242 | 4120 |

## 21.7 Pesticide Use

Table 21.7 above indicates the performance of Bt cotton, relative to Non-Bt cotton in the reduction of pesticide use. The information shows that pesticides are still used by farmers after shifting to Bt cotton. However, there is a significant reduction in the number of sprays that are applied as well as the cost of pesticides. The average number of sprays is reduced from 5.28 to 3.37, a 36 % reduction. The cost per hectare reduces by 21 %. Thus, even though pesticide spraying is not eliminated, there is a substantial reduction in the pesticide use and cost under Bt cotton.

# Chapter 22
# Economics of Bt Cotton vs. Non-Bt Cotton in Maharashtra

## 22.1 Introduction

This chapter examines the relationship of various factors to the yield, value of output and profit of cotton cultivation reported by the sample farmer households. It includes average Bt and non-Bt cotton yields, cost of production, value of output and profit, and econometric analysis of the performance of Bt cotton vs. non-Bt cotton.

## 22.2 Average Bt and Non-Bt Cotton Yields

Most sample farmers, over 90 %, report at least five cotton fibre pickings and the rest had up to six pickings. The average yield per picking under Bt cotton was invariably higher (Table 22.1 and Figs. 22.1 and 22.2). However, the percentage distribution pattern of quantity of cotton realized under each picking did not differ much for both Bt and Non-Bt cotton. Thus, there is not much difference between Bt and Non-Bt in the number of pickings or the distribution across pickings. The main difference is the quantity obtained in each picking especially from the second picking onwards.

The Table 22.2 provides a summary comparison of the performance of Bt and Non-Bt cotton in terms of their yield and value of output under irrigated and unirrigated conditions. The Table shows that in all cases, the yields of Bt cotton are higher than the yields of Non-Bt cotton. This is found to be true under irrigated as well as unirrigated conditions. The yields obtained with irrigations are typically higher than those without irrigations . The results indicate a sizeable impact of Bt cotton on the yield and value of output under both irrigated and unirrigated conditions.


V.P. Gandhi, D. Jain, *Introduction of Biotechnology in India's Agriculture*, DOI 10.1007/978-981-10-1091-0_22

**Table 22.1** Yield per hectare in kilogram per picking among Bt and non-Bt cotton farmer households

| | Pickings | | | | | | Total |
|---|---|---|---|---|---|---|---|
| | First | Second | Third | Fourth | Fifth | Sixth | |
| **Bt cotton** | | | | | | | |
| **Small** | | | | | | | |
| Irrigated | 256 | 525 | 710 | 571 | 216 | 185 | 2465 |
| Unirrigated | 343 | 633 | 529 | 402 | 193 | 46 | 2146 |
| Total | 302 | 582 | 614 | 481 | 204 | 112 | 2296 |
| **Medium** | | | | | | | |
| Irrigated | 316 | 750 | 733 | 485 | 309 | 88 | 2681 |
| Unirrigated | 448 | 696 | 626 | 417 | 232 | 31 | 2449 |
| Total | 368 | 729 | 691 | 459 | 279 | 66 | 2590 |
| **Large** | | | | | | | |
| Irrigated | 333 | 708 | 813 | 623 | 347 | 62 | 2885 |
| Unirrigated | 351 | 663 | 780 | 475 | 221 | 26 | 2517 |
| Total | 340 | 689 | 799 | 560 | 294 | 47 | 2729 |
| **Overall** | | | | | | | |
| Irrigated | 315 | 690 | 772 | 574 | 314 | 90 | 2755 |
| Unirrigated | 373 | 664 | 682 | 443 | 217 | 32 | 2410 |
| Total | 340 | 679 | 733 | 517 | 272 | 65 | 2605 |
| **Non-Bt cotton** | | | | | | | |
| **Small** | | | | | | | |
| Irrigated | 350 | 480 | 494 | 343 | 170 | 27 | 1866 |
| Unirrigated | 270 | 438 | 358 | 327 | 189 | 18 | 1609 |
| Total | 301 | 455 | 411 | 333 | 182 | 21 | 1709 |
| **Medium** | | | | | | | |
| Irrigated | 242 | 371 | 494 | 482 | 128 | 25 | 1742 |
| Unirrigated | 204 | 475 | 532 | 319 | 185 | 23 | 1739 |
| Total | 215 | 446 | 522 | 364 | 170 | 23 | 1740 |
| **Large** | | | | | | | |
| Irrigated | 282 | 530 | 530 | 406 | 159 | 18 | 1924 |
| Unirrigated | 181 | 470 | 565 | 397 | 200 | 32 | 1844 |
| Total | 206 | 485 | 556 | 399 | 190 | 29 | 1864 |
| **Overall** | | | | | | | |
| Irrigated | 302 | 471 | 506 | 397 | 157 | 24 | 1856 |
| Unirrigated | 213 | 462 | 496 | 355 | 193 | 25 | 1747 |
| Total | 240 | 465 | 499 | 368 | 182 | 25 | 1780 |

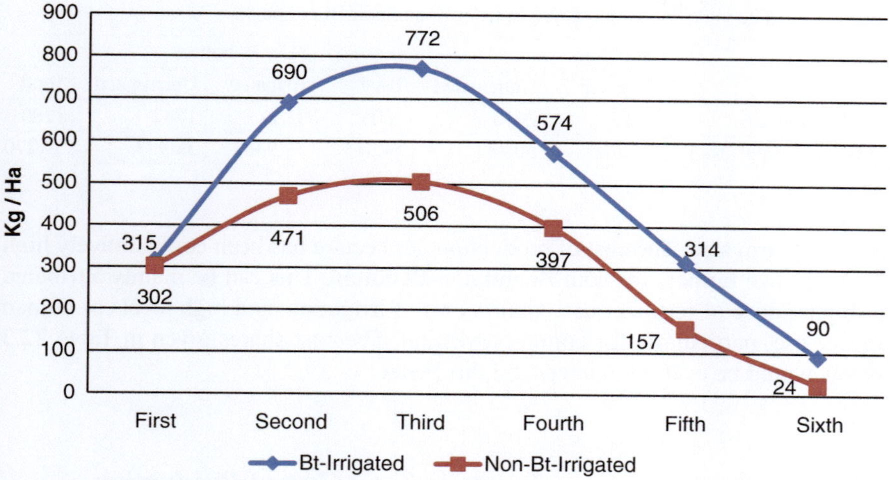


**Fig. 22.1** Average yield per picking: Bt and non-Bt irrigated

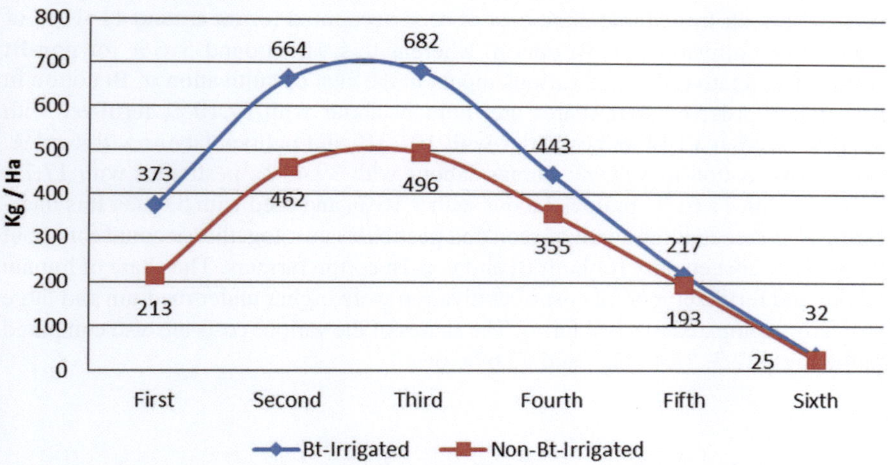


**Fig. 22.2** Average yield per picking: Bt and non-Bt unirrigated

## 22.3 Cost of Production, Value of Output and Profit

In this section we examine the input use pattern under Bt and non-Bt cotton. This has been examined by farm size under irrigated and unirrigated conditions. The total cost of cultivation per hectare, including the marketing cost, were respectively Rs.32368 for Bt cotton and Rs.24102 for non-Bt cotton under irrigated conditions (Table 22.3). The corresponding figures under the unirrigated conditions were respectively Rs.30783 and Rs.22815. Thus the total cost of cultivation of Bt cotton was about 30 % higher compared to non-Bt cotton irrespective of the irrigation status. There was marked difference in the cost of production per hectare under

**Table 22.2** The yield and value of output from Bt and non-Bt cotton

| | Bt cotton | | | Non-Bt cotton | | |
|---|---|---|---|---|---|---|
| | Irrigated | Unirrigated | Total | Irrigated | Unirrigated | Total |
| Yield (Kg/Ha) | 2755 | 2410 | 2605 | 1856 | 1747 | 1780 |
| Value of output (Rs.) | 57,262 | 50,487 | 54,313 | 39,948 | 38,973 | 39,270 |

different farm size. The cost of production per hectare has been comparatively high for large size farmers for both Bt and non-Bt cotton. This can be mainly attributed to higher dose of fertilizer use, greater cost of irrigation, and high levels of human labour use, particularly for cotton harvesting. The cost shares given in Table 22.3 below would be useful to understand this better.

### 22.3.1 Share of Various Inputs in Total Cost of Production

As can be seen from Table 22.4, cost of seed accounted for on around 11–13 % of the cost of cultivation for Bt cotton, whereas this was around 5–6 % for non-Bt cotton. The relative share of various inputs in the cost of cultivation of Bt cotton in descending order of their shares are: human labour with 37.10 %, fertilizer with 22.46 %, seed at 12.18 %, pesticides with 10.23 % and bullock labour with 6.04 %. For non-Bt cotton they were: human labour with 39.43 %, pesticides with 17.75, fertilizer with 17.61 %, bullock labour with 9.16 %, and seed with 5.68 %. It is interesting to notice that the share of seed and pesticides cost together account for about 23 % of the cost equally for both Bt and non-Bt cotton farmers. The share of human labour and fertilizer in total cost of cultivation were higher under medium and large size farms compared to small farms. The shares of the various costs are also compared in the Figs. 22.3, 22.4, 22.5, and 22.6 below.

### 22.3.2 Value of Output and Net Profit

The Table 22.5 and Figs. 22.7, 22.8, 22.9, and 22.10 below give the findings on the value of production, cost of production, and net profit per hectare for Bt and Non-Bt farmer households. The net profit per hectare under Bt cotton was Rs.24894 under irrigated and Rs.19702 under unirrigated conditions. The net profit per hectare under Non-Bt cotton was Rs.14871 under irrigated and Rs.14075 under unirrigated. Thus, both under irrigated and unirrigated conditions the net profits are found to be substantially higher with Bt cotton. There is some positive association with the farm size but even small farmers are able to realize substantial gains in net profits. In percentage terms, under Bt cotton as compared to non-Bt cotton for all sample farmers together, the total cost of production of Bt cotton was higher by 34.3 %

**Table 22.3** Cost of production of Bt and non-Bt cotton among sample farmer households per hectare in rupees

| | Bt cotton | | | Non-Bt cotton | | |
|---|---|---|---|---|---|---|
| | I | UI | Total | I | UI | Total |
| **Small** | | | | | | |
| 1. Seed | 3631 | 3656 | 3644 | 1222 | 1299 | 1269 |
| 2. Human labour | 8515 | 9429 | 8999 | 7098 | 8540 | 7976 |
| 3. Bullock labour | 2427 | 1344 | 1854 | 2433 | 1906 | 2112 |
| 4. Tractor | 958 | 988 | 974 | 536 | 582 | 564 |
| 5. Fertilizer | 5535 | 6789 | 6199 | 2893 | 3469 | 3244 |
| 6. Pesticides | 2859 | 3920 | 3421 | 4471 | 3459 | 3855 |
| 7. Irrigation | 1520 | 0 | 715 | 1400 | 0 | 548 |
| 8. Other operational costs | 247 | 380 | 317 | 55 | 71 | 65 |
| 9. Total operational cost | 25,692 | 26,506 | 26,123 | 20,108 | 19,326 | 19,632 |
| 10. Total marketing cost | 572 | 1245 | 928 | 645 | 1200 | 983 |
| Total cost | 26,264 | 27,750 | 27,051 | 20,753 | 20,528 | 20,616 |
| **Medium** | | | | | | |
| 1. Seed | 3800 | 4338 | 4011 | 1200 | 1365 | 1319 |
| 2. Human labour | 10,998 | 11,810 | 11,316 | 9513 | 9126 | 9234 |
| 3. Bullock labour | 1752 | 2158 | 1911 | 2506 | 2281 | 2344 |
| 4. Tractor | 1183 | 1019 | 1119 | 1597 | 532 | 828 |
| 5. Fertilizer | 7122 | 6589 | 6913 | 4095 | 4753 | 4570 |
| 6. Pesticides | 2626 | 3637 | 3022 | 4074 | 3969 | 3998 |
| 7. Irrigation | 1935 | 0 | 1178 | 1620 | 0 | 450 |
| 8. Other operational costs | 300 | 310 | 304 | 346 | 57 | 137 |
| 9. Total operational cost | 29,716 | 29,861 | 29,773 | 24,951 | 22,083 | 22,880 |
| 10. Total marketing cost | 1245 | 1282 | 1259 | 841 | 1330 | 1194 |
| Total cost | 30,959 | 31,142 | 31,031 | 25,791 | 23,414 | 24,074 |
| **Large** | | | | | | |
| 1. Seed | 3808 | 3932 | 3860 | 1293 | 1383 | 1361 |
| 2. Human labour | 13,244 | 12,712 | 13,019 | 10,840 | 9801 | 10,061 |
| 3. Bullock labour | 2010 | 1835 | 1936 | 2900 | 1694 | 1996 |
| 4. Tractor | 984 | 975 | 980 | 738 | 885 | 848 |
| 5. Fertilizer | 7741 | 7327 | 7566 | 4259 | 4150 | 4177 |
| 6. Pesticides | 3127 | 3504 | 3286 | 4014 | 4548 | 4415 |
| 7. Irrigation | 2203 | 0 | 1273 | 1710 | 0 | 428 |
| 8. Other operational costs | 390 | 299 | 352 | 388 | 82 | 159 |
| 9. Total operational cost | 33,507 | 30,584 | 32,273 | 26,142 | 22,543 | 23,443 |
| 10. Total marketing cost | 1141 | 1131 | 1137 | 741 | 1059 | 980 |
| Total cost | 35,006 | 32,053 | 33,759 | 27,200 | 23,969 | 24,777 |
| **Overall** | | | | | | |
| 1. Seed | 3776 | 3963 | 3857 | 1240 | 1354 | 1319 |
| 2. Human labour | 11,801 | 11,694 | 11,754 | 8920 | 9251 | 9150 |
| 3. Bullock labour | 2004 | 1794 | 1913 | 2606 | 1915 | 2125 |

(continued)

**Table 22.3** (continued)

| | Bt cotton | | | Non-Bt cotton | | |
|---|---|---|---|---|---|---|
| | I | UI | Total | I | UI | Total |
| 4. Tractor | 1037 | 989 | 1016 | 856 | 701 | 748 |
| 5. Fertilizer | 7193 | 7017 | 7116 | 3984 | 4131 | 4086 |
| 6. Pesticides | 2936 | 3638 | 3242 | 4224 | 4074 | 4120 |
| 7. Irrigation | 2011 | 0 | 1136 | 1556 | 0 | 474 |
| 8. Other operational costs | 340 | 321 | 332 | 235 | 72 | 122 |
| 9. Total operational cost | 31,098 | 29,416 | 30,366 | 23,621 | 21,498 | 22,144 |
| 10. Total marketing cost | 1271 | 1369 | 1314 | 830 | 1334 | 1181 |
| Total cost | 32,368 | 30,785 | 31,679 | 24,102 | 22,815 | 23,207 |

**Table 22.4** Percentage share of various input costs in the total cost of production

| | Bt cotton | | | Non-Bt cotton | | |
|---|---|---|---|---|---|---|
| | I | UI | Total | I | UI | Total |
| **Small** | | | | | | |
| 1. Seed | 13.83 | 13.17 | 13.47 | 5.89 | 6.33 | 6.16 |
| 2. Human labour | 32.42 | 33.98 | 33.27 | 34.20 | 41.60 | 38.69 |
| 3. Bullock labour | 9.24 | 4.84 | 6.85 | 11.72 | 9.28 | 10.24 |
| 4. Tractor | 3.65 | 3.56 | 3.60 | 2.58 | 2.84 | 2.74 |
| 5. Fertilizer | 21.07 | 24.46 | 22.92 | 13.94 | 16.90 | 15.74 |
| 6. Pesticides | 10.89 | 14.13 | 12.65 | 21.54 | 16.85 | 18.70 |
| 7. Irrigation | 5.79 | 0.00 | 2.64 | 6.75 | 0.00 | 2.66 |
| 8. Other operational costs | 0.94 | 1.37 | 1.17 | 0.27 | 0.35 | 0.32 |
| 9. Total operational cost | 97.82 | 95.52 | 96.57 | 96.89 | 94.14 | 95.23 |
| 10. Total marketing cost | 2.18 | 4.49 | 3.43 | 3.11 | 5.85 | 4.77 |
| Total cost | 100 | 100 | 100 | 100 | 100 | 100 |
| **Medium** | | | | | | |
| 1. Seed | 12.27 | 13.93 | 12.93 | 4.65 | 5.83 | 5.48 |
| 2. Human labour | 35.52 | 37.92 | 36.47 | 36.88 | 38.98 | 38.36 |
| 3. Bullock labour | 5.66 | 6.93 | 6.16 | 9.72 | 9.74 | 9.74 |
| 4. Tractor | 3.82 | 3.27 | 3.61 | 6.19 | 2.27 | 3.44 |
| 5. Fertilizer | 23.00 | 21.16 | 22.28 | 15.88 | 20.30 | 18.98 |
| 6. Pesticides | 8.48 | 11.68 | 9.74 | 15.80 | 16.95 | 16.61 |
| 7. Irrigation | 6.25 | 0.00 | 3.80 | 6.28 | 0.00 | 1.87 |
| 8. Other operational costs | 0.97 | 1.00 | 0.98 | 1.34 | 0.24 | 0.57 |
| 9. Total operational cost | 95.99 | 95.89 | 95.95 | 96.74 | 94.32 | 95.04 |
| 10. Total marketing cost | 4.02 | 4.12 | 4.06 | 3.26 | 5.68 | 4.96 |
| Total cost | 100 | 100 | 100 | 100 | 100 | 100 |
| **Large** | | | | | | |
| 1. Seed | 10.88 | 12.27 | 11.43 | 4.75 | 5.77 | 5.49 |
| 2. Human labour | 37.83 | 39.66 | 38.56 | 39.85 | 40.89 | 40.61 |
| 3. Bullock labour | 5.74 | 5.72 | 5.73 | 10.66 | 7.07 | 8.06 |
| 4. Tractor | 2.81 | 3.04 | 2.90 | 2.71 | 3.69 | 3.42 |

(continued)

**Table 22.4** (continued)

| | Bt cotton | | | Non-Bt cotton | | |
|---|---|---|---|---|---|---|
| | I | UI | Total | I | UI | Total |
| 5. Fertilizer | 22.11 | 22.86 | 22.41 | 15.66 | 17.31 | 16.86 |
| 6. Pesticides | 8.93 | 10.93 | 9.73 | 14.76 | 18.97 | 17.82 |
| 7. Irrigation | 6.29 | 0.00 | 3.77 | 6.29 | 0.00 | 1.73 |
| 8. Other operational costs | 1.11 | 0.93 | 1.04 | 1.43 | 0.34 | 0.64 |
| 9. Total operational cost | 95.72 | 95.42 | 95.60 | 96.11 | 94.05 | 94.62 |
| 10. Total marketing cost | 3.26 | 3.53 | 3.37 | 2.72 | 4.42 | 3.96 |
| Total cost | 100 | 100 | 100 | 100 | 100 | 100 |
| **Overall** | | | | | | |
| 1. Seed | 11.67 | 12.87 | 12.18 | 5.14 | 5.93 | 5.68 |
| 2. Human labour | 36.46 | 37.99 | 37.10 | 37.01 | 40.55 | 39.43 |
| 3. Bullock labour | 6.19 | 5.83 | 6.04 | 10.81 | 8.39 | 9.16 |
| 4. Tractor | 3.20 | 3.21 | 3.21 | 3.55 | 3.07 | 3.22 |
| 5. Fertilizer | 22.22 | 22.79 | 22.46 | 16.53 | 18.11 | 17.61 |
| 6. Pesticides | 9.07 | 11.82 | 10.23 | 17.53 | 17.86 | 17.75 |
| 7. Irrigation | 6.21 | 0.00 | 3.59 | 6.46 | 0.00 | 2.04 |
| 8. Other operational costs | 1.05 | 1.04 | 1.05 | 0.98 | 0.32 | 0.53 |
| 9. Total operational cost | 96.08 | 95.55 | 95.86 | 98.00 | 94.23 | 95.42 |
| 10. Total marketing cost | 3.93 | 4.45 | 4.15 | 3.44 | 5.85 | 5.09 |
| Total cost | 100 | 100 | 100 | 100 | 100 | 100 |

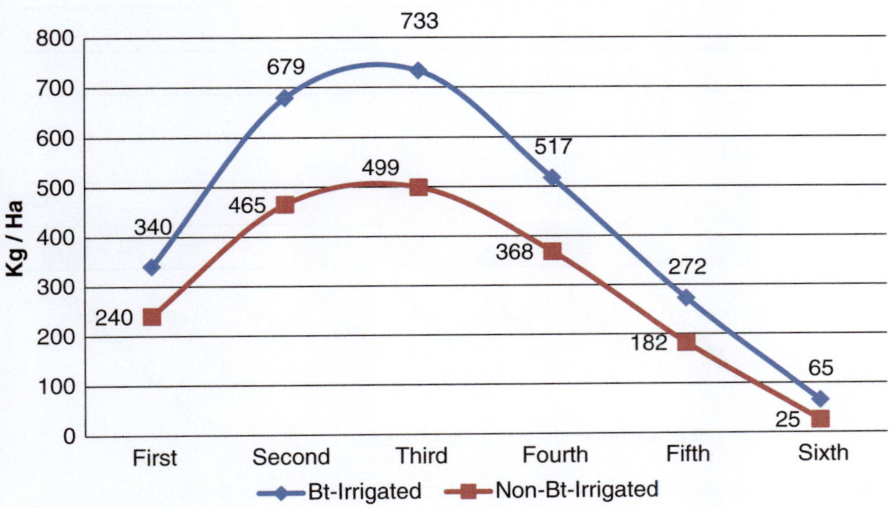


**Fig. 22.3** Average yield per picking: Bt and non-Bt overall

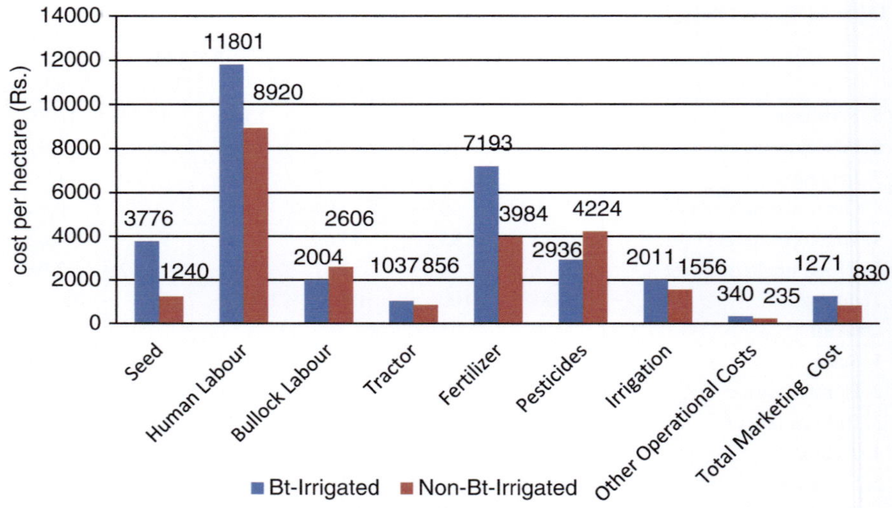


**Fig. 22.4** Per hectare cost of production: Bt and non-Bt under irrigated conditions

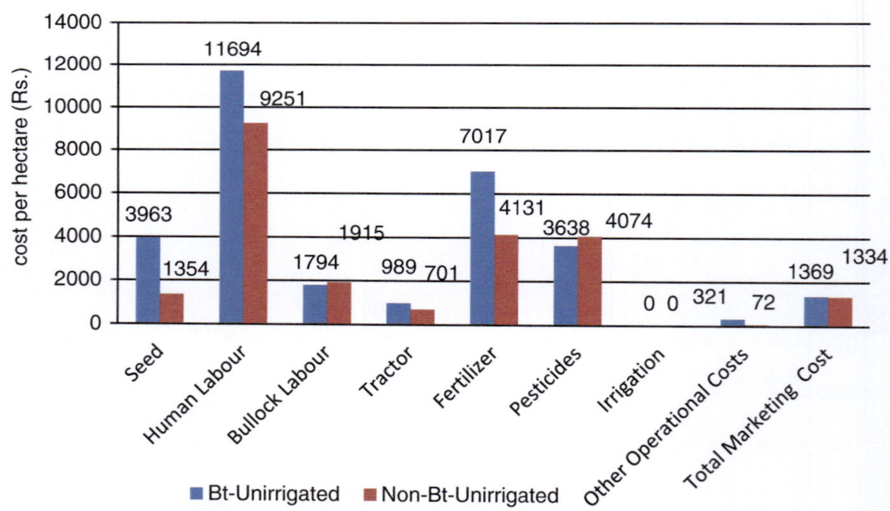


**Fig. 22.5** Per hectare cost of production: Bt and non-Bt under unirrigated conditions

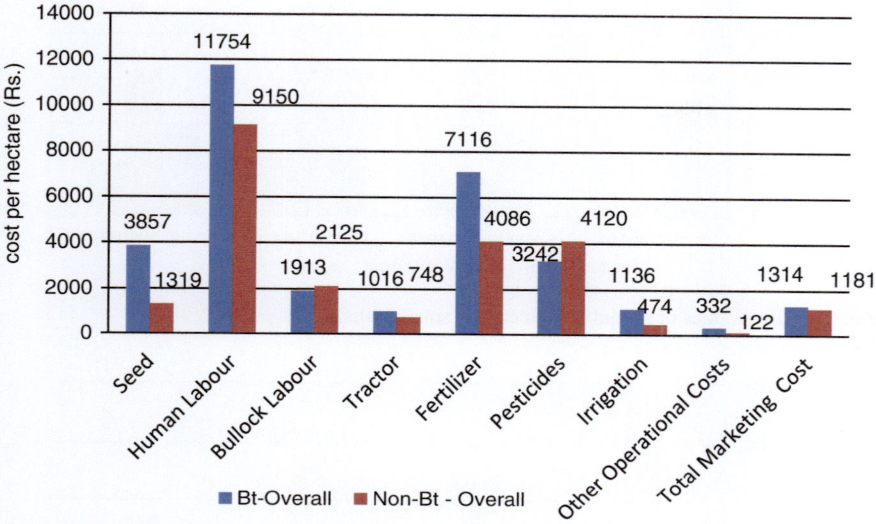


**Fig. 22.6** Per hectare cost of production: Bt and non-Bt overall

**Table 22.5** Economics of Bt cotton over non-Bt cotton among sample farmer households (rupees per hectare)

| | Bt cotton | | | Non-Bt cotton | | | Bt over non-Bt cotton (%) | | |
|---|---|---|---|---|---|---|---|---|---|
| | I | UI | Total | I | UI | Total | I | UI | Total |
| **Small** | | | | | | | | | |
| Total cost | 26,264 | 27,750 | 27,051 | 20,753 | 20,528 | 20,616 | 26.6 | 35.2 | 31.2 |
| Value of output | 49,210 | 42,206 | 45,502 | 36,276 | 32,118 | 33,745 | 35.7 | 31.4 | 34.8 |
| Net profit | 22,946 | 14,456 | 18,451 | 15,523 | 11,590 | 13,129 | 47.8 | 24.7 | 40.5 |
| **Medium** | | | | | | | | | |
| Total cost | 30,959 | 31,142 | 31,031 | 25,791 | 23,414 | 24,074 | 20.0 | 33.0 | 28.9 |
| Value of output | 56,160 | 47,955 | 52,949 | 39,158 | 37,090 | 37,664 | 43.4 | 29.3 | 40.6 |
| Net profit | 25,201 | 16,813 | 21,918 | 13,367 | 13,676 | 13,590 | 88.5 | 22.9 | 61.3 |
| **Large** | | | | | | | | | |
| Total cost | 35,006 | 32,053 | 33,759 | 27,200 | 23,969 | 24,777 | 28.7 | 33.7 | 36.3 |
| Value of output | 60,333 | 55,609 | 58,338 | 42,309 | 39,948 | 40,538 | 42.6 | 39.2 | 43.9 |
| Net profit | 25,327 | 23,556 | 24,579 | 15,109 | 15,979 | 15,761 | 67.6 | 47.4 | 55.9 |
| **Overall** | | | | | | | | | |
| Total cost | 32,368 | 30,785 | 31,679 | 24,102 | 22,815 | 23,207 | 34.3 | 34.9 | 36.5 |
| Value of output | 57,262 | 50,487 | 54,313 | 38,973 | 36,890 | 37,524 | 46.9 | 36.9 | 44.7 |
| Net profit | 24,894 | 19,702 | 22,634 | 14,871 | 14,075 | 14,317 | 67.4 | 40.0 | 58.1 |

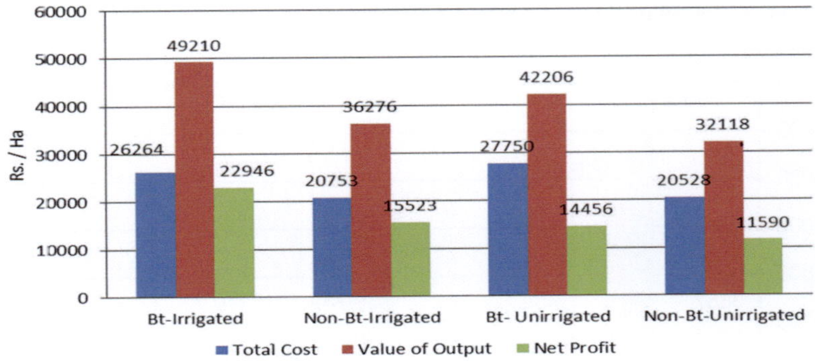


**Fig. 22.7** Economics of Bt and non-Bt cotton: small farms

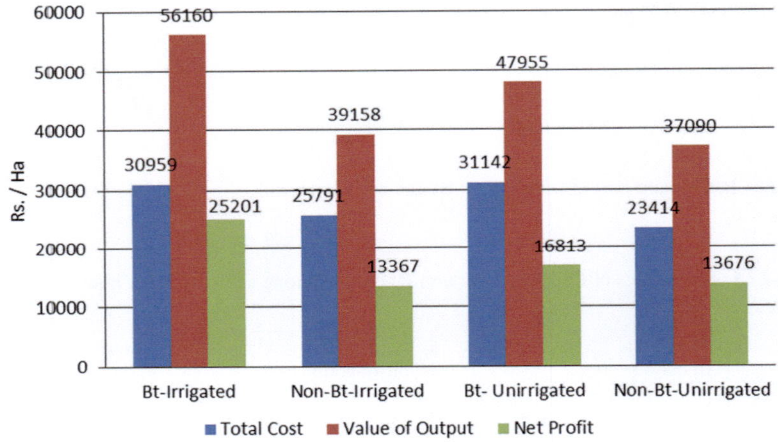


**Fig. 22.8** Economics of Bt and non-Bt cotton: medium farms

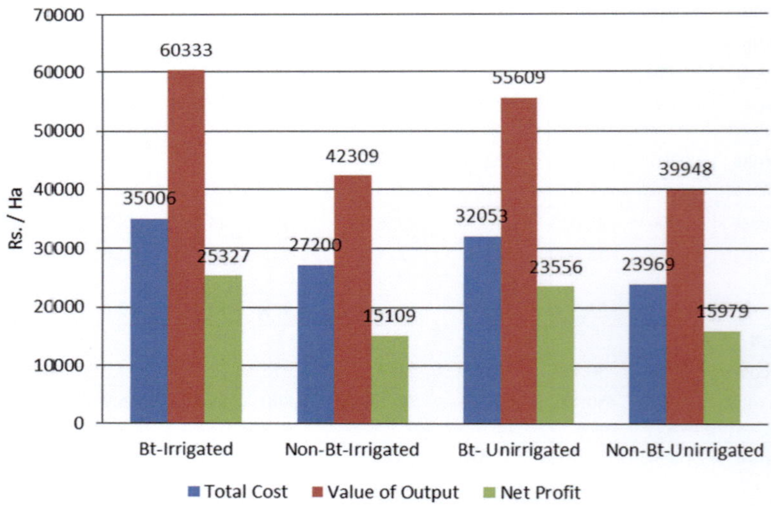


**Fig. 22.9** Economics of Bt and non-Bt cotton: large farms

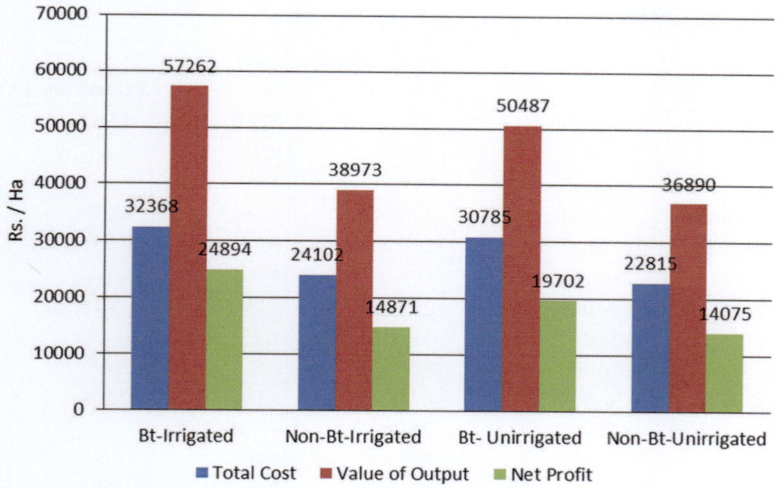


**Fig. 22.10** Economics of Bt and non-Bt cotton: overall

under irrigated and 34.9 % under unirrigated conditions. The value of output of Bt cotton was higher by 46.9 % under irrigated area and 36.9 % under unirrigated area. The net profit of Bt cotton under irrigated and unirrigated area as compared to those under non-Bt cotton was 67.4 % and 40 % higher respectively. Thus, even though the cost of production is higher with Bt cotton, the value of production and net profits are substantially higher in Bt cotton as compared to non-Bt.

## 22.4 Econometric Analysis of the Performance of Bt Cotton vs. Non-Bt Cotton

Given the variation in the sample, it is important to examine the statistical significance of impact of Bt on the economics of cotton cultivation. This has first been examined through a regression approach relating yield with a dummy variable for Bt cotton, which is 1 for Bt cotton and 0 for Non-Bt cotton. The results would be identical to those obtained through analysis of variance (ANOVA). The results (Table 22.6) below indicate that Bt cotton clearly has a statistically significant impact on the yield, significant at the 99 % level, and the mean impact is estimated to be 43 %. The seed cost increases and the pesticide cost reduces, but the impact on the total cost is relatively low and not statistically significant. Pesticide cost is reduces by 22.38 %, and the profit increase is 120.08 % and statistically highly significant at 99 %. This indicates that the technology is highly profitable.

Even though Bt appears to have a dominant effect, perhaps it is also pulling in other inputs to boost the profitability. The performance can also be examined through a function including various inputs and factors.

**Table 22.6** Regression results: impact of Bt cotton N = 154

| Dependent variable | | Independent variables | | Percent impact of Bt |
|---|---|---|---|---|
| | | Constant | Bt | |
| Yield | Coefficient | 1821 | 777.01 | 42.67 |
| | t-stat | 29.04 | 9.21 | |
| | Signifi. | *** | *** | |
| Value of output | Coefficient | 38,944 | 16,663 | 42.79 |
| | t-stat | 25.29 | 8.04 | |
| | Signifi. | *** | *** | |
| Total cost | Coefficient | 26,198 | 1357.71 | 5.18 |
| | t-Stat | 31.63 | 1.22 | |
| | Signifi. | *** | NS | |
| Pesticide cost | Coefficient | 8241.22 | −1844.21 | −22.38 |
| | t-Stat | 22.96 | −3.82 | |
| | Signifi. | *** | *** | |
| Seed cost | Coefficient | 1319.28 | 2487.20 | 188.53 |
| | t-Stat | 38.42 | 53.81 | |
| | Signifi. | *** | *** | |
| Price | Coefficient | 21.36 | −0.0415 | −0.1943 |
| | t-Stat | 89.92 | −0.13 | |
| | Signifi. | *** | NS | |
| Profit | Coefficient | 12,746 | 15,305 | 120.08 |
| | t-Stat | 8.18 | 7.29 | |
| | Signifi. | *** | *** | |

*NS* not significant
*significant at 90 %, **significant at 95 %, ***significant at 99 %

### 22.4.1 Econometric Analysis of the Determinants of Yield, Value of Output and Profit

This is examined by using regression models of both linear and logarithmic forms. The determinant variables used based on the information obtained in the survey includes the costs of seeds, fertilizers, pesticides, human labour, and farm power (including tractor, bullock, and irrigation). The irrigation costs could not be considered separately due to data limitations. The results of the linear and logarithmic models are presented in Tables 22.7 and 22.8 below respectively.

**Table 22.7** Contribution of various factors to yield, value of production and profit of cotton: linear model

| Dependent variable | Particulars | Constant | Bt cotton dummy | Seed cost | Fertilizer cost | Pesticide cost | Human labour cost | Farm power cost[a] |
|---|---|---|---|---|---|---|---|---|
| | | All sample farmer households | | | | | | |
| Yield | Coefficient | 761.09 | 948.39 | −0.109 | 0.136 | 0.003 | 0.062 | 0.009 |
| | t-Value | 3.97 | 3.55 | −1.04 | 9.31 | 0.33 | 3.26 | 0.28 |
| | Significance | *** | *** | NS | *** | NS | *** | NS |
| | N | 154 | | | | | | |
| | Adjusted $R^2$ | 0.71 | | | | | | |
| | F-Statistics | 62.01 | | | | | | |
| Value of output | Coefficient | 9296.5 | 15,235.89 | −0.38 | 3.125 | −0.005 | 1.778 | 0.584 |
| | t-Value | 1.99 | 2.34 | −0.15 | 8.77 | −0.02 | 3.82 | 0.786 |
| | Significance | ** | ** | NS | *** | NS | *** | NS |
| | N | 154 | | | | | | |
| | Adjusted $R^2$ | 0.68 | | | | | | |
| | F-Statistics | 55.81 | | | | | | |
| Profit | Coefficient | 9601.96 | 14,914.28 | −1.336 | 2.061 | −1.007 | 0.648 | −0.583 |
| | t-Value | 2.07 | 2.31 | −0.524 | 5.83 | −4.19 | 1.39 | −0.79 |
| | Significance | ** | ** | NS | *** | *** | NS | NS |
| | N | 154 | | | | | | |
| | Adjusted $R^2$ | 0.57 | | | | | | |
| | F-Statistics | 34.76 | | | | | | |

*90 %, **95 %, ***99 %
[a]Includes cost of irrigation

The results indicate that Bt even by itself has a positive and statistically significant impact on the yield. The impacts of fertilizers and human labour are also positive and statistically significant. Though the cost of seed is negatively associated with the yield, it was statistically not significant – perhaps due to multi-collinearity with Bt. Although, the cost of pesticides and farm power are positive associated, but they are statistically non-significant. These results remain similar for the value of output. While the cost of pesticides has a strong and negative influence in determining the profit levels, fertilizer and Bt variety have a strong and positive influence. The overall results based on the linear and logarithmic models are very similar – both confirm the statistically significant and positive association of Bt cotton with the yield, revenue and profits in cotton.

**Table 22.8** Contribution of various factors to yield, value of production and profit of cotton: logarithmic model

| Dependent variable | Particulars | Constant | Bt cotton dummy | Seed cost | Fertilizer cost | Pesticide cost | Human labour cost | Farm power cost # |
|---|---|---|---|---|---|---|---|---|
| | | All sample farmer households | | | | | | |
| Yield | Coefficient | 3.87 | 0.45 | −0.11 | 0.33 | 0.01 | 0.16 | 0.01 |
| | t-Value | 4.95 | 4.18 | −1.15 | 7.02 | 0.24 | 3.56 | 0.19 |
| | Significance | *** | *** | NS | *** | NS | *** | NS |
| | N | 154 | | | | | | |
| | Adjusted R2 | 0.66 | | | | | | |
| | F-Statistics | 50.83 | | | | | | |
| Value of output | Coefficient | 6.04 | 0.43 | −0.10 | 0.36 | −0.01 | 0.22 | 0.04 |
| | t-Value | 6.84 | 3.58 | −0.95 | 6.69 | −0.33 | 4.29 | 0.96 |
| | Significance | *** | *** | NS | *** | NS | *** | NS |
| | N | 154 | | | | | | |
| | Adjusted R2 | 0.64 | | | | | | |
| | F-Statistics | 45.80 | | | | | | |
| Profit | Coefficient | 11.70 | 1.35 | −0.50 | 0.45 | −0.37 | 0.15 | −0.10 |
| | t-Value | 3.14 | 2.64 | −1.08 | 1.98 | −2.59 | 0.69 | −0.52 |
| | Significance | *** | *** | NS | ** | *** | NS | NS |
| | N | 154 | | | | | | |
| | Adjusted R2 | 0.32 | | | | | | |
| | F-Statistics | 12.80 | | | | | | |

*90%, **95%, ***99%
#Includes cost of irrigation

# Chapter 23
# Farmers' Perceptions on Various Features of Bt Cotton in Maharashtra

This chapter examines the findings on various perceptions of the sample farmers' households growing Bt cotton on various aspects including economic benefits, availability of Bt seeds, their characteristic features, environmental impact, extension activities, and so on.

## 23.1 Some Basic Features of Bt Cotton: Views of Bt Cotton Sample Farmer Households

### *23.1.1 Adoption of Bt Cotton: Some Determining Factors*

In this section we have examined the determinants of adoption of Bt Cotton. The stratified nature of the sample is a limitation to this analysis. We have related different characteristics of the sample farmers to the adoption/non-adoption of Bt Cotton. The characteristics include farm size, age, education, experience in cotton cultivation and the area irrigated. A simple linear regression model is used here. The results presented in Table 23.1 could be summarized as follows. The farm size variable is found to be not significant and in fact it has a negative sign, which indicates that the adoption is scale neutral and not biased towards large farm size. Age of the head of household is also found not significant. However, education, experience with cotton, and irrigated area show positive and significant associations, indicating that that they are positively associated with the adoption of Bt cotton.

Table 23.2 presents responses of sample farmers' households about various factors that help determine the adoption. The responses indicate that Bt cotton growers were not approached or pressured by private sales agents for the promotion of Bt cotton. As regards to plant size and boll size, the respondents did not indicate any major difference compared to non-Bt cotton. About 96.5 % of the sample farmer


V.P. Gandhi, D. Jain, *Introduction of Biotechnology in India's Agriculture*, DOI 10.1007/978-981-10-1091-0_23

**Table 23.1** Adoption of Bt cotton in relation to characteristics of farmer households

| Independent variables | Coefficient | t-Value | Level of significance |
|---|---|---|---|
| Constant | 0.1182 | 0.56 | NS |
| Farm size | –0.0054 | –0.56 | NS |
| Age of head of household | 0.0009 | 0.03 | NS |
| Experience in cotton cultivation | 0.0116 | 2.20 | ** |
| Level of education | 0.1073 | 2.84 | *** |
| Irrigated area | 0.0306 | 2.19 | ** |

*NS* non-significant
** 90 %, *** 99 %

**Table 23.2** Response of Bt cotton sample farmer households on various aspects

| Particulars | Yes | No | No opinion |
|---|---|---|---|
| 1. Did any private sales agent approach you for promoting Bt cotton? | 0 | 100 | 0 |
| 2. Is Bt cotton plant shorter? | 0 | 100 | 0 |
| 3. Does Bt cotton have smaller bolls? | 0 | 100 | 0 |
| 4. Does Bt cotton give lesser number of cotton pickings? | 0 | 96.5 | 3.5 |
| 5. Did any government agency approach you for inspecting the cotton variety you have sown? | 0 | 100 | 0 |
| 6. Do you face any problem in marketing of Bt cotton? | 0 | 100 | 0 |
| 7. In your opinion is Bt cotton is more pest resistant than non-Bt? | 82.4 | 14.1 | 3.5 |
| 8. Do you need to buy Bt cotton seed every year? | 100 | 0 | 0 |
| 9. Is Bt cotton seed easily available | 97.6 | 2.4 | 0 |
| 10. Will you continue with Bt cotton cultivation? | 94.1 | 2.4 | 2.4 |
| 11. Do you feel that the pest/insect attack on other crop is higher or lower, when Bt cotton is cultivated? | 0 | 0 | 100 |
| 12. Have you observed any adverse effect on the environment due to Bt cotton cultivation? | 0 | 100 | 0 |

households indicated that the number of pickings under Bt and non-Bt cotton is same. No government agencies had approached them for the inspection of Bt cotton, and none of the sample households felt specific problems with respect to the marketing of Bt cotton fibre. None of the Bt growers had observed any adverse environmental impact as a result of the cultivation of Bt cotton. They also did not indicate any increase in the pest attack on other crops as a result of the cultivation of Bt cotton. All the sample farmers indicated that they need to buy Bt seed every year for cultivation. Almost 98 % of the sample farmers did not face any difficulty in getting quality Bt seeds in time. As high as 94.1 % of the sample farmers were positive on continuing with Bt cultivation in the future.

**Table 23.3** Bt cotton farmer households' response to "Who Recommended" in growing Bt cotton (percentage)

| | |
|---|---|
| Per cent of farmers responding | 95.30 |
| 1. Extension worker | 2.47 |
| 2. Fellow farmer | 50.62 |
| 3. Village leader | 7.41 |
| 4. Village cooperative | 7.41 |
| 5. Seed company | 20.99 |
| 6. Seed dealer | 11.11 |

**Table 23.4** Advantages of Bt seed as opposed to traditional cotton conveyed by agents referred in above table (percentage)

| | |
|---|---|
| Per cent of farmers responding | 83.53 |
| 1. More profit | 73.24 |
| 2. Less pesticides spraying | 66.20 |
| 3. No boll shedding | 15.49 |
| 4. Comparatively more bolls | 59.15 |

#### 23.1.1.1 Awareness and Adoption of Bt Cotton

Over 50.6 % of the sample households adopted Bt seeds with the recommendation of fellow farmers, and another major sources of information was seed company/dealer (Table 23.3). Some farmers came to know about the Bt technology from the village cooperatives and the village leaders. But government extension agencies did not play much role in creating farmer awareness of Bt cotton. The main communication was about its superiority in terms of better profits and lesser amount of pesticides sprays, more bolls per plant, no bolls shedding etc. (Table 23.4).

### *23.1.2 Agronomic Features of Bt Cotton: Response of Bt Growers*

All the sample farmer households reported early flowering of Bt cotton compared to non-Bt cotton. About 50.6 % of them reported that Bt cotton flowers 1–10 days early, 36.5 % reported it at 10–20 days earlier, and 12.9 % reported that Bt cotton flowers more than 20 days earlier than non-Bt cotton (Table 23.5).

The sample farmers were asked about the reasons behind their preference and comparison of Bt cotton over non-Bt cotton. Over 74 % of the households responded to this question (Table 23.6). Better yield, more bolls, and less pest attack were the major reasons expressed by them. On the other hand, 83 % of them expressed that the cost of Bt cotton seed was very high.

**Table 23.5** Farmers response on early flowering of Bt cotton

| | |
|---|---|
| Per cent of farmers responding early flowering | 100 |
| 1–10 Days | 50.59 |
| 10–20 Days | 36.47 |
| Above 20 Days | 12.94 |

**Table 23.6** Response on comparison of Bt vis-à-vis traditional varieties by sample farmer households

| | |
|---|---|
| Per cent of farmers responding | 74.12 |
| 1. More bolls | 66.23 |
| 2. Better yield | 71.56 |
| 3. Less pest attack | 30.00 |
| 4. High cost of seeds | 82.56 |

## 23.2 Farmers' Perception on Advantages or Disadvantages of Bt Cotton vis-à-vis Non-Bt Cotton

The Bt cotton sample farmer households were specifically asked about their opinion on advantages and disadvantages of Bt cotton. The major advantages of Bt cotton that are expressed by majority of the sample farmer households are yield superiority, more profit, lesser need of pesticides, better quality, and its suitability for early sowing (Table 23.7). On the other hand a common disadvantage expressed was the high cost of seed. Other issues expressed include higher fertilizer and irrigation cost, and higher harvest cost. No major differences were observed in other matters.

## 23.3 Measures Suggested by Sample Farmers to Improve the Acceptance of Bt Cotton

The sample farmers were asked about their suggestion to improve the Bt technology and 74.12 % of the sample households responded to it (Table 23.8). The most important suggestion given was to reduce the cost of Bt cotton seed. Other suggestions in order of their importance are: arranging field demonstrations, seed packages with smaller quantities, and assurance of seed quality.

**Table 23.7** Advantages or disadvantages of Bt Cotton-G vis-à-vis non-Bt cotton reported by Bt cotton-growers: percentage

| | | Strong advantage | Advantage | No difference | Disadvantage | Strong disadvantage |
|---|---|---|---|---|---|---|
| 1 | Availability of seeds | 0.0 | 11.0 | 80.5 | 8.5 | 0.0 |
| 2 | Seed cost/price | 0.0 | 3.7 | 2.5 | 35.8 | 58.0 |
| 3 | Quality of avail. Seeds | 4.8 | 46.4 | 44.0 | 4.8 | 0.0 |
| 4 | Pest Incidence/ problem | 3.6 | 58.3 | 23.8 | 14.3 | 0.0 |
| 5 | Pesticide need/ cost | 8.3 | 60.7 | 23.8 | 6.0 | 1.2 |
| 6 | Fertilizer need/ cost | 1.2 | 15.9 | 47.6 | 35.4 | 0.0 |
| 7 | Labour need/ cost | 2.4 | 8.4 | 73.5 | 15.7 | 0.0 |
| 8 | Machine need/ cost | 2.4 | 6.0 | 89.3 | 1.2 | 1.2 |
| 9 | Irrigation need/ cost | 4.8 | 12.0 | 45.8 | 37.3 | 0.0 |
| 10 | Harvesting cost | 2.4 | 6.0 | 56.0 | 34.5 | 1.2 |
| 11 | Cotton quality | 4.8 | 51.8 | 41.0 | 2.4 | 0.0 |
| 12 | Market preference | 2.4 | 16.7 | 73.8 | 6.0 | 1.2 |
| 13 | Staple length | 2.4 | 34.9 | 50.6 | 12.0 | 0.0 |
| 14 | Fibre colour | 2.4 | 16.7 | 73.8 | 6.0 | 1.2 |
| 15 | Cotton price | 2.4 | 2.4 | 88.0 | 7.2 | 0.0 |
| 16 | Easy marketing | 2.4 | 15.7 | 74.7 | 7.2 | 0.0 |
| 17 | By-product output | 2.4 | 9.6 | 88.0 | 0.0 | 0.0 |
| 18 | Yield | 14.5 | 81.9 | 2.4 | 1.2 | 0.0 |
| 19 | Profit | 12.0 | 77.1 | 9.6 | 1.2 | 0.0 |
| 20 | Livestock feeding | 2.4 | 7.2 | 90.4 | 0.0 | 0.0 |
| 21 | Water saving | 8.5 | 20.7 | 53.7 | 17.1 | 0.0 |
| 22 | Suitable for early sowing | 14.6 | 58.5 | 26.8 | 0.0 | 0.0 |
| 23 | Suitable for late sowing | 2.4 | 2.4 | 79.3 | 15.9 | 0.0 |

**Table 23.8** Suggestion by Bt cotton farmer households for improving the acceptance of Bt cotton

| | |
|---|---|
| Per cent of farmers responding | 74.12 |
| 1. Reduce seed cost | 53.97 |
| 2. Seed packages with less quantity seeds | 14.29 |
| 3. Field demonstration | 33.33 |
| 4. Assurance of seed quality | 6.35 |

# Part VI
# State Study 4: Tamil Nadu

The part is based on research study done by A Pushpavalli, Agro-Economic Research Centre University of Madras, Chennai, under the guidance of Vasant P. Gandhi CMA-IIMA, and the Head of the Agro-Economic Research Centre.

# Chapter 24
# Cotton Cultivation in Tamil Nadu

## 24.1 Area Under Cotton in Tamil Nadu

During 1980–1981 the area under cotton was 220,524 ha which represented 3.41 % of the total cropped area in the State. It was 254,345 ha in 1985–1986 which was 3.57 % of the total cropped area. The area under cotton in the year 1990–1991 was 239,282 ha which formed 3.61 % of the total cropped area. The area further increased to 252,197 ha in 1995–1996 and drastically decreased to 164,169 ha in 2000–2001 which represented 3.91 % and 2.64 % of the total cropped area. The maximum area under cotton was witnessed in the year 1989–1990 (281,000 ha) and the minimum area was 164,169 ha in 2001–2002 during the period under study.

## 24.2 Production of Cotton

The production of cotton was 260,240 bales in 1980–1981 and it had increased to 484,710 bales in 1985–1986. During 1990–1991 the production had decreased to 408,710 bales which further had decreased to 339,350 ha in 1995–1996 and 229,730 bales in 2001–2002. The maximum production was seen in the year 1984–1985 (585,000 bales) and the minimum production was noticed in the year 2001–2002 (229,730 bales) during the period under study.


V.P. Gandhi, D. Jain, *Introduction of Biotechnology in India's Agriculture*, DOI 10.1007/978-981-10-1091-0_24

## 24.3 Productivity of Cotton

The productivity of cotton was 201 kg in 1980–1981 and it had increased to 324 kg in 1985–1986. During 1990–1991 it had decreased to 290 kg and further decreased to 221 kg in 1995–1996. It was 238 kg in the year 2001–2002. The maximum yield was observed in the year 1984–1985 (393 kg) and the minimum yield was seen in the year 1980–1981 during the study period.

## 24.4 Growth Rates of Cotton in Tamil Nadu

The growth rates were calculated for the data on area, production and productivity of cotton in Tamil Nadu during 1980–1981 to 2001–2002. They were analysed and growth rates were calculated for three periods. Period I referred to 1980–1981 to 1989–1990; period II referred to 1990–1991 to 2001–2002 and period III referred to 1980–1981 to 2001–2002.

Table 24.1 gives the growth rates of cotton for past decades. The Table shows that the growth rates of area, production, and productivity were found to be positive during 1980–1981 to 1989–1990. But on the other hand, during 1990–1991 to 2001–2002 the growth rates of area, production and productivity were found to be negative. For the whole period under study, i.e., 1980–1981 to 2001–2002, it was observed that except area, the growth rates of production and productivity were found to be positive. It was 0.26% and 0.70% for production and productivity respectively (see Table 24.1).

## 24.5 Decomposition of Change of Output

In order to find out the important factors for the change of output, decomposition analysis was used. The change of cotton production has been decomposed using Minhas Seven Factor Model.

**Table 24.1** Growth rates (%) of cotton in Tamil Nadu

| | Period I | Period II | Period III |
|---|---|---|---|
| Area | 3.00 | –3.89 | –0.44 |
| Production | 7.64 | –3.97 | 0.26 |
| Productivity | 4.50 | –0.08 | 0.70 |

The model is

$$
\begin{aligned}
P_t - P_o = {} & (A_t - A_o)\Sigma w_i C_{io} Y_{io} + A_o \Sigma w_i (Y_{it} - Yio) \\
& + A_o \Sigma w_i Yio(C_{it} - C_{io}) \\
& + (A_t - A_o)\Sigma w_i C_{io}(Y_{it} - Yio) \\
& + (A_t - A_o)\Sigma w_i Yio(C_{it} - C_{io}) \\
& + A_o \Sigma wi(Y_{it} - Yio)(C_{it} - C_{io}) \\
& + (A_t - A_o)\Sigma wi(Y_{it} - Yio)(C_{it} - C_{io})
\end{aligned}
$$

Where,

$P_t = A_t\ \Sigma w_i\ Y_{it}\ C_{it}$ (Total value of output at time 't')
$P_o = A_o\ \Sigma w_i Y_{io}\ C_{io}$ (Total value of output at to base period)
$A_t$ = Cropped area under cotton at time 't'
$A_o$ = Cropped area under cotton at the base period
$Y_{it}$ = Yield of cotton at time 't'
Yio = Yield of cotton at base period
$C_{it}$ = Cropping pattern of cotton at time 't'
$C_{io}$ = Cropping pattern of cotton at base period
$\Sigma wi$ = Farm harvest period of cotton at 95–96 level

The decomposition was done for the period 1980–1981 to 2001–2002.

The Table 24.2 gives the results of decomposition of growth of output into area, yield, cropping pattern and their interaction effects using Minhas Seven Factor Analysis.

From the Table 24.2 it is observed that in Tamil Nadu State during 1980–1983 to 1990–1993 the change of output of cotton was due to yield, cropping pattern and area – the main and effects and the interaction effects of yield and cropping pattern; yield and Area, Area and cropping pattern and Area, yield and cropping pattern.

About 55 % change of output was mainly due to Yield and 22 % change was due to Cropping Pattern and nearly 9 % charge was due to Area – the main effects. The inter-

**Table 24.2** Decomposition of growth of output into area, yield, cropping pattern and their interactions using Minhas Seven Factors Model (in percentage)

| | Main effects | | | Interaction effects | | | |
|---|---|---|---|---|---|---|---|
| Period | Area | Yield | Cropping pattern | Area × yield | Area × cropping pattern | Yield × cropping pattern | Area × yield × cropping pattern |
| 1980–1983 to 1990–1993 | 9.10 | 54.95 | 21.79 | 3.59 | 1.42 | 8.59 | 0.56 |
| 1990–1993 to 1999–1902 | 271.90 | −124.75 | 9.64 | 9.61 | −74.22 | 34.05 | −26.23 |
| 1980–1983 to 1999–1902 | 7.4 | 195.7 | −70.35 | −3.28 | 1.18 | −31.19 | 0.54 |

action effects such as Yield and Cropping Pattern contributed nearly 8.59 %; Area and Yield contributed 3.59 %; Area and Cropping Pattern contributed 1.42 % and 0.56 % Change of output was contributed by Area, Yield and Cropping pattern respectively.

During 1990–1993 to 1999–2002, the changes of output was due to Area and cropping pattern the main effects and yield and cropping pattern and area and yield – the interaction effects. About 2.72 % change of output was due to Area effect and 9.64 % charge was due to Area effect and 9.64 % change was due to Cropping Pattern – the main effects.

The interaction effects such as Yield and Cropping pattern and Area and Yield contributed 34.05 % and 9.61 % respectively. Yield – the main effect and the interaction effects such as Area and Cropping Pattern contributed negative impacts. Nearly – 125 % change of output was due to Yield effect and 74 % change of output was due to negative impact of Area and Cropping Pattern and 26.23 % was due to the negative impact of Area, Yield and Cropping Pattern.

During the period 1980–1983 to 1990–2002, the change of output was due to Yield and Area – the main effects and Area and Cropping pattern and Area and Yield and Cropping Pattern the interaction effects. Nearly 195.70 % change of cotton output was due to Yield and 7.4 % change was due to Area effect – the main effects. The interaction effects such as Area and Cropping Pattern and Area Cropping Pattern and Yield contributed 1.18 % and 0.54 % which was the minimal. Nearly 70 % change of output was due to negative impact of Cropping Pattern – the main effect. Nearly 3.28 and 31.19 percentage change of output was due to negative impact of Area and Yield and yield and Cropping Pattern respectively.

## 24.6 District Wise Number of Textile Mills and Their Approximate Lint

There are about 732 cotton spinning mills located all over Tamil Nadu except a few districts. Among the total number of mills 60 % of them are situated in Coimbatore district. The total ring spindles account to 82 lakhs and rotors to 62,000 in Tamil Nadu. The yarn spun from these mills range from 10s to 120s counts.

## 24.7 Varieties of Cotton Used in Tamil Nadu Mills

Varieties of cotton cultivated by the farmers in Tamil Nadu are MCU 5, MCU 5(VT) MCU 9, Suvin, TCHB 213 (Hybrid) Savitha (Hybrid) MCU 12, Surabhi, Surriangala, RCH2 (Hybrid, Bunny (Hybrid), DCH 32 (Hybrid) LRA 5166, SVPR2, Supriya, K11, KC2 and Anjali. Apart from these varieties the mills import cotton from other countries also. They are T/Manbo-s, T/Kaba-S, T/Bola-s, T/Sarama, T/Jubli-s T/ Jrma, T/Chad, Us cotton Guna 86, Pima, Sudan, Egyptian Giza 45, Giza 70, Us Mot/ Texas/Califormia/SJV, Syrian Sudan etc.

# Chapter 25
# Sampling and Methodology

## 25.1 Sampling Design

### 25.1.1 District Selection

Tamil Nadu has been classified into five agro climatic zones. The zone wise data on area under irrigated as well as unirrigated cotton were collected for 5 years (1998–2003). Then the zones were ranked on the basis of area under cotton under irrigated and unirrigated condition. Under irrigated cotton, zone V was chosen. Under unirrigated cotton, zone III high intensity of irrigation was selected. Among the districts in the selected zones, the district with maximum area under cotton was chosen for the in-depth study. Accordingly Salem district in Zone V and Perambalur district in Zone III were selected. Moreover, the data on the distribution of Bt cotton in each district of the State were also collected and ranked. Based on the concentration of Bt cotton area, Salem and Perambalur districts occupied I & II position. Therefore, these districts were chosen for the in-depth study.

### 25.1.2 Selection of Villages

Considering the concentration of Bt cotton, Pethanaicken Palayam block in Attur taluk of Salem district was purposively selected. Applying the same criteria, three villages viz, Mettupalayam, Olappadi Erumai Samudram have been chosen. In Mettupalayam and Olappadi both Bt and Non-Bt Cotton growers were available but in Erumai Samudram only Bt cotton was cultivated.

In this block, the farmers had almost abandoned the cultivation of cotton for some years. After the introduction of Bt and hybrid cotton varieties they resumed cotton cultivation. Therefore it was difficult to select farmers cultivating non Bt




V.P. Gandhi, D. Jain, *Introduction of Biotechnology in India's Agriculture*, DOI 10.1007/978-981-10-1091-0_25

**Table 25.1** Methodology adopted for the study

| Bt cotton | District I (High intensity of irrigation) | | | | | | District II low intensity of irrigation) | | | | | |
|---|---|---|---|---|---|---|---|---|---|---|---|---|
| | VI | | V2 | | V3 | | VI | | V2 | | V3 | |
| | I | UI | I | UI | I | UI | I | UI | I | UI | I | UI |
| Small | 4 | 3 | 4 | 3 | 4 | 3 | – | 7 | – | 7 | – | 7 |
| Medium | 3 | 2 | 3 | 2 | 3 | 2 | 1 | 4 | 1 | 4 | 0 | 5 |
| Large | 2 | 1 | 2 | 1 | 2 | 1 | 1 | 2 | 1 | 2 | 1 | 2 |
| Overall | 9 | 6 | 9 | 6 | 9 | 6 | 2 | 13 | 2 | 13 | 1 | 14 |
| **Non-Bt cotton** | | | | | | | | | | | | |
| Small | 4 | 3 | 4 | 3 | 4 | 3 | – | 7 | – | 7 | – | 7 |
| Medium | 3 | 2 | 3 | 2 | 3 | 2 | 1 | 4 | 1 | 4 | 0 | 5 |
| Large | 2 | 1 | 2 | 1 | 2 | 1 | 1 | 2 | 1 | 2 | 1 | 2 |
| Overall | 9 | 6 | 9 | 6 | 9 | 6 | 2 | 13 | 2 | 13 | 1 | 14 |

V1, V2 and V3 = Village 1, 2 and 3 respectively
*I* irrigated, *UI* unirrigated

cotton in Erumai Samudram village. Therefore, we had to select another village where non Bt cotton was largely cultivated. Hence Kurumapatti village had been selected for Non Bt in Salem district.

Following same criteria in Perambalur district, three villages viz Kurumbalur, Anukoor and Siruganpur villages have been chosen for the in-depth study.

### *25.1.3 Selection of Samples*

The samples were selected at random in both the districts. In Perambalur district, 99 % of the cotton cultivation was under rainfed condition. Only handfuls of farmers having wells were raising cotton under irrigated condition. Under such circumstances, it was not possible to follow fully the methodology adopted in other states and the modified methodology is as given in Table 25.1.

The primary data were collected with the help of a structured questionnaire and survey method was followed for collection of data.

The secondary data were collected from the Season and Crop Reports of various issues. The secondary data on Bt cotton were collected from the Rasi Hybrid Seed Co through the Department of Agriculture, Tamil Nadu.

### *25.1.4 Area Under Cotton in Salem District*

During the year 1980–1981, the area under cotton was 13,577 ha which represented 6.16 % of the cotton area in the State. There were fluctuations in the cotton area in this district. It was 24,988 ha in the year 1989–1990 which was the maximum area under cotton in the district which represented 8.89 % of area under cotton in the State during the study period. In the year 1994–1995 the area under cotton was 21,825 ha representing 8.55 % of the State's area under cotton. It was 15,898 ha in the year 1999–2000 and decreased and reached the minimum of 13,396 ha in the year 2001–2002 representing 8.16 % of the State's area under cotton.

### 25.1.5 Production of Cotton in Salem District

The production of cotton in Salem district was 18,260 bales of 170 kg in the year 1980–1981 representing 8.28 % of the State's cotton production. It had increased to 48,050 bales representing 17.1 % in the year 1989–1990 which was considered as peak production period. Thereafter, the production had fluctuated and it had come down to 37,010 bales in the year 1994–1995 representing 14.5 % of the State's cotton production. In the year 2001–2002 the production of cotton further declined to 24,140 bales which was 14.7 % of the State's cotton production.

### 25.1.6 Yield of Cotton in Salem District

The yield of cotton in Salem District was 229 kg in the year 1980–1981 while the State average was 201 kg. During the year 1984–1985 the yield became very high as 537 kg which was much higher than the while the State's yield average yields of 393 kg. There were fluctuations every year and the yield was 324 kg in the year 1989–1990 while it was 307 for the State. There were fluctuations year after year and yield declined to 288 kg in 1994–1995 which was lower than the State average yield of 293 kg. Again yield increased to 306 in the year 2001–2002 while it was 238 kg in the State.

### 25.1.7 Area Under Cotton in Perambalur District

As this district was a newly created district, data were available from the year 1997 only. The area under cotton was 28,508 ha which was 11.30 % of the State's total cotton area. The area decreased to 26,017 ha in the year 1999–2000 representing 14.59 % of the State's area. The area further declined to 18,396 ha in 2001–2002 representing 11.20 % of the State's area under cotton.

### *25.1.8 Production of Cotton in Perambalur District*

The production of cotton was 22,130 bales in 1996–1997 and increased to 54,160 bales in 1999–2000. But in the subsequent years production had decreased to 44,060 bales in 2000–2001 and again declined to 22,910 bales in 2001–2002.

### *25.1.9 Yield of Cotton in Perambalur District*

The cotton in Perambalur district yielded 132 kg/ha in 1996–1997 and reached the maximum of 354 in 1999–2000. In the subsequent years the yield decreased and reached the low level of 212 kg in 2001–2002.

### *25.1.10 Statistical Tools Used*

1. Percentage and Tabular Analysis.
2. De composition Analysis.
3. Growth Rate analysis.
4. Students 't' test.

### *25.1.11 Scope of the Study*

The cotton cultivation in India has been plagued with rising cost of cultivation, ineffective pesticides and other inputs leading to frequent crop failures. The transgenic Bt cotton with in-built protein toxin is the key to control the important pest i.e., boll worm and sustain the higher cotton productivity. Therefore, the study will throw the merits and demerits of Bt cotton over Non Bt cotton.

### *25.1.12 Limitation*

The study was undertaken with the available primary and secondary data. Care was taken to scrutinize the data given by the respondents.

# Chapter 26
# Nature and Performance of Bt Cotton vs. Non-Bt Cotton in Tamil Nadu

## 26.1 Introduction

As mentioned in the methodology, the sample households consisted of 90 Bt cotton cultivators and 90 Non Bt cotton cultivators. The Bt as well as Non Bt cotton cultivators had been selected from seven villages in Salem and Perambalur districts as mentioned in the methodology.

## 26.2 Cropping Pattern

### *26.2.1 Bt Cotton Growers*

The details of cropping pattern are presented in Tables 26.1 and 26.2. In case of small farmers, cotton occupied 7.9 % of total cultivated area followed by groundnut (5.7 %) cholam (3.1 %), maize (2.3 %), paddy (1.6 %) pulses (1.2 %), cumbu (0.9 %), sugarcane (0.5 %), and vegetables. In case of medium farmers, cotton was cultivated in 12.9 % of the total area sown by the sample farmers followed by groundnut (6.2 %), maize (5.4 %), cholam (4.0 %), cumbu (2.2 %), pulses (2.0 %) and vegetables. In case of large farmers cotton occupied (10.5 %) followed by groundnut (5.6 %), (maize 5.0 %), cholam (4.4 %), pulses (2.5 %) sugarcane (2.1 %) and other crops.

On the whole, Bt Cotton growers had brought nearly 31 % of the total cropped under cotton; nearly 18 % under groundnut; 11 % under maize, nearly 6 % under pulses; 5 % under cumbu and 4 % under sugar cane; nearly 4 % under sunflower. The remaining areas were cultivated with turmeric (1 %) and tapioca (0.8 %).


V.P. Gandhi, D. Jain, *Introduction of Biotechnology in India's Agriculture*, DOI 10.1007/978-981-10-1091-0_26

**Table 26.1** Cropping pattern in percentage – Bt cotton

| | Small | | | Medium | | | Large | | | Overall | | |
|---|---|---|---|---|---|---|---|---|---|---|---|---|
| Particulars | I | UI | T | I | UI | T | I | UI | T | I | UI | T |
| **Summer** | | | | | | | | | | | | |
| Pulses | 0.8 | – | 0.8 | 0.7 | 0.9 | 1.6 | 1.4 | 0.3 | 1.7 | 2.9 | 1.2 | 4.1 |
| Groundnut | 0.5 | 1.6 | 2.1 | 0.2 | – | 0.2 | – | – | – | 0.7 | 1.6 | 2.3 |
| Total | 1.3 | 1.6 | 2.9 | 0.9 | 0.9 | 1.8 | 1.4 | 0.3 | 1.7 | 3.6 | 2.8 | 6.4 |
| **Kharif 2004** | | | | | | | | | | | | |
| Sunflower | – | – | – | 0.7 | – | 0.7 | 0.7 | – | 0.7 | 1.4 | – | 1.4 |
| Cotton | 3.2 | 4.7 | 7.9 | 7.2 | 5.7 | 12.9 | 6.4 | 4.1 | 10.5 | 16.8 | 14.5 | 31.3 |
| Cholam | 1.1 | 2.0 | 3.1 | 1.8 | 2.2 | 4.0 | 1.5 | 2.9 | 4.4 | 4.4 | 7.1 | 11.5 |
| Maize | 0.9 | 1.2 | 2.1 | 1.0 | 2.5 | 3.5 | 1.4 | 2.6 | 4.0 | 3.3 | 6.3 | 9.6 |
| Cumbu | 0.8 | 0.1 | 0.9 | 0.5 | 1.7 | 2.2 | 0.3 | 1.5 | 1.8 | 1.6 | 3.3 | 4.9 |
| Groundnut | 0.3 | 2.1 | 2.4 | 1.5 | 2.6 | 4.1 | 1.9 | 2.7 | 4.6 | 3.7 | 7.4 | 11.4 |
| Sugarcane | 0.5 | – | 0.5 | 1.5 | – | 1.5 | 2.1 | – | 2.1 | 4.1 | – | 4.1 |
| Tapioca | – | – | – | 0.2 | – | 0.2 | 0.6 | _ | 0.6 | 0.8 | – | 0.8 |
| Pulses | – | – | – | – | – | – | – | – | – | – | – | – |
| Paddy | – | – | – | 0.5 | – | 0.5 | 1.5 | – | 1.5 | 2.0 | – | 2.0 |
| Total | 6.8 | 10.1 | 16.9 | 14.9 | 14.7 | 29.6 | 16.4 | 13.8 | 30.2 | 38.1 | 38.6 | 76.7 |
| **Rabi 2004–2005** | | | | | | | | | | | | |
| Paddy | 1.6 | – | 1.6 | 2.4 | – | 2.4 | 1.9 | – | 1.9 | 5.9 | – | 5.9 |
| Sunflower | 0.6 | – | 0.6 | 1.0 | – | 1.0 | 0.9 | – | 0.9 | 2.5 | – | 2.5 |
| Maize | 0.1 | 0.1 | 0.2 | 0.3 | – | 0.3 | 1 | – | 1.0 | 1.4 | 0.1 | 1.5 |
| Groundnut | 0.8 | 0.4 | 1.2 | 1.8 | _ | 1.9 | 1 | _ | 1.0 | 3.6 | 0.4 | 4.0 |
| Vegetables | 0.1 | – | 0.1 | 0.3 | – | 0.3 | – | – | – | 0.4 | – | 0.4 |
| Turmeric | – | – | – | Neg | – | neg | 1.0 | – | 1.0 | 1.0 | . | 1.0 |
| Pulses | 0.3 | 0.1 | 0.4 | 0.4 | – | 0.4 | 0.8 | – | 0.8 | 1.5 | 0.1 | 1.6 |
| Chillies | – | – | – | – | – | – | – | – | – | – | – | – |
| Total | –3.5 | 0.6 | 4.1 | 6.2 | – | 6.2 | 6.6 | – | 6.6 | 16.5 | 0.6 | 16.9 |
| Overall | – | 12.3 | 23.9 | 22 | 15.6 | 37.6 | 24.4 | 14.1 | 38.5 | 58.0 | 42.0 | 100.0 |

*I* irrigated, *UI* unirrigated, *T* total

### 26.2.2 *Non Bt Cotton Growers*

In case of small farmers the important crops like cotton occupied nearly 13 % followed by cholam (4.6 %) maize (3.4 %) cumbu (1.8 %) and groundnut (1.7 %). In case of medium farmers, the important crop cotton was cultivated under 17.4 % of total cropped area. The other important crops were cholam, maize, groundnut etc. In case of large farmers the important crops grown by sample farmers were cotton (12.6 %), groundnut (8.4 %), maize (5.4 %), pulses (2.9 %) cholam (2.9 %) followed by other crops like cumbu, chillies etc., On the whole the important crops grown by non Bt Cotton growers were cotton (42.7 %) groundnut (15.9 %), maize (15.2 %) and cholam (12.6 %) followed by other crops.

**Table 26.2** Cropping pattern in percentage – non-Bt cotton

| | Small | | | Medium | | | Large | | | Overall | | |
|---|---|---|---|---|---|---|---|---|---|---|---|---|
| Particulars | I | UI | T | I | UI | T | I | UI | T | I | UI | T |
| **Summer** | | | | | | | | | | | | |
| Pulses | – | 0.2 | 0.2 | 0.4 | – | 0.4 | 1.1 | 0.4 | 1.5 | 1.5 | 0.6 | 2.1 |
| Groundnut | 0.3 | – | 0.3 | 1 | – | 1.0 | 3.2 | – | 3.2 | 4.5 | – | 4.5 |
| Maize | – | 1.0 | 1.0 | 0.2 | 0.7 | 0.9 | – | 0.4 | 0.4 | 0.2 | 2.1 | 2.3 |
| Total | 0.3 | 1.2 | 1.5 | 1.6 | 0.7 | 2.3 | 4.3 | 0.8 | 5.1 | 6.2 | 2.7 | 8.9 |
| **Kharif 2004** | | | | | | | | | | | | |
| Cotton | – | 5.7 | 5.7 | 1.2 | 7.4 | 18.6 | 2.6 | 4.7 | 7.3 | 3.8 | 17.8 | 21.6 |
| Cholam | 0.6 | 1.5 | 2.1 | 1.5 | 2.2 | 3.7 | 0.9 | 1.6 | 2.5 | 3.0 | 5.3 | 8.3 |
| Maize | 0.5 | 1.6 | 2.1 | 0.6 | 4.9 | 5.5 | 0.7 | 4.3 | 5.0 | 1.8 | 10.8 | 12.6 |
| Cumbu | – | 1.8 | 1.8 | – | 0.5 | 0.5 | – | 0.9 | 0.9 | – | 3.2 | 3.2 |
| Groundnut | 0.3 | 1.1 | 1.4 | 0.3 | 4.2 | 4.5 | 0.5 | 3.9 | 4.4 | 1.1 | 9.2 | 10.3 |
| Sugarcane | 0.7 | – | 0.7 | – | – | – | – | – | – | 0.7 | – | 0.7 |
| Pulses | – | – | – | 0.4 | – | 0.4 | 0.2 | 0.4 | 0.6 | 0.6 | 0.4 | 1.0 |
| Paddy | – | – | – | 1.9 | 0.2 | 2.1 | 1.6 | 0.5 | 2.2 | 3.5 | 0.7 | 4.2 |
| Chillies | – | – | – | – | – | – | 0.4 | – | 0.4 | 0.4 | – | 0.4 |
| Sunflower | – | – | – | – | – | – | 0.5 | – | 0.5 | 0.5 | – | 0.5 |
| Total | 2.1 | 11.7 | 13.8 | 5.9 | 19.4 | 25.3 | 7.5 | 16.3 | 23.8 | 15.5 | 47.4 | 62.9 |
| **Rabi 2004–2005** | | | | | | | | | | | | |
| Paddy | – | – | – | – | – | – | – | – | – | – | – | – |
| Maize | – | 0.3 | 0.3 | – | – | – | – | – | – | – | 0.3 | 0.3 |
| Groundnut | – | – | – | 0.2 | – | 0.2 | – | – | – | 0.2 | – | 0.2 |
| Pulses | – | – | – | 0.4 | – | 0.4 | 0.8 | – | 0.8 | 1.2 | – | 1.2 |
| Chillies | – | – | – | 0.2 | – | 0.2 | – | – | – | 0.2 | – | 0.2 |
| Cumbu | – | 0.4 | 0.4 | – | 0.1 | 0.1 | – | 0.4 | 0.4 | – | 0.9 | 0.9 |
| Cholam | – | 2.5 | 2.5 | – | 1.1 | 1.1 | – | 0.7 | 0.7 | – | 4.3 | –4.3 |
| Cotton | 4.5 | 2.5 | 7.0 | 6.3 | 2.5 | 8.8 | 3.5 | 1.8 | 5.3 | 14.3 | 6.8 | 21.1 |
| Total | 4.5 | 5.7 | 10.2 | 7.1 | 3.7 | 10.8 | 4.3 | 2.9 | 7.2 | 15.9 | 12.3 | 28.2 |
| Grand total | 6.9 | 18.6 | 25.5 | 14.6 | 23.8 | 38.4 | 16.1 | 20.0 | 36.1 | 37.6 | 62.4 | 100 |

## 26.3 Area Under Cotton

Table 26.3 furnishes the details on area under cotton in the study area.

It is seen from the above Table that, in the study area, nearly 54 % of 74.27 ha of Bt cotton was cultivated under irrigated condition, whereas 46.3 % was raised under unirrigated condition. On the other hand nearly 42 % of Non Bt cotton was grown under irrigated condition and 57.7 % of 99.11 ha was cultivated under unirrigated condition.

In case of Bt cotton 40 % and 60 % of cotton areas were brought under irrigated and unirrigated cotton cultivation by the small farmers nearly 56 % and 61 % of total area under cotton was raised under irrigated condition by the medium and large fanners respectively.

**Table 26.3** Area under cotton

| Farm size | Bt cotton | | | Non Bt cotton | | |
|---|---|---|---|---|---|---|
| | I | UI | Total | I | UI | Total |
| Small | 7.59 (40.5) | 11.15 (59.5) | 18.74 (100.0) | 10.32 (35.2) | 18.99 (64.8) | 29.31 (100.0) |
| Medium | 17.11 (55.9) | 13.52 (44.1) | 30.63 (100.0) | 17.41 (43.0) | 23.04 (57.0) | 40.45 (100.0) |
| Large | 15.18 (61.0) | 9.72 (39.0) | 24.90 (100.0) | 14.17 (48.3) | 15.18 (51.7) | 29.35 (100.0) |
| Overall | 39.88 (53.7) | 34.39 (46.3) | 74.27 (100.0) | 41.90 (42.3) | 57.21 (57.7) | 99.11 (100.0) |

Figures in brackets give percentage to total. *I* irrigated, *UI* unirrigated

It can also be seen that the percentage of irrigated cotton area had increased with the size group and under unirrigated condition the percentage of area under cotton had decreased with the size group.

With regard to Non Bt Cotton, the percentage share of irrigated Non Bt cotton had increased with the size group and that of unirrigated Non Bt cotton had decreased with the size group in the study area. The percentage share of irrigated non Bt cotton was the highest for the large size group (48.3 %) followed by medium (43.0 %) and small (35.2 %). On the other hand the percentage share of unirrigated Non Bt cotton was the highest for small farm size (64.8 %) followed by medium (57.0 %) and large size group (51.7 %).

### *26.3.1 Area Under Cotton Per Farm*

The Table 26.4 gives the details of area under per farm.

It is noticed from the above Table that the average area under cotton per farm was 1.25 ha for irrigated cotton and 0.59 ha for unirrigated Bt cotton. It is also noticed that the average area under irrigated Non Bt cotton was 1.31 ha and that under unirrigated Non Bt cotton was 0.99 ha. On the whole it was found that the area under cotton per farm was 0.83 ha and 1.10 ha for Bt as well as Non Bt cotton growers respectively.

Under irrigated Bt cotton the average per farm area was the highest (1.56 ha) for the medium size group followed by large (1.38 ha) and small (0.63 ha) respectively.

Under unirrigated Bt cotton, the area per farm was the highest for large (1.08 ha) followed by medium (0.71 ha) and small (0.37 ha) respectively. In other words, it can be said that the average area under unirrigated Bt cotton had increased with the increase in the size group.

In case of Non Bt Cotton, it is observed that the per farm area under irrigated cotton was the highest for medium size (1.58 ha) followed by large (1.57 ha) and

**Table 26.4** Area under cotton per farm (ha)

| Size group | Bt cotton | | | Non Bt | | |
|---|---|---|---|---|---|---|
| | I | UI | Total | I | UI | Total |
| Small | 0.63 | 0.37 | 0.45 | 0.86 | 0.63 | 0.70 |
| Medium | 1.56 | 0.71 | 1.02 | 1.58 | 1.21 | 1.35 |
| Large | 1.38 | 1.08 | 1.38 | 1.57 | 1.69 | 1.63 |
| Overall | 1.25 | 0.59 | 0.83 | 1.31 | 0.99 | 1.10 |

small (0.86 ha). In case of unirrigated non Bt cotton the area under per farm had increased with the increase in the size group. It was 0.63 ha; 1.21 ha and 1.69 ha in case of small, medium and large farms respectively.

On the whole it can be seen that the area under cotton both Bt and Non Bt had increased with the increase in the size group. It was 0.45 ha; 1.02 ha and 1.38 ha for small, medium and large farmers respectively in case of Bt cotton while it was 0.70 ha; 1.35 ha and 1.63 ha for small, medium and large farmers respectively under Non Bt cotton.

## 26.4 Cotton Varieties Used by the Sample Households

The farmers in the study area were about to abandon the cotton cultivation due to loss caused by pests, excessive expenditure on pesticides, and price fluctuations which is also risky. The sample farmers had started cultivating cotton after the introduction of hybrid varieties. Therefore, the respondents were asked about the important factors motivating them to take up cotton cultivation. The important reasons mentioned by them were "remunerative returns due to Bt and hybrid varieties"; "Other crops were not so remunerative" and experience of other farmers in the village" was that they were motivated by seed companies and Seed, Fertilizer and Pesticides dealers who encouraged them to cultivate Bt and hybrid varieties of cotton. Majority of the respondents had taken up cotton cultivation with the hope that the returns would be remunerative.

### *26.4.1 Relative Preference Among Different Varieties*

An attempt had been made to elicit the preference of respondents towards the varieties and the reasons therefore. It is observed during the discussion with the farmers that a majority of the respondents expressed their preferences to RCH2 (Rasi cotton Hybrid) Bt variety and RCH2 Non Bt variety only. The Maharashtra Hybrid Cotton seed varieties (MAHYCO) were also available in the study area. Though this variety is also popular in the study area, the sample respondents had preferred only RCH2 Bt & RCH2 Non Bt cotton varieties only.

**Table 26.5** Varieties of cotton grown by the sample farmers

| Farm size | Year | Bt cotton | | | | Non Bt cotton | | | |
|---|---|---|---|---|---|---|---|---|---|
| | | I | | UI | | I | | UI | |
| | | Number | Variety | Number | Variety | Number | Variety | Number | Variety |
| Small | 2004–2005 | 12 | $RCH_2$Bt | 30 | $RCH_2$Bt | 12 | $RCH_2$ | 30 | $RCH_2$ |
| | 2003–2004 | 8 | $RCH_2$Bt | 17 | $RCH_2$Bt | 6 | 5 MAHYCO<br>1 $RCH_2$ | 19 | $RCH_2$ |
| | 2002–2003 | – | – | 1 | MAHYCO | 1 | MAHYCO | – | $RCH_2$ |
| Medium | 2004–2005 | 11 | $RCH_2$Bt | 19 | $RCH_2$ | 11 | $RCH_2$ | 19 | $RCH_2$ |
| | 2003–2004 | 8 | $RCH_2$Bt | 6 | 4 $RCH_2$ 2<br>MAHYCO | 5 | 4 MAHYCO<br>1 $RCH_2$ | 4 | $RCH_2$ |
| | 2002–2003 | 1 | MAHYCO | 1 | MAHYCO | 2 | $RCH_2$ | 5 | $RCH_2$ |
| Large | 2004–2005 | 9 | $RCH_2$ | 9 | $RCH_2$ | 9 | $RCH_2$ | 9 | $RCH_2$ |
| | 2003–2004 | 8 | $RCH_2$Bt | 2 | $RCH_2$ | 5 | MAHYCO | 7 | $RCH_2$ |
| | 2002–2003 | – | – | 2 | MAHYCO | 3 | MAHYCO | 4 | $RCH_2$ |
| Overall | 2004–2005 | 32 | $RCH_2$ | 58 | $RCH_2$ | 32 | $RCH_2$ | 58 | $RCH_2$ |
| | 2003–2004 | 24 | $RCH_2$ | 25 | 23 $RCH_2$<br>MAHYCO | 16 | 14 MAHYCO 2 $RCH_2$ | 30 | $RCH_2$ |
| | 2002–2003 | 1 | MAHYCO | 4 | MAHYCO | 6 | 14 MAHYCO 2 RCH2 | 9 | $RCH_2$ |

The Table 26.5 gives the details of cotton variety grown by the sample respondents. The varieties largely grown were $RCH_2$ Bt and $RCH_2$.

## 26.5 Average Cost of Seed Used

The sample farmers had purchased seed from various sources like Rasi Seed Co; and Seed Fertiliser and Pesticide dealers. The Table 26.6 shows the details of sources from where the cotton seeds were purchased by them.

It is seen from the Table that more than 70 % of the sample respondents had purchased seed from the seed, fertilizer and pesticide dealers. The reason gives by the sample respondent was that the seeds can be bought on credit under the precondition that the produce should be sold to them. Not only seed, the sample respondents purchased other inputs such as pesticides, and fertilizers on credit from the dealers.

The sample respondents were enquired about the quantity of seed used and the purchase rate of the cotton seeds. Table 26.7 furnishes the details. The seed cost per kg of Bt cotton had ranged from Rs. 3604 to Rs 3694. On an average, the seeds were purchased at the rate of Rs. 3620 per kg by the Bt cotton growers to grow cotton under irrigated condition and Rs. 3696/kg under unirrigated condition.

The quantity of seed used by the medium farmers was the lowest (1119 g). The large farmers had used the highest (1171 g) quantity of seed among irrigated Bt cotton growers. In case of unirrigated Bt cotton the lowest quantity of (1111 g) was used by the large farmers and the highest quantity (1132 g) by the medium farmers. On an average, the farmers used more seed under irrigated cotton (1148 g) than that under unirrigated cotton (1,125g).

Considering Non Bt Cotton, it was observed that the seed quantity used by the sample farmers under both irrigated and unirrigated was almost the same. The seed quantity ranged from 1104 to 1110 g in case of irrigated Bt cotton while it ranged from 1107 to 1110 g in case of unirrigated Non Bt cotton. In case of irrigated non

**Table 26.6** Purchase of cotton seeds from various sources

| Farm size | Bt cotton | | | Non Bt cotton | | |
|---|---|---|---|---|---|---|
| | Rasi seed Co | Seed, fertiliser, pesticide dealers | Total | Rasi seed Co | Seed, fertiliser, pesticide dealers | Total |
| Small | 10 (23.8) | 32 (76.2) | 42 (100.0) | 11 (26.2) | 31 (73.8) | 42 (100.0) |
| Medium | 9 (30.0) | 21 (70.0) | 30 (100.0) | 6 (20.0) | 24 (80.0) | 30 (100.0) |
| Large | 2 (11.1) | 16 (88.9) | 18 (100.0) | 2 (11.1) | 16 (88.9) | 18 (100.0) |
| Overall | 21 (23.3) | 69 (76.7) | 90 (100.0) | 19 (21.1) | 71 (78.9) | 90 (100.0) |

Figures in brackets give percentage to total

**Table 26.7** Purchase, cost and seed rate of cotton

| Farm size | Bt cotton | | | | | | | | Non Bt cotton | | | | | | | |
|---|---|---|---|---|---|---|---|---|---|---|---|---|---|---|---|---|
| | Area planted under cotton (hectare) | | Quantity of seed purchase (gm) | | Cost per kg | | Seed rate/ hectare (gm) | | Area planted under cotton (hectare) | | Quantity of seed purchase (gm) | | Cost per kg | | Rate/hectare (gm) | |
| | I | UI | I | UI | I | UI | I | UI | I | UI | I | UI | I | UI | I | UI |
| Small | 7.59 | 11.15 | 8775 | 12,600 | 3604 | 3604 | 1156 | 1130 | 10.32 | 18.99 | 11,400 | 21,100 | 1000 | 1010 | 1104 | 1110 |
| Medium | 17.11 | 13.52 | 19,150 | 15,300 | 3615 | 3694 | 1119 | 1132 | 17.41 | 23.04 | 19,250 | 25,500 | 1008 | 1007 | 1106 | 1107 |
| Large | 15.18 | 9.72 | 17,850 | 10,800 | 3941 | 3919 | 1176 | 1111 | 14.17 | 15.18 | 15,740 | 16,600 | 1010 | 1009 | 1110 | 1094 |
| Overall | 39.88 | 34.39 | 45,775 | 38,700 | 3620 | 3696 | 1148 | 1125 | 41.9 | 57.21 | 46,390 | 63,200 | 1006 | 1009 | 1107 | 1105 |

Bt cotton, the seed rate had increased with increase in the size group. On the other hand, in case of irrigated Non Bt cotton it had decreased with increase in the size group.

The cost of Non Bt Cotton seed ranged from Rs. 1000 to 1010 in all the size groups. On an average it was Rs 1006 and Rs. 1009 for irrigated and unirrigated cotton. The Table 26.7 presents the details of seed rate, cost per kg of seed.

### *26.5.1 Seed Rate of Bt Cotton vis-à-vis Non Bt Cotton*

The seed rate of 450 g/per acre or 1111.5 g per hectare was recommended for the cultivation of Bt and Non Bt cotton. The price of Bt cotton seed ranged between Rs 1600 and 1550 per 450 g. The Non Bt Cotton Seed was Rs 450 per 450 g. Some farmers felt that seed rate was lower and for some it was normal. The Table 26.8 shows the details of seed rates.

It is observed from the Table that majority of the farmers felt that the seed rate was normal since company had released the seeds after experiments and trials in their fields. Only 6 of the farmers it was low for their field.

## 26.6 Application of Pesticides

### *26.6.1 Rating on Insect or Pest Attack on Cotton*

The sample respondents were asked whether the pest attack was heavy, moderate, light or none. The opinions given by them were analysed and presented in Table 26.9.

A very few number of farmers have reported mild or very low Pink/Spotted bollworm attack, whereas a very large number of farmers growing Non Bt Cotton have reported severe bollworm attack. This result clearly supports the claim that BT technology in cotton is highly effective in providing strong resistance to American bollworm for virtually Bt cotton growers.

Both, Bt cotton and non-Bt cotton suffered attack of soil pests, sucking pests and foliage feeding pests. However, intensity of attack of these pests was found slightly lower in Bt cotton as compared to non-Bt cotton. This clearly suggests that Bt cot-

**Table 26.8** The details of seed rates

| | Bt | | | NonBt | | |
|---|---|---|---|---|---|---|
| Particulars | I | UI | Total | I | UI | Total |
| Lower | 2 | 4 | 6 (6.7) | – | – | – |
| Normal | 30 | 54 | 84 (93.3) | 32 | 58 | 90 (100.0) |
| Higher | – | – | – | – | – | – |
| Total | 32 | 58 | 90 (100.0) | 32 | 58 | 90 (100.0) |

**Table 26.9** Rating on insect or pest attack on cotton

| | | Bt Cotton | | | | | | | | Non Bt Cotton | | | | | | | |
|---|---|---|---|---|---|---|---|---|---|---|---|---|---|---|---|---|---|
| | | I | | | | UI | | | | I | | | | UI | | | |
| | | 4 | 3 | 2 | 1 | 4 | 3 | 2 | 1 | 4 | 3 | 2 | 1 | 4 | 3 | 2 | 1 |
| (A) Boll worm | | | | | | | | | | | | | | | | | |
| American | Small | – | – | 12 | – | – | – | 9 | 21 | 12 | – | – | – | 30 | – | – | – |
| | Marginal | – | – | 9 | 2 | – | – | 6 | 13 | 11 | – | – | – | 19 | – | – | – |
| | Large | – | – | 6 | 3 | – | – | 3 | 6 | 9 | – | – | – | 9 | – | – | – |
| Pink | Small | 5 | 7 | – | – | 3 | 4 | 1 | 22 | 12 | – | – | – | 30 | – | – | – |
| | Marginal | 2 | 2 | 4 | 3 | 1 | 1 | 3 | 14 | 11 | – | – | – | 19 | – | – | – |
| | Large | 1 | – | 3 | 5 | 2 | – | – | 7 | 9 | – | – | – | 9 | – | – | – |
| Spotted | Small | 2 | 9 | 1 | – | 1 | 5 | 11 | 13 | 12 | – | – | – | 30 | – | – | – |
| | Marginal | 1 | 1 | 6 | 3 | – | 2 | 9 | 8 | 11 | – | – | – | 19 | – | – | – |
| | Large | – | – | 4 | 5 | 1 | 1 | 2 | 5 | 9 | – | – | – | 9 | – | – | – |
| (B) Sucking pest | | | | | | | | | | | | | | | | | |
| Thrips | Small | 5 | 5 | 2 | – | 3 | 11 | 6 | – | 12 | – | – | – | 30 | – | – | – |
| | Marginal | 2 | 8 | 1 | – | 2 | 16 | 1 | – | 11 | – | – | – | 19 | – | – | – |
| | Large | 2 | 3 | 3 | – | 2 | 5 | 1 | 1 | 9 | – | – | – | 9 | – | – | – |
| Leaf hopper | Small | 3 | 8 | 1 | – | 2 | 11 | 17 | – | 12 | – | – | – | 30 | – | – | – |
| | Marginal | 2 | 4 | 5 | – | 1 | 8 | 10 | – | 11 | – | – | – | 19 | – | – | – |
| | Large | 1 | 4 | 3 | 1 | – | 5 | 4 | – | 9 | – | – | – | 9 | – | – | – |
| Whitefly | Small | 4 | 6 | 2 | – | 5 | 11 | 12 | 2 | 12 | – | – | – | 30 | – | – | – |
| | Marginal | 2 | 8 | – | 1 | 4 | 4 | 9 | 2 | 11 | – | – | – | 19 | – | – | – |
| | Large | 2 | 4 | 3 | – | 2 | 5 | 2 | – | 9 | – | – | – | 9 | – | – | – |
| (C) Foliage feeding pest | | | | | | | | | | | | | | | | | |
| Leaf roller | Small | – | 1 | 8 | 3 | – | 10 | 16 | 4 | – | 12 | – | – | 29 | – | – | – |
| | Marginal | – | 2 | 8 | 1 | – | 4 | 10 | 5 | 2 | 9 | – | – | 19 | – | – | – |
| | Large | – | – | 5 | 4 | – | 1 | 6 | 2 | 3 | 6 | – | – | 9 | – | – | – |
| Caterpillar | Small | – | 7 | 5 | – | 1 | 17 | 12 | - | - | 12 | – | – | 22 | – | – | – |
| | Marginal | – | 5 | 6 | – | – | 12 | 6 | 1 | 2 | 9 | – | – | 19 | – | – | – |
| | Large | – | 4 | 3 | 2 | – | 2 | 7 | – | 3 | 6 | – | – | 9 | – | – | – |
| (D) Soil pests termites | | | | | | | | | | | | | | | | | |
| | Small | 1 | 9 | 2 | – | 4 | 21 | 5 | – | – | 12 | – | – | 30 | – | – | – |
| | Marginal | 3 | 8 | – | – | 4 | 11 | 4 | – | 2 | 9 | – | – | 19 | – | – | – |
| | Large | 1 | 6 | 2 | – | 2 | 5 | 2 | – | 3 | 6 | – | – | 9 | – | – | – |

4 – Heavy, 3 – Moderate, 2 – Light, 1 – None

ton is substantially effective in protecting cotton plants from only bollworm infestation.

### 26.6.2 Pesticides Spraying

The farmers had used pesticides to control boll worm and other pests. The sample households had given details of total sprays and the cost on pesticide sprays. It was very difficult to get the information regarding the pesticide spraying for boll worm and other pests separately. The Table 26.10 gives the details regarding pesticide spraying for bollworm and other pests together.

The pesticides used by the sample farmers are Imida Gold for Thrips; Monophos, Planofix, Confidor, Cyber Methodine, Larvin, Ecolux and Manik. All the pesticides are in liquid form except Larvin and Confidor which are in powder form.

On an average, for Bt Cotton (both I and UI) four sprays were used while four sprays we used for irrigated Non Bt and six sprays for unirrigated Non Bt cotton. In other words unirrigated Non Bt required more sprays than irrigated Non Bt Cotton. On an average, the area sprayed was 0.40 ha for irrigated Bt cotton while it was 0.15 ha for unirrigated Bt Cotton. In case of Non Bt Cotton, the average area sprayed was 0.29 ha irrigated cotton and 0.17 ha for unirrigated cotton. It can also be inferred that the average area sprayed under irrigated cotton both Bt and Non Bt was more than unirrigated Bt and Non Bt Cotton.

The sample farmers reported that in irrigated Bt cotton farms on an average 618 ml of pesticide was sprayed per spray while under unirrigated Bt Cotton farms, it was 321 ml Among the size groups, under irrigated Bt Cotton, large farmers sprayed more quantity than other two size groups. In case of unirrigated Bt farms, the same pattern was seen. Under irrigated Non Bt Cotton, the quantity of pesticide used per spray was found to be 1663 ml while that of unirrigated Non Bt farms it was 141 ml.

The quantity of pesticide used per hectare was the highest by small farms (2717 ml) followed by large (2090 ml) and medium (1779 ml) farms under irrigated Bt cotton. Under unirrigated Bt cotton, the quantity of pesticides was the highest (2447 ml) by the small farmers followed by medium (2254 ml) and large farmers (1790 ml) respectively. On the whole unirrigated Bt growers had used more pesticide (2186 ml) than the irrigated Bt cotton growers (2076 ml) (see Table 26.10).

In case of Non Bt cotton, the quantity used per hectare was the highest in medium farm (6395 ml) followed by small (5797 ml) and large farms in case of irrigated cotton. While for unirrigated Non Bt Cotton, small farmers used the highest quantity of pesticide (903 ml) followed by large (666 ml) and medium (599 ml) farmers respectively (see Table 26.10).

On an average the cost of pesticide used per hectare was Rs. 887 for Bt and Rs. 1934 for Bt unirrigated cotton. The highest amount was spent by small farmers (Rs 1991) followed by medium farmers (Rs 1913) and large farmers (Rs. 1807) in case

**Table 26.10** Practices followed by sample farmers in pesticides spraying for boll worm and other pests

| | | Bt Cotton | | | | | | | | Non Bt Cotton | | | | | | | |
|---|---|---|---|---|---|---|---|---|---|---|---|---|---|---|---|---|---|
| | | I | | | | UI | | | | I | | | | UI | | | |
| | | S | M | L | T | S | M | L | T | S | M | L | T | S | M | L | T |
| 1 | Average no. of spray | 5 | 4 | 4 | 4 | 4 | 4 | 4 | 4 | 3 | 6 | 6 | 4 | 6 | 6 | 6 | 6 |
| 2 | Average area sprayed | 0.14 | 0.40 | 0.40 | 0.40 | 0.09 | 0.17 | 0.26 | 0.15 | 0.29 | 0.21 | 0.33 | 0.29 | 0.1 | 0.22 | 0.3 | 0.17 |
| 3 | Quantity of pesticides sprayed /spray/ml | 389 | 708 | 835 | 618 | 231 | 386 | 470 | 321 | 1661 | 1392 | 1861 | 1663 | 92 | 130 | 202 | 141 |
| 4 | Quantity sprayed/ ha(ml) | 2117 | 1779 | 2090 | 2076 | 2447 | 2254 | 1790 | 2186 | 5797 | 6395 | 5519 | 5798 | 903 | 599 | 666 | 720 |
| 5 | Cost/ha. (Rs) | 1991 | 1913 | 1807 | 1887 | 2122 | 1880 | 1795 | 1934 | 5279 | 5439 | 3077 | 5316 | 3937 | 3003 | 4573 | 3074 |

of Bt irrigated cotton. The same pattern was observed in case of Bt unirrigated cotton also.

In case of Non Bt Cotton, the amount spent on pesticides by Non Bt irrigated cotton growers was Rs. 5316 while Rs. 3074 by unirrigated growers. Among the farm size, medium farmers under irrigated cotton spent the highest amount Rs. 54,397 followed by small farmers (Rs. 5439) and large farmers (Rs 3077). Under unirrigated Non Bt cotton, the highest amount of Rs. 4573 was spent by large farmers followed by small (Rs. 3937) and medium farmers (Rs. 3003).

On the whole irrigated Non Bt cotton growers spent more (Rs. 5316) than the unirrigated cotton farmers.

# Chapter 27
# Economics of Bt Cotton vs. Non-Bt Cotton in Tamil Nadu

## 27.1 Introduction

In this chapter, an attempt has been made to examine economics of Bt cotton vis-à-vis non-Bt cotton by examining the relationship of various factors to the yield, value of output and profit of cotton cultivation reported by the sample farmer households. It includes average Bt and non-Bt cotton yields, cost of production, value of output and profit, and econometric analysis of the performance of Bt cotton vs. non-Bt cotton.

## 27.2 Average Bt and Non-Bt Cotton Yields

Tables 27.1 gives the average yield of Bt cotton It is observed from the Table that the yield per hectare of irrigated cotton was 2424 kg, 2384 kg and 2340 kg for small, medium and large farms respectively. It is also noticed that the quantity of cotton per picking was increasing up to third picking and reached the maximum in all the three size groups. Thereafter the quantity started reducing up to sixth picking. The left over cotton was plucked for the seventh time.

It is observed from the Table 27.2 that nearly 8 % of the yield was obtained during first picking in all the farms. During second picking nearly 23 % of the yield was procured. During third picking, 25–27 % was secured. Small farmers plucked 27.3 % of the yield followed by large (26.1 %) and medium (25.5 %) farmers during this period.

During fourth picking, small farmers got 21.5 % medium farmers got (19.4 %) and the large farmers got (19.6 %) respectively. During fifth time the yield ranged between 11.4 % and 12.7 %. During sixth picking the highest quantity was obtained


V.P. Gandhi, D. Jain, *Introduction of Biotechnology in India's Agriculture*, DOI 10.1007/978-981-10-1091-0_27

**Table 27.1** Harvesting of xotton – Bt yield in Kg/Ha

| | I | | | | UI | | | |
|---|---|---|---|---|---|---|---|---|
| Picking | S | M | L | T | S | M | L | T |
| First | 206 (8.5) | 191 (8) | 186 (8) | 192 (8.1) | 190 (14.8) | 150 (11.1) | 139 (10.1) | 160 (12) |
| Second | 572 (23.6) | 541 (22.7) | 531 (22.7) | 544 (22.9) | 280 (21.8) | 263 (19.5) | 293 (21.3) | 276 (20.7) |
| Third | 662 (27.3) | 608 (25.5) | 611 (26.1) | 620 (26.1) | 293 (22.8) | 287 (21.3) | 318 (23.1) | 298 (22.3) |
| Fourth | 521 (215) | 462 (19.4) | 459 (19.6) | 473 (19.9) | 239 (18.6) | 280 (20.8) | 292 (21.2) | 270 (20.2) |
| Fifth | 308 (12.7) | 272 (11.4) | 295 (12.6) | 287 (12.1) | 173 (13.4) | 206 (15.3) | 196 (14.3) | 192 (14.4) |
| Sixth | 145 (6) | 279 (11.7) | 164 (7) | 211 (8.9) | 110 (8.6) | 143 (10.6) | 138 (10) | 131 (9.8) |
| Seventh | 10 (0.4) | 31 (1.3) | 94 (4) | 48 (2) | – | 18 (1.4) | – | 8 (0.6) |
| Total (yield/ Ha.) | 2424 (100) | 2384 (100) | 2340 (100) | 2375 (100) | 1285 (100) | 1347 (100) | 1376 (100) | 1335 (100) |

Figures in brackets give percentage to total

by medium farmers (11.7 %) followed by large (7 %) and small farmers (6.0 %) respectively.

The left over cotton was harvested for the seventh time. It was the highest for large farmers (4.0 %) followed by medium (1.3 %) and small farmers (0.4 %).

The average of Bt Cotton Unirrigated yield per hectare was found to be 1285 kg, 1347 kg and 1376 kg for small, medium and large farms respectively. The maximum picking of cotton took place in third picking in all the size groups it is noticed that in all the size groups the picking of cotton was increasing up to third picking and started decreasing after third picking. During first picking, small farms had the highest quantity (14.8 %) followed by medium (11.1 %) and large farms (10.1 %). During second picking the per cent was the highest for small farms (21.8 %) followed by large (21.3 %) and medium (19.5 %). In the third picking it was the highest in large farms (23.1 %) followed by small (22.8 %) and medium farms (21.3 %). During fourth picking, large farms obtained the highest quantity (21.2 %) followed by medium (20.8 %) and small farms (18.6 %). During fifth picking, the percentage was the highest for medium farms (15.3 %) followed by large (14.3 %) and medium farms (13.4 %) and small (13.4 %) farms. During sixth picking, the highest percentage was obtained by medium farms (10.6 %) followed by large (10.0) and small (8.6) farms respectively (Table 27.2).

The yield of Non Bt Cotton Irrigated per hectare of cotton was 1734 kg, 1678 and 1693 kg for small, medium and large farmers respectively. The sample farmers had only six pickings. The yield of cotton had increased up to fourth picking and started declining thereafter. The highest picking by the small farmers was 22.4 % followed by medium (22.1 %) and large farmers (21.6 %). During first picking, the highest

**Table 27.2** Harvesting of non-Bt cotton yield in kg/ha

| | I | | | | UI | | | |
|---|---|---|---|---|---|---|---|---|
| Picking | S | M | L | T | S | M | L | T |
| First | 175 (10.1) | 166 (9.9) | 174 (10.3) | 173 (10.2) | 169 (13.0) | 149 (11.3) | 131 (10) | 149 (114) |
| Second | 264 (15.2) | 290 (17.3) | 223 (13.2) | 258 (15.2) | 263 (20.2) | 246 (18.7) | 201 (15.3) | 237 (18.1) |
| Third | 340 (19.6) | 352 (21) | 296 (17.5) | 334 (19.7) | 348 (26.7) | 299 (22.7) | 255 (19.4) | 300 (22.9) |
| Fourth | 388 (22.4) | 371 (22.1) | 366 (21.6) | 373 (22) | 307 (23.6) | 313 (23.8) | 303 (23.1) | 308 (23.5) |
| Fifth | 335 (19.3) | 281 (16.7) | 325 (19.2) | 314 (18.5) | 172 (13.2) | 184 (14) | 248 (18.9) | 202 (15.4) |
| Sixth | 232 (13.4) | 218 (13) | 309 (18.2) | 245 (14.4) | 42 (3–3) | 117 (8.9) | 153 (117) | 100 (7.6) |
| Seventh | – | – | – | – | – | 8 (0.6) | 21 (1.6) | 14 (1.1) |
| Total (yield/ha) | 1734 (100) | 1678 (100) | 1693 (100) | 1697 (100) | 1301 (100) | 1316 (100) | 1312 (100) | 1310 (100) |

quantity was obtained by large farmers (10.3 %) followed by small (10.1 %) and medium farmers (9.9 %) respectively. During second picking, the medium farms obtained 17.3 % followed by small (15.2 %) and large farmers (13.2 %) respectively. During third picking, it was the highest for medium (21.0 %) followed by small (19.6 %) and large (17.5 %). During fifth picking it was the highest for small (19.3 %) followed by large 19.2 % and medium farms (16.7 %) respectively. In the last picking, large farms obtained the highest quantity (18.2 %) followed by small (13.4 %) and medium farms (13.0 %) respectively.

The yield of Bt Cotton Unirrigated per hectare of cotton was 1301 kg, 1316 kg and 1312 kg for small, medium and large farms respectively. The maximum was obtained during third picking for small and during fourth picking for medium and large farms. The small farms had only six picking while medium and large had seven pickings.

During first picking, the quantity per hectare was the highest for small farms (13 %) while it was (11.3 %) and (10.0 %) for medium and large farms respectively. During second picking small farms obtained the highest quantity (20.2 %) followed by medium (18.7 %) and large farms respectively. During third picking small had the maximum of 26.7 % followed by medium (22.7 %) and large farms (19.4 %) respectively. In the fourth picking, the highest quantity was obtained by medium farms (23.8 %) followed by small (23.6 %) and medium farms (23.1 %) respectively. In the fifth picking large farm (18.9 %) had the maximum followed by medium (14.0 %) and small (13.2 %) respectively. In the sixth picking large farms had the maximum (11.7 %) followed by medium (9.9 %) and small (3.3 %) respectively. During seventh picking large farms had higher (1.6 %) quantity than medium farms.

On the whole, comparing the yield of Bt irrigated and unirrigated it was higher Bt irrigated in than Bt unirrigated cotton by 43.8 %. It was 1335 kg for unirrigated Bt cotton. Similarly comparing Non Bt irrigated and unirrigated yield, it was found that the yield was higher by 22.8 %. It was 1697 kg for irrigated Non Bt cotton while it was 1310 kg for unirrigated Non Bt cotton.

## 27.3 Cost of Production, Value of Output and Profit

### *27.3.1 Overall Cost of Cultivation*

Table 27.3 gives the cost of cultivation of BT and Non Bt Cotton in irrigated and unirrigated conditions. The cost of cultivation per hectare of irrigated Bt cotton was Rs. 25,672 while that of Non Bt irrigated cotton was Rs 22,073. It was Rs. 19,991 for Bt unirrigated cotton while Rs 18,454 for Non Bt unirrigated cotton in the study.

### *27.3.2 Bt Cotton Irrigated*

Of the total cost of Rs. 25,672, per hectare of irrigated Bt cotton cultivation, the cost towards human labour was the highest (25.1 %) followed by harvesting cost (18.1). Seed accounted for 15.6 % of the total cost. These three items formed 58.8 % of the total cost of cultivation. The cost towards tractor and Farm Yard Manure accounted for 20.6 %. The cost of fertilizer was 14.5 % while that of pesticide was 9.7 % of the total cost. Very meagre amount (0.4 %) was spent on irrigation. Nearly 97.5 % was the operational cost the total cost. The total cost of marketing was found to be 2.5 % of the total cost.

### *27.3.3 Bt Cotton: Unirrigated*

The cost of cultivation per hectare was Rs. 19,991. Of this, human labour was the important cost item (21.1 % followed by seed (19.7 %) and cost towards harvesting (12.9 %). These three cost items formed 53.7 % of the total cost. The cost of tractor and FYM was 20.6 % and that of fertilizer and pesticides was 14.5 % and 9.7 % respectively. The total operational cost formed about 98 % of total cost. Of this cost, total marketing costs formed about 1.5 %. The transport cost alone formed 0.6 % of total cost.

**Table 27.3** Cost of cultivation per hectare in rupees

| Sr no | Particulars | Bt cotton I | Bt cotton UI | Bt cotton Total | Non Bt cotton I | Non Bt cotton UI | Non Bt cotton Total |
|---|---|---|---|---|---|---|---|
| Small | | | | | | | |
| 1 | Seed | 4167 (15.6) | 3908 (19.0) | 4013 (17.2) | 1181 (5.6) | 1144 (6.6) | 1160 (6.0) |
| 2 | Human labour | 6967 (26.2) | 4466 (21–7) | 5478 (23.5) | 5945 (28.5) | 4830 (27.8) | 5321 (27.7) |
| 3 | Bullock labour | – | – | – | – | – | – |
| 4 | Tractor | 2555 (9.6) | 2472 (12.0) | 2506 (10.08) | 1876 (9–0) | 1974 (11.4) | 1931 (10.1) |
| 5 | FYM | 2819 (10.6) | (12.1) | 2623 (11.3) | 1266 (6.1) | 1742 (10.01 | 1532 (8.0) |
| 6 | Fertilizer | 2568 (9.7) | 2428 (11.8) | 2782 (11.9) | 2673 (12.8) | 910 (5.2) | 1790 (9.3) |
| 7 | Pesticides | 1991 (7.5) | 2122 (10.3) | 2069 (8.9) | 5279 (25.2) | 3937 (22.6) | 4608 (24.0) |
| 8 | Irrigation | 178 (0.7) | – | 72 (0.3) | 151 (0–6) | – | 66 (0–3) |
| 9 | Harvesting | 4625 (17.4) | 2439 (11.9) | 3324 (14–2) | 2447 (11.7) | 2555 (14.7) | 2507 (13) |
| 10 | Other costs | – | – | – | – | – | – |
| 11 | Total operational cost (1–10) | 25,870 (97.3) | 20,323 (98.8) | 22,867 (98.1) | 20,518 (98.5) | 17,092 (98.3) | 18,915 (98.4) |
| 12 | Transport | 145 (0.5) | 123 (0.6) | 132 (0.6) | 106 (0.5) | 96 (0.6) | 101 (0.5) |
| 13 | Marketing cost | 576 (2.2) | 132 (0.6) | 312 (1.3) | 212 (1.0) | 198 (1.1) | 204 (1.1) |
| 14 | Other mktg. cost | – | – | – | – | – | – |
| 15 | Total marketing cost (12–14) | 721 (2.7) | 255 (1.2) | 444 (1.9) | 318 (1.5) | 294 (1.7) | 305 (1.6) |
| 16 | Total cost (11 + 15) | 26,591 (100.0) | 20,578 (100.0) | 23,311 (100.0) | 20,836 (100.0) | 17,386 (100.0) | 19,220 (100.0) |
| Medium | | | | | | | |
| 1 | Seed | 4041 (15.4) | 3998 (20.2) | 4022 (17.2) | 1219 (5.2) | 1123 (6.2) | 1137 (5.5) |
| 2 | Human labour | 6248 (23.9) | 4255 (21.5) | 5368 (23.0) | 5797 (24.5) | 4982 (27.6) | 5097 (24.9) |
| 3 | Bullock | – | – | – | – | | – |
| 4 | Tractor | 2477 (9.5) | 2146 (10.9) | 2331 (10.0) | 1936 (8.2) | 1190 (6.6) | 1540 (7.5) |
| 5 | FYM | 2595 (10.0) | 1668 (8.4) | 2186 (9.4) | 1307 (5.5) | 132 (6.3) | 1220 (6.0) |
| 6 | Fertilizer | 3248 (12.4) | 2815 (14.2) | 3057 (13.1) | 4823 (20.4) | 3662 (20.3) | 4242 (20.8) |

(continued)

**Table 27.3** (continued)

| Sr no | Particulars | Bt cotton | | | Non Bt cotton | | |
|---|---|---|---|---|---|---|---|
| | | I | UI | Total | I | UI | Total |
| 7 | Pesticides | 1913 (7.2) | 1880 (9.6) | 1898 (8.1) | 5439 (23.0) | 3003 (16.6) | 4226 (20.7) |
| 8 | Irrigation | 104 (0.4) | – | 58 (0.2) | 96 (0.4) | – | 48 (0.2) |
| 9 | Harvesting | 4662 (17.8) | 2720 (13.8) | 3805 (16.3) | 2562 (10.8) | 2571 (14.3) | 2569 (12.5) |
| 10 | Other cost | – | – | – | – | | – |
| 11 | Total op. cost | 25,288 (96.6) | 19,482 (98.6) | 22,725 (97.3) | 23,179 (98.0) | 17,633 (97.9) | 20,079 (98.1) |
| 12 | Transport | 83 (0.3) | 82 (0.4) | 83 (0.4) | 95 (0.4) | 89 (0.5) | 93 (0.4) |
| 13 | Marketing cost | 805 (3.1) | 196 (1.0) | 536 (2.3) | 346 (1.6) | 292 (1.6) | 319 (1–5) |
| 14 | Other mktg. cost | – | | | – | | – |
| 15 | Total marketing cost (12–14) | 888 (3.4) | 278 (1.4) | 619 (2.7) | 441 (2.0) | 381 (2.1) | 412 (1.9) |
| 16 | Total cost (11 + 15) | 26,176 (100.0) | 19,760 (100.0) | 23,344 (100.0) | 23,620 (100.0) | 18,044 (100.0) | 20,491 (100.0) |
| Large | | | | | | | |
| 1 | Seed | 3883 (15.7) | 3915 (20.5) | 3896 (17.3) | 1203 (5.4) | 1235 (7.0) | 1223 (6.1) |
| 2 | Human labour | 6433 (26.0) | 3846 (20.2) | 5423 (24.1) | 5559 (25.1) | 4137 (23.3) | 4848 (24.4) |
| 3 | Bullock | – | – | – | – | – | – |
| 4 | Tractor | 2404 (9.7) | 2200 (11.5) | 2324 (10.3) | 1779 (8.1) | 1674 (9.4) | 1711 (8.6) |
| 5 | FYM | 2372 (9.6) | 1379 (7.2) | 1984 (8.8) | 2224 (10.0) | 1177 (6.6) | 1700 (8.6) |
| 6 | Fertilizer | 2768 (11.2) | 3004 (15.8) | 2860 (12.7) | 3077 (13.9) | 4573 (25.8) | 3820 (19.2) |
| 7 | Pesticides | 1807 (7.3) | 1795 (9.4) | 1802 (8.0) | 5214 (23.5) | 2283 (12.9) | 3748 (18.9) |
| 8 | Irrigation | 164 (0.7) | – | 100 (0.4) | 120 (0.5) | – | 65 (0.3) |
| 9 | Harvesting | 4562 (18.4) | 2558 (13.4) | 3780 (16.8) | 2733 (12.3) | 2455 (13.8) | 2554 (12.8) |
| 10 | Other cost | – | – | – | – | – | |
| 11 | Total op. cost | 24,393 (98.6) | 18,697 (98.0) | 22,169 (98.4) | 21,909 (98.8) | 17,534 (98.8) | 19,669 (98.9) |
| 12 | Transport | 36 (0.1) | 163 (0.9) | 86 (0.4) | 78 (0.4) | 56 (0.3) | 69 (0.3) |
| 13 | Marketing cost | 313 (1.3) | 208 (1–1) | 272 (1–2) | 184 (0.8) | 152 (0.9) | 169 (0.8) |
| 14 | Other mktg. cost | | – | – | | | – |

(continued)

**Table 27.3** (continued)

| Sr no | Particulars | Bt cotton | | | Non Bt cotton | | |
|---|---|---|---|---|---|---|---|
| | | I | UI | Total | I | UI | Total |
| 15 | Total marketing cost (12–14) | 349 (1.4) | 371 (2.0) | 358 (1.6) | 262 (1.2) | 208 (1.2) | 238 (1.1) |
| 16 | Total cost (11 + 15) | 24,742 (100.0) | 19,068 (100.0) | 22,527 (100.0) | 22,171 (100.0) | 17,742 (100.0) | 19,907 (100.0) |
| Overall | | | | | | | |
| 1 | Seed | 4005 (15.6) | 3945 (19.7) | 3977 (17.3) | 1194 (5.4) | 1167 (6.3) | 1180 (5.8) |
| 2 | Human labour | 6455 (25.1) | 4208 (21.1) | 5414 (23.5) | 5669 (25.7) | 4650 (25.2) | 5160 (25.5) |
| 3 | Bullock | – | – | – | – | – | – |
| 4 | Tractor | 2464 (9.6) | 2267 (11.3) | 2373 (10.3) | 1854 (8.4) | 1613 (8.7) | 1734 (8.6) |
| 5 | FYM | 2553 (9.9) | 1852 (9.3) | 2228 (9.7) | 1575 (7.1) | 1075 (5.8) | 1325 (6.5) |
| 6 | Fertilizer | 2936 (11.4) | 2905 (14.5) | 2922 (12.7) | 3423 (15.5) | 4057 (22.0) | 3740 (18.5) |
| 7 | Pesticides | 1887 (7.4) | 1934 (9.7) | 1909 (8.3) | 5316 (24.1) | 3074 (16.7) | 4195 (20.7) |
| 8 | Irrigation | 103 (0.4) | – | 55 (0.2) | 122 (0.6) | – | 60 (0.3) |
| 9 | Harvesting | 4618 (18.1) | 2583 (12.9) | 3675 (16.0) | 2581 (11.7) | 2527 (13.7) | 2554 (12.6) |
| 10 | Other cost | – | – | – | – | – | – |
| 11 | Total op. cost | 25,021 (97.5) | 19,694 (98.5) | 22,553 (98.0) | 21,734 (98.5) | 12,163 (98.4) | 19,948 (98.5) |
| 12 | Transport | 77 (0.3) | 118 (0.6) | 96 (0.3) | 94 (0.4) | 81 (0.4) | 87 (0.4) |
| 13 | Marketing cost | 574 (2.2) | 179 (0.6) | 391 (1.7) | 245 (1.1) | 210 (1.2) | 225 (1.1) |
| 14 | Other mktg. cost | – | – | – | – | – | – |
| 15 | Total marketing cost (12–14) | 651 (2.5) | 297 (1.5) | 487 (2.0) | 339 (15) | 291 (1.60) | 312 (1.5) |
| 16 | Total cost (11 + 15) | 25,672 (100.0) | 19,991 (100.0) | 23,040 (100.0) | 22,073 (100.0) | 18,454 (100.0) | 20,260 (100.0) |

### *27.3.4 Non Bt Cotton Irrigated*

The cost towards cultivation of irrigated Non Bt Cotton was found to be Rs. 22,073. Of this, the cost towards labour was 25.7 % followed by pesticides (24.1 %), fertilizer (15.5 %) and cost towards harvesting (11.7 %). The cost on tractor and Farm Yard Manure formed 15.5 % of the total cost. The cost of seed formed 5.4 % of total cost. The cost of irrigation was only 0.6 % of the total cost. The operational cost formed about 98.5 %. The cost of marketing was 1.5 %. Of which, 0.4 % was towards transport and 1.1 % was towards marketing cost.

### 27.3.5 Non Bt Cotton Unirrigated

The cost of cultivation of unirrigated Non Bt Cotton was Rs. 18,454. the important cost item was human labour (25.2 %) followed by fertilizer (22.0 %) pesticides (16.7 %) and the cost towards harvesting (13.7 %). These items formed 77.5 % of the total cost. The cost towards tractor was 8.7 % while seed was 6.3 % and FYM was 5.8 % of the total cost. These operational costs formed 98.4 % of the total cost. The total marketing cost was 1.6 % of the total cost of cultivation. Of total marketing cost 0.4 % was towards transport and 1.2 % was the marketing cost.

### 27.3.6 Value of Output

The value of Cotton depended on the price of the Cotton. During the period of survey it was Rs. 1600–1800 per quintal.

For Bt cotton in irrigated farms the value of Cotton per ha ranged between Rs. 46,324 and Rs, 44,551. It was the highest for small farms (Rs 46,324) and the lowest for the large farms (Rs. 44,551). The medium farms obtained Rs. 46,207 per hectare. In unirrigated farms, the value of Cotton per hectare ranged between 25,259 and Rs. 35,915. It was the lowest for the small farmers (Rs. 25,259) and the highest for the medium farmers (Rs. 35,915). The value of Cotton per hectare of large farmers was Rs. 26,494 per hectare. On the whole per hectare value of Cotton was Rs. 45,599 and Rs. 29,797 under irrigated and unirrigated cotton respectively.

For non Bt Cotton in irrigated farms, it is seen that value of Cotton per hectare ranged between Rs 28,713 and Rs. 29,120. It was the lowest for the small farmers (Rs.28713) and the highest for the large farmers (Rs. 29,971). The value of Cotton for the medium farmers was Rs. 29,120. In unirrigated farms, the value of Cotton per hectare ranged between Rs. 23,374 and Rs. 23,814. It was the lowest for the small farmer (Rs 23,374) and the highest for the medium farmers (Rs. 23,814). The value of Cotton was Rs. 23,679 for large farmers.

On the whole, it was seen that value of Cotton under irrigated and unirrigated farms was Rs. 29,308 and Rs. 23,632 respectively.

### 27.3.7 Net Income

Net income 1 was arrived at as the difference between value of the Cotton and operational cost of cultivation. Net income 2 was arrived at as the difference between the value of Cotton and the total cost of cultivation.

**Table 27.4** Net income per hectare in rupees

| | | Bt cotton | | Non Bt cotton | |
|---|---|---|---|---|---|
| | | I | UI | I | UI |
| Small | N1 | 20,454 | 4936 | 8195 | 6282 |
| | N2 | 19,803 | 4681 | 7877 | 5988 |
| Medium | N1 | 20,919 | 16,433 | 5941 | 6181 |
| | N2 | 20,031 | 16,155 | 5500 | 5770 |
| Large | N1 | 20,158 | 7797 | 8062 | 6145 |
| | N2 | 19,809 | 7426 | 7800 | 5937 |
| Overall | N1 | 20,578 | 10,103 | 7574 | 5469 |
| | N2 | 19,927 | 9806 | 7235 | 5178 |

The following Table 27.4 gives the details of net income obtained by the sample farmers.

It is observed from the Table that Bt cotton growers are benefited better than their counter parts.

The net returns received by irrigated and unirrigated Bt cotton growers were the highest for medium farms followed by large farms and small farms respectively. On the whole the net returns of Bt irrigated cotton were Rs. 19,927 and that of Bt unirrigated cotton were Rs. 9806.

Normally the sample farmers reported that flowering start from 90 days of sowing and picking of bolls starts from 120 days of sowing. Depending upon the availability of labourers, the picking would take place at the interval of 1 week or 10 days. The sample farmers picked the bolls for six to seven times. The Tables 27.1 and 27.2 give the details of harvesting of cotton.

For non Bt Cotton the net returns were the highest for small farmers under irrigated and unirrigated cotton followed by large and medium farmers. On the whole, the net returns of Non Bt irrigated cotton were Rs. 7235/- while that of Non Bt Unirrigated cotton were Rs 7235 while that of Non Bt Unirrigated were Rs. 5178.

## 27.4 Econometric Analysis of the Performance of Bt Cotton vs. Non-Bt Cotton

### *27.4.1 Students' 't' Test*

In order to find out whether there was any significant difference between the yields of Bt cotton and Non Bt cotton. Keeping the other factors remaining the same, the students 't' test was done.

The hypothesis was that there was no significant difference between the yields of

- Irrigated Bt cotton and Unirrigated Bt cotton
- Irrigated Non Bt cotton and Unirrigated Non Bt cotton

- Irrigated Bt cotton and irrigated Non Bt cotton
- Unirrigated Bt cotton and Unirrigated Non Bt cotton

The test confirms that there was significant difference in the yields of irrigated Bt cotton and unirrigated Bt cotton. The estimated 't' value (16.5461) was greater than the tabulated value (2.63). Therefore it is concluded that the hypothesis was rejected i.e. there was significant difference in the yields of irrigated Bt cotton and unirrigated Bt cotton.

Similarly it is also observed that there was significant difference in the yields of Non Bt irrigated and Non Bt unirrigated Cotton. The calculated value of 't' (23.1867) was greater than the tabulated value (2.62). Therefore it is concluded that the hypothesis was rejected. Therefore, the test confirms that there was significant difference in the yields of irrigated Non Bt cotton and unirrigated Non Bt Cotton.

In case of irrigated Bt cotton and irrigated Non Bt cotton, the test was done. The test confirms that there was significant difference in the yields of irrigated Bt cotton and irrigated Non Bt cotton since the calculated value of 't' (9.622) was greater than the tabulated value (2.66). Therefore the hypothesis was rejected and it is confirmed that there was significant difference in the yields of irrigated Bt cotton and irrigated Non Bt cotton.

Considering unirrigated Bt cotton and unirrigated Non Bt cotton, the 't' test confirms that there was no significant difference in the yields of unirrigated Bt cotton and unirrigated Non Bt cotton as the calculated value of 't' (2.2856) was lesser that the Table value (2.62). Therefore the hypothesis was accepted and it was confirmed that there was no significant difference in the yields of unirrigated Bt cotton and unirrigated Non Bt cotton.

# Chapter 28
# Farmers' Perceptions on Various Features of Bt Cotton in Tamil Nadu

In this chapter, the practice followed by the sample farmers, attitude and opinions of sample farmers have been presented.

## 28.1 Some Basic Features of Bt Cotton: Views of Bt Cotton Sample Farmer Households

It was reported that government agencies inspected the cotton field and helped the farmers in technical knowhow. If they are approached by the farmers they recommend the dosages for pesticide, fertilizer etc. The seed dealers also helped them. They advised them in eradicating the pests by supplying pesticide etc.

The seed company also visited the fields and recommended the pesticide, fertilizer and supervises the plant growth, boll size and pest attack etc. The information provided by the farmers are analysed and provided in the following Table 28.1.

It is observed from the above Table that nearly 20 % of the sample farmers, needed no guidance and the rest needed guidance from the seed company, government officials and seed dealers. Of them 54.4 got the guidance from seed dealers and 16.7 % of them from seed companies, and the rest from the department of agriculture.


V.P. Gandhi, D. Jain, *Introduction of Biotechnology in India's Agriculture*, DOI 10.1007/978-981-10-1091-0_28

**Table 28.1** Inspection of cotton by government agencies

| Inspected by | Bt cotton | | |
|---|---|---|---|
| | I | UI | Total |
| Seed company | 10 | 5 | 15 (16.7) |
| Department of Agriculture Officials | 5 | 3 | 8 (8.9) |
| Seed dealers | 3 | 46 | 49 (54.4) |
| None of the above | 14 | 4 | 18 (20.0) |
| Total | 32 | 58 | 90 (100.0) |

Figure in brackets give percentage to total

**Table 28.2** Experience of sample farmers in the Bt cotton cultivation

| Year | Bt cotton growers | Percentage |
|---|---|---|
| 2004–2005 | 90 | 100 |
| 2003–2004 | 67 | 74.4 |
| 2002–2003 | 32 | 35.6 |

## 28.2 Experience of Sample Farms in the Bt Cotton Cultivation

In the study area, the Bt cotton growers started cultivation of Bt cotton from 2002 to 2003. The information provided by the sample farmers are given in Table 28.2 above.

Thirty-two sample farmers started cultivating Bt cotton during the year 2002–2003 and 67 of them cultivated Bt cotton during 2003–2004. Ninety sample farmers of the study cultivated Bt cotton during the study period under irrigated and unirrigated condition. They used RCH2BT RCH2 Hybrid varieties of cotton seeds of their farms.

### *28.2.1 Agency/Person Recommended for Growing Bt Cotton*

The Rasi Seed Company and seed dealers who are dealers of pesticide and Fertiliser encouraged the farmers and recommended for growing Bt cotton. They also exhibited Bt Cotton plant in their shops and motivated the farmers on the basis of size of bolls, plant height etc. Some farmers were also motivated by the fellow farmers. The Table 28.3 gives the details of agencies recommended for showing Bt cotton.

It is observed form the Table that of the 90 sample Bt growers, 11.1 % of them were motivated by fellow farmers and 40 % were by Rasi seed company and the rest by seed, fertilizer and pesticide dealers.

Among the Bt cotton growers raining irrigated crop, 9.4 % were motivated by fellow farmers, 40.6 % by Rasi Seed Co and the rest by seed, fertilizer and pesticide dealers.

**Table 28.3** Agency/person recommended the growing of Bt cotton

| | Small | | Medium | | Large | | L | | Grand | |
|---|---|---|---|---|---|---|---|---|---|---|
| Particulars | I | UI | I | UI | I | UI | I | UI | total | Percentage |
| Fellow farmer | 1 | 2 | 1 | 3 | 1 | 2 | 3 (9.4) | 7 (12.1) | 10 | 11.1 |
| Seed co | 5 | 17 | 5 | 4 | 3 | 2 | 13 (40.6) | 23 (39.7) | 36 | 40.0 |
| Seed, fertiliser and pesticide dealer | 6 | 11 | 5 | 12 | 5 | 5 | 16 (50.0) | 28 (48.2) | 44 | 48.9 |
| Total | 12 | 30 | 11 | 19 | 9 | 9 | 32 (100.0) | 58 (100.0) | 90 | 100 |

**Table 28.4** Sample households approached by private agencies for promoting Bt cotton

| | Bt | % | Non Bt | % |
|---|---|---|---|---|
| Rasi seeds company | 49 | 54.4 | 63 | 70 |
| Seed dealers | 41 | 45.6 | 27 | 30 |
| Total | 90 | 100 | 90 | 100 |

Among unirrigated Bt cotton growers, 12.1 % were motivated by fellow farmers; 39.7 % by Rasi Seed and the rest by seed dealers.

### 28.2.2 *Roll of Private Agencies*

The Rasi seed company approached the farmers since 1999–2000 for promoting Bt Cotton cultivation. The seed dealers also encouraged the farmers to start Bt cotton cultivation. The Table 28.4 gives the details of role of private agencies in promoting the Bt cotton cultivation.

It is observed that Rasi Seed Company approached the farmers to use $RCH_2Bt$ and RCH2 Hybrid seed for cultivation. Out of 90 farmers, 54.4 % were approached and motivated by Rasi Seeds for growing Bt cotton and the rest were approached by the seed dealers. In case of Non Bt also Rasi Seed approached the farmers to spread the cultivation of Non Bt cotton. The seed dealers also approached the 30 % of the farmers for $RCH_2$ Hybrid cotton cultivation.

## 28.3 Agronomic Features of Bt Cotton: Response of Bt Growers

### 28.3.1 *Flowering of Bt Cotton*

The flowering of Bt cotton took place in more than 60–65 days after sowing. The information regarding flowering was collected from the farmers and presented in the Table 28.5.

**Table 28.5** Flowering of Bt cotton

| | Does the Bt cotton flower earlier or not? | | |
|---|---|---|---|
| | I | UI | Total |
| Yes | 2 (6.3) | 5 (8.6) | 7 (7.8) |
| No difference | 30 (93.7) | 53 (91.4) | 83 (92.2) |
| Total | 32 (100.00) | 58 (100.00) | 90 (100.0) |

**Table 28.6** Plant and boll size and number of picking of Bt cotton as compared to non-Bt cotton reported by Bt cotton growers

| Particulars | Number | % |
|---|---|---|
| **Percentage of sample households reporting on Bt cotton plant size** | | |
| Shorter | 84 | 93.4 |
| Taller | 2 | 2.2 |
| Same | 4 | 4.4 |
| **Percentage of sample households reporting on Bt cotton plant boll size** | | |
| Bigger | 79 | 87.8 |
| Smaller | 9 | 10 |
| Same | 2 | 2.2 |
| **Percentage of sample households reporting on Bt cotton picking** | | |
| More | 3 | 3.3 |
| Less | 4 | 4.4 |
| Same | 8.3 | 92.3 |

It is seen from the Table that 92.2 % of the sample farmers reported that there is no difference between the flowering time of Bt Cotton and non Bt cotton. But very few farmers (7) reported that flowering occurred 2 or 3 days earlier in Bt cotton then No, Bt cotton.

### *28.3.2 Plant and Boll Size, Number of Picking of Bt Cotton as Compared to Non Bt Cotton as Reported by Bt Cotton Growers*

The sample Bt Cotton Growers were asked about plant size of Bt cotton, boll size and the number of pickings as compared to Non Bt cotton. The replies given by them were analysed and presented in the Table 28.6.

It is seen from the Table that 93.4 % of the sample farmers reported that the Bt cotton plant was shorter in size when compared with Non Bt cotton plant. Four of them reported that it was of the same size while the rest reported that it was slightly taller. Nearly 88 % of the sample households had reported that the boll size was bigger and 10 % of them had reported that it was smaller and the rest told that it was same as Non Bt Cotton.

Normally picking of cotton will be done for six times. Eighty-three farmers (92.3 %) reported that the number of picking was same as that of Non Bt cotton while four of them reported that less number of picking was found while others reported that there were more pickings compared to non Bt cotton.

## 28.4 Farmers' Perception on Advantages or Disadvantages of Bt Cotton vis-à-vis Non-Bt Cotton

The seed dealers explained various advantages of Bt cotton such as less boll worm attack less pesticide usage etc. to the farmers. The advantages of Bt Cotton conveyed by the seed dealers are given below in Table 28.7.

It is observed from the Table that majority of the farmers i.e., nearly 81 % were told about the less boll worm attack and pesticide usage. Only 10 % reported that they were told about less pesticide use and others about less boll worm attack.

The opinion of Bt cotton growers were obtained on the advantages and disadvantages of Bt cotton over Non Bt cotton. The opinions provided by them were classified as strong advantage, advantage, no difference, disadvantage and strong disadvantage. Their opinions were presented in the following Table 28.8.

The respondents were asked about the availability of seed. Nearly 98 % reported that the seeds were available in sufficient quantity during the time of sowing. The entire sample farmers reported that seed cost was very high and they found it difficult to buy the seed. Hence it was disadvantageous to them as compared to the seed cost of non Bt cotton. Nearly 77 % of the sample households reported that there was no difference in the quality of seeds. Only 23 % of them reported that the quality of Bt Cotton seed was good and advantageous to them when they were asked about the quality of available Bt Cotton seed.

When the sample household were asked about the pest incidence problem 84.4 % of them reported that pest attack was less and the seeds were strongly advantageous of them. Only 15.6 % reported that it was advantageous to them. Regarding the cost of pesticide/need it was observed that strong advantage was reported by 72.2 % and advantage was reported by 15.6 % and 12.2 % reported that there was no difference in the cost of pesticide. All the sample households reported that there was no differ-

**Table 28.7** Advantages conveyed by the seed dealers

| Advantages | Sample farmers | % |
|---|---|---|
| Less boll worm attack | 8 | 8.9 |
| Less pesticide use | 9 | 10.0 |
| Both | 73 | 81.1 |
| Total | 90 | 100 |

**Table 28.8** Advantages or disadvantages of Bt cotton vis-à-vis non-Bt cotton as reported by Bt cotton growers (percentage)

| Sl. No. | Particulars | Average rating (%) | | | | |
|---|---|---|---|---|---|---|
| | | 2 | 1 | 0 | −1 | −2 |
| 1 | Availability of seeds | – | 2.2 | 97.8 | – | – |
| 2 | Seed cost/price | – | – | – | – | 100 |
| 3 | Quality of avail. Seeds | – | 23.3 | 76.7 | – | – |
| 4 | Pest incidence/problem | 84.4 | 15.6 | – | – | – |
| 5 | Pesticide need/cost | 72.2 | 15.6 | 12.2 | – | – |
| 6 | Fertilizer need/cost | – | – | 100.0 | – | – |
| 7 | Labour need/cost | – | 6.7 | 93.3 | – | – |
| 8 | Machine need/ cost | – | – | 100.0 | – | – |
| 9 | Irrigation need/cost | – | – | 100.0 | – | – |
| 10 | Harvesting cost | – | 43.3 | 56.7 | – | – |
| 11 | Cotton quality | 13.3 | 86.7 | – | – | – |
| 12 | Market preference | 6.7 | 37.8 | 55.5 | – | – |
| 13 | Staple length | 13.3 | 68.9 | 17.8 | – | – |
| 14 | Fibre colour | 27.8 | 70.0 | 2.2 | – | – |
| 15 | Cotton price | 12.2 | 17.8 | 70 | – | |
| 16 | Easy marketing | – | – | 100.0 | – | – |
| 17 | By product output | – | – | 100.0 | – | – |
| 18 | Yield | 30.0 | 70.0 | – | – | – |
| 19 | Profit | 18.7 | 58.9 | 22.4 | – | – |
| 20 | Livestock feeding | – | – | 100.0 | – | – |
| 21 | Water Saving | – | – | 100.0 | – | – |
| 22 | Suitable for early sowing | – | – | 100.0 | – | – |
| 23 | Suitable for late sowing | – | – | 100.0 | – | – |
| | Overall | 14.4 | 85..6 | – | – | – |

2 Strong advantage, 1 Advantage, 0 no difference, −1 Disadvantage, −2 Strong disadvantage

ence in the cost of fertilizer need when asked about the cost of fertilizer need (see Table 28.8).

When the sample household were asked about the pest incidence problem 84.4 % of them reported that pest attack was less and the seeds were strongly advantageous of them. Only 15.6 % reported that it was advantageous to them. Regarding the cost of pesticide/need it was observed that strong advantage was reported by 72.2 % and advantage was reported by 15.6 % and 12.2 % reported that there was no difference in the cost of pesticide. All the sample households reported that there was no difference in the cost of fertilizer need when asked about the cost of fertilizer need.

It was reported by the 93.3 % of the sample household that there was no difference in the cost of labour need. All the Bt cotton growers reported that there was no difference in the cost and machine and irrigation need. All of them reported that there was no difference in marketing the cotton and the marketing was easy for them. There was no difference in the by-product **as** reported by all the sample households.

Seventy per cent of the sample households reported that there was advantage of Bt cotton and rest of 30 % reported that there was strong advantage of Bt cotton regarding yield. 58.9 % of the sample households reported that the profit was advantageous to them while 18.7 % of them reported that the profit was strong advantageous to them but 22.4 % reported that there was no difference in profit.

All the respondents reported that there is no difference in livestock feeding, water saving and also suitable for late and early sowing. Nearly 57 % of the sample households reported that there was no difference in the harvesting cost of the crop but 43.3 % of them reported the harvesting cost was advantageous to them as compared to non Bt cotton. Nearly 87 % of the sample households accepted that it was advantageous to them and 13.3 % of them reported that cotton quality was strongly advantageous to them. It was reported by 55.5 % of the sample households that there was no difference in the market preference between Bt cotton and non-Bt cotton whereas 37.8 % reported that there was strong market preference for Bt cotton. Only 6.7 % of them reported that market preference was very strong for Bt Cotton. About 69 % of the sample households reported that it was strong and 17.8 % reported that there was no difference and the rest reported that there was strong advantage of staple length which brought good price to them. Regarding fibre colour 70 % of them reported that it was good and fetched good price. 27.8 % of them reported that fibre colour was advantageous to them and 2.2 % reported that there was no difference in the fibre colour of Bt cotton and Non Bt cotton.

On the whole, 85.6 % of the Bt growers reported that Bt cotton was advantageous to them and 14.4 % reported that it was strongly advantageous to them.

### *28.4.1 Willingness/Non Willingness to Continue with Bt Cotton*

The sample respondents were asked whether they would continue to cultivate Bt cotton in the years to come. All the sample households expressed their willingness to continue to cultivate the Bt cotton crop since Bt cotton is resistant to American Boll worm to some extent thereby pesticide cost can be reduced to that extent. But they expressed the concerns that the cost of the seed is very expensive. All farmers in addition to the above replied that other crops were not so remunerative. They would like to continue Bt cotton cultivation. Table 28.9 gives the details regarding willingness to continue with Bt cotton.

**Table 28.9** Reasons for cultivating Bt cotton

| Sl. No. | Reasons | Number | % |
|---|---|---|---|
| 1 | Less American boll worm and less pesticide use | 90 | 100 |
| 2 | Other crops were not so remunerative | 90 | 100 |

**Table 28.10** Direct/indirect impact of Bt cotton technology on the economic development of the following reported by Bt cotton growers: percentage

| Sl. No | Particulars | Average rating (%) | | | | |
|---|---|---|---|---|---|---|
| | | 2 | 1 | 0 | –1 | –2 |
| 1 | Village as a whole | 26.7 | 73.3 | – | – | |
| 2 | Any particular religious group (specify) | – | – | 100 | – | – |
| 3 | Any particular caste (specify) | – | – | 100 | – | – |
| 4 | Any other group (specify) | – | – | 100 | – | – |
| 5 | Women | 13.3 | 27.8 | 58.9 | – | – |
| 6 | Poor | 5.6 | 37.8 | 56.6 | – | – |
| 7 | Middle income | – | 71.1 | 28.9 | – | – |
| 8 | Upper income | 7.8 | 61.1 | 31.1 | – | – |
| 9 | Large/medium farmers | 16.7 | 83.3 | – | – | – |
| 10 | Small/marginal farmers | 3.3 | 96.7 | – | – | – |
| 11 | Landless | – | 10 | 90 | – | – |
| 12 | Labour/wage earners | 1.1 | 98.9 | – | – | – |
| 13 | Livestock owners | – | – | 100 | – | – |
| 14 | Tribals | – | – | 100 | – | – |
| 15 | Upper caste | – | – | 100 | – | – |
| 16 | Lower caste | – | – | 100 | – | – |
| 17 | Scheduled castes | – | – | 100 | – | – |
| 18 | Youth | – | – | 100 | – | – |
| 19 | Any other occupational group (specify) | – | – | 100 | – | – |
| 20 | Any other (specify) | – | – | – | – | – |
| | Overall | – | – | 100 | – | – |

2 Strong advantage, 1 Advantage, 0 no difference, –1 Disadvantage, –2 Strong disadvantage

### *28.4.2 Direct or Indirect Impact of Bt Cotton Technology on the Economic Development*

The Direct Indirect impact of Bt cotton technology on the economic development was analysed and presented in Table 28.10.

About 73 % of the sample respondents reported that impact of Bt cotton technology had favourable impact on the economic development of village and the rest reported it had very strong advantage over non Bt cotton. All of them reported that there was no difference on any particular religious group because of impact of Bt cotton technology in the study area. 58.9 of sample households reported that there was no difference on women due to Bt cotton technology and 27.8 of them reported that the impact was advantage and the rest reported that it was strong.

57.6 % of s sample households reported that there was no difference on poor due to impact of Bt cotton technology and 37.8 % of them reported that the impact was strong and the rest reported that it was very strong. 71.1 % of the sample households reported that the impact of Bt cotton technology on middle income was strong while the rest reported that there was no difference on middle income due to the impact of Bt cotton technology. The impact on upper income was strongly reported by 61.1 % of sample households 31.1 % reported that there was no difference on upper income due to the impact of Bt cotton technology.

83.3 % of the farmers reported that the impact of Bt Cotton technology was advantageous and rest reported that it was very strong advantageous to middle and large farmers. 96.7 % of the fanners reported that the impact of Bt cotton technology was advantageous to small marginal farmers and the rest reported that it was strong advantageous to small/marginal farmers. Ninety percent of farmers reported that there was no difference on the landless due to the impact of Bt Cotton technology and the rest reported the impact on landless was advantage to them.

98.9 % of the reported that the impact was advantage to labour and wage earners and the rest reported that it was a strong advantage to labour/wage earners. All the farmers reported that there was no difference on livestock owners tribals, upper caste, lower caste, schedule caste, youth and any other occupational group due to the impact of Bt cotton technology.

On the whole there was no difference on the economic development due to impact of Bt cotton technology.

## 28.5 Measures Suggested by Sample Farmers to Improve the Acceptance of Bt Cotton

The sample farmers were asked to give their suggestions to facilitate Bt cotton cultivated by different agencies like government, extension agencies, private traders and seed company. All of them unanimously replied that the government should take steps to reduce the price of the cotton seed and extension workers should visit the field and give guidance on application of fertilizer and pesticides. They also demanded that the private traders like seed fertilizers pesticide dealers should not sell spurious seeds. Labelled, good quality seeds must be sold to the farmers. The seed company should supervise the sale of seeds whether there are spurious seeds for sale and also they should provide good quality seeds with high termination rate. Moreover, the seeds companies should take steps to reduce the prices of Bt cotton seeds.

## 28.6 Impact of Bt Cotton Cultivation on Environment

None of the farmers reported any adverse impact on environment due to Bt cotton cultivation (see Table 28.11).

From the above analysis it was found that the Bt cotton technology was a welcome feature among the farming community as it had created favourable impact on cotton cultivation.

**Table 28.11** The impact of Bt cotton on environment

| | Adverse environment effect | |
|---|---|---|
| | Bt | Non Bt |
| Yes | – | – |
| No | 90 | 90 |
| Total | 90 | 90 |

# Bibliography


Bose, A. (2000). From population to pests in Punjab: American boll worm and suicides in cotton belt. *Economic and Political Weekly, 35*(38), 3375–3378.

Central Institute for Cotton Research. (2001). *Cotton: March towards millennium*. Nagpur: CICR.

Chaturvedi, S. (2004). Bio-safety regulation: Need for fine balancing. *Economic and Political Weekly, 39*(33), 3693–3697.

David, G. S., & Sai, Y. V. S. T. (2002). Bt cotton: Farmers' reactions. *Economic and Political Weekly, 37*(46), 4601–4602.

Deshpande, R. S. (2002). Suicide by farmers in Karnataka: Agrarian distress and possible alleviatory steps. *Economic and Political Weekly, 37*(26), 2601–2610.

Financial Express. (2003). Regulator says no to Bt cotton cultivation. *Financial Express*, 25 April 2003.

Gandhi, V. P., & Namboodiri, N. V. (2006, August). *The adoption and economics of Bt cotton in India: preliminary results from a study*. Paper presented at the conference of the International Association of Agricultural Economists (IAAE), Gold Coast.

Gandhi, V. P., & Namboodiri, N. V. (2007a). Cotton revolution in Gujarat: The impact of Bt technology. In R. H. Dholakia (Ed.), *Frontiers of agricultural development in Gujarat* (pp. 56–76). Ahmedabad: Centre for Management in Agriculture, Indian Institute of Management.

Gandhi, V. P., & Namboodiri, N. V. (2007b). *Returns and economics of Bt cotton vis-à-vis traditional cotton varieties in the state of Maharashtra in India*. Ahmedabad: Centre for Management in Agriculture, Indian Institute of Management, Ahmedabad.

Ghosh, P. K. (2000). *Genetically modified crops in India with special reference to Cotton*. Mumbai: Indian Society of Cotton Improvement, Central Institute of Research on Cotton Technology.

Huang, J., Rozelle, S., Pray, C., & Wang, Q. (2002). Plant biotechnology in China. *Science, 295*, 674–677.

Indian Express (2003, January 15). Government defends Bt cotton. *Indian Express Bureau, New Delhi*.

Joseph, R. K. (2005). Is GM technology desirable? *Economic and Political Weekly, 40*(49), 5210.

Kranthi, K. R. (2005, September 5). Is Bt cotton unsustainable? *The Hindu*. Retrieved from: http://www.thehindu.com/2005/09/05/stories/2005090505731100.htm.

Mayee, C. D. (2002). Cotton scenario in India. Challenges and approaches for increasing productivity. In Workshop, TNAU, Coimbatore.

Naik, G., Qaim, M., Subramanian, A., & Zilberman, D. (2005). Bt cotton controversy: Some paradoxes explained. *Economic & Political Weekly, 40*, 1514–17.

Narayanamoorthy, A., & Kalamkar, S. S. (2005). *Economics of Bt cotton cultivation in Maharashtra*. Pune: Gokhale Institute of Politics and Economics.





V.P. Gandhi, D. Jain, *Introduction of Biotechnology in India's Agriculture*, DOI 10.1007/978-981-10-1091-0

Pathak, A. (2005, January 8). Bumper Bt cotton harvest at the root of glut in Gujarat. *Times of India*.

Pray, C. E., Huang, J., Hu, R., & Rozelle, S. (2002). Five years of Bt cotton in China – the benefits continue. *The Plant Journal, 31*(4), 423–430. doi:10.1046/j.1365-313X.2002.01401.x.

Purcell, J. P., & Perlak, F. J. (2004). Global impact of insect resistant (Bt) cotton. *AgBioforum, 17*(1 & 2), 27–30.

Pushpavalli, A. (2004). *Returns to Bt cotton vis-à-vis traditional cotton in Tamil Nadu, Agro-Economic Research Centre*. Chennai: University of Madras.

Qayum, A., & Sakkhari, K. (2003). *Did Bt cotton save farmers in Warangal?* Hyderabad: Deccan Development Society.

Ramgopal, N. (2006). *Economics of Bt cotton vis-à-vis traditional cotton varieties: Study in Andhra Pradesh*. Visakhapatnam: Agro-Economic Research Centre, Andhra University.

Shah, V. D. (2007). *Returns to Bt cotton vis-à-vis traditional cotton varieties in Gujarat State*. Gujarat: Agro-Economic Research Centre, Sardar Patel University.

Sharma, D. C. (2010, March 6). *Bt cotton has failed admits Monsanto*. India Today. Retrieved from: http://indiatoday.intoday.in/story/Bt+cotton+has+failed+admits+Monsanto/1/86939.html

Sharma, H. C., Crouch, J. H., Sharma, K. K., Seetharama, N., & Hash, C. T. (2002). Applications of biotechnology for crop improvement: Prospects and constraints. *Plant Science, 163*(2002), 381–395.

Shiva, V., & Jafri, A. (2003, May 31). *Failure of the GMO's in India*. Retrieved September 2014, from, Mindfully website:http://www.mindfully.org/GE/2003/India-GMO-Failure-Shiva31may03.htm

Shiva, V., Emani, A., & Jafri, A. H. (1999). Globalisation and threat to seed security: Case of transgenic cotton trials in India. *Economic and Political Weekly, 34*(10/11), 601–613.

Smetacek, R. (2003). Performance of Bt cotton [letter to the editor]. *Economic and Political Weekly, 38*(33), 3434.

Times of India. (2006, June 1). *Three states fix price of Monsanto's Bt cotton seeds*. Retrieved 2010 October, from The Times of India: http://timesofindia.indiatimes.com/india/3-states-fix-price-of-Monsantos-Bt-cottonseeds/articleshow/1606491.cms

Thomas, E. C. (2002). India adopts Bt cotton. *Yojana, 46*(6), 14–17.

USEPA. (2001, October 15). *Bt plant-incorporated protectants – Biopesticides registration action document*. Washington, DC: U.S. Environmental Protection Agency.

Venkateshwarlu, K. (2002). *The Hindu*, India, December 30.

# Index





V.P. Gandhi, D. Jain, *Introduction of Biotechnology in India's Agriculture*, DOI 10.1007/978-981-10-1091-0

MIX
Papier aus verantwortungsvollen Quellen
Paper from responsible sources
FSC® C105338

If you have any concerns about our products,
you can contact us on
**ProductSafety@springernature.com**

In case Publisher is established outside the EU,
the EU authorized representative is:
**Springer Nature Customer Service Center GmbH**
**Europaplatz 3, 69115 Heidelberg, Germany**

Printed by Libri Plureos GmbH
in Hamburg, Germany